Research Reports in Physics

Research Reports in Physics

Nuclear Structure of the Zirconium Region
Editors: J. Eberth, R. A. Meyer, and K. Sistemich

Ecodynamics Contributions to Theoretical Ecology
Editors: W. Wolff, C.-J. Soeder, and F. R. Drepper

Nonlinear Waves 1 Dynamics and Evolution
Editors: A. V. Gaponov-Grekhov, M. I. Rabinovich, and J. Engelbrecht

Nonlinear Waves 2 Dynamics and Evolution
Editors: A. V. Gaponov-Grekhov, M. I. Rabinovich, and J. Engelbrecht

Nuclear Astrophysics
Editors: M. Lozano, M. I. Gallardo, and J. M. Arias

Optimized LCAO Method and the Electronic Structure of Extended Systems By H. Eschrig

Nonlinear Waves in Active Media
Editor: J. Engelbrecht

Problems of Modern Quantum Field Theory
Editors: A.A. Belavin, A.U. Klimyk, and A.B. Zamolodchikov

Fluctuational Superconductivity of Magnetic Systems
By M.A. Savchenko and A.V. Stefanovich

Nonlinear Evolution Equations and Dynamical Systems
Editors: S. Carillo and O. Ragnisco

Nonlinear Physics
Editors: Gu Chaohao, Li Yishen, and Tu Guizhang

M.A. Savchenko A.V. Stefanovich

Fluctuational Superconductivity of Magnetic Systems

With 63 Figures

Springer-Verlag Berlin Heidelberg GmbH

Professor Maxim A. Savchenko
Paustovskogo Street, d.8, korpus 3, apt. 653, SU-117463 Moscow, USSR

Dr. Alexei V. Stefanovich
Volokolamskoe shosse, d.16b, korpus 2, apt. 37, SU-125080 Moscow, USSR

Translators

Dr. Ram S. Wadhwa
Natalie Wadhwa
28-3-224 Menzhinsky Street, SU-129281 Moscow, USSR

Editor of the English Edition

Dr. Alan R. Bishop
Los Alamos National Laboratory, Los Alamos, NM 87545, USA

Title of the original Russian edition:
Fluktuatsionnaia sverkhprovodimost' magnitnykh sistem

ISBN 978-3-540-50561-7 ISBN 978-3-642-74287-3 (eBook)
DOI 10.1007/978-3-642-74287-3

Library of Congress Cataloging-in-Publication Data. Savchenko, M. A. [Fluktuatsionnaia sverkhprovodimost' magnitnykh sistem. English] Fluctuational superconductivity of magnetic systems / M. A. Savchenko, A. V. Stefanovich. p. cm. – (Research reports in physics) Translation of: Fluktuatsionnaia sverkhprovodimost' magnitnykh sistem. Includes bibliographical references. ISBN 0-387-50561-X (U.S.) 1. Superconducting magnets – Fluctuations. 2. Rare earth metals-Magnetic properties. I. Stefanovich, A. V. (Alekseĭ Viacheslavovich) II. Title. III. Series.
QC761.3.S2813 1990 538'.3--dc20 89-26081

2157/3150-543210 – Printed on acid-free paper

Preface

This is a monograph on the fluctuational theory of superconductivity. The theory was originally developed by M. A. Savchenko in 1964 in response to the work of B. T. Matthias, the discoverer of superconductive compounds. Further development of the theory led to the *prediction* of the existence of high-temperature superconductors among magnetic and nonmagnetic compounds of rare-earth metals, ceramics, and polymers. In 1987 this prediction was experimentally verified by the discovery of high-T_c superconducting rare-earth metal oxides by J. Bednorz and K. Müller. To date, this is the only account that explains consistently all the available data.

The theory of high-temperature superconductivity is based on the concept of an enhanced electron-phonon interaction which leads to an attraction between electrons forming superconducting pairs. This interaction is due to the exchange spin fluctuations (exchange enhancement effect). In compounds in which there is no magnetic ordering except at very low temperatures, such as in rare-earth metal oxides, the electron-phonon interaction is strengthened due to fluctuations in the spins of the conducting electrons. If there is magnetic ordering in a superconductor at a temperature higher than or of the same order as the critical superconducting temperature T_c, then the attraction in the electron pairs will be further increased because the Coulomb repulsion is overwhelmed by fluctuations in the spins forming the long-range antiferromagnetic order.

On the basis of this theory the methods of synthesizing new high-temperature superconductors (including polymers) with improved critical parameters, that is, increased critical superconducting transition temperature, lower and upper fields H_{c1}, H_{c2} and critical current density, were developed.

The monograph begins with an introduction to the exchange enhancement effect at phase transitions in superconductor and magnetic crystals. It continues with a discussion of the phenomenological fluctuation theory of superconductivity in compounds of rare-earth metals, ceramic oxides, and spin glasses, and the microscopic theory of high-temperature superconductivity. In the final chapter the possibility of finding high-temperature superconductive polymer systems is considered.

The book is intended for specialists in physics and chemistry of high-temperature superconductivity, under- and (post)graduate students of physics, chemistry, and mathematics.

Moscow, 1989 *M. A. Savchenko · A. V. Stefanovich*

Contents

1. Introduction

Phase transitions in the condensed state of matter have been the subject of intensive studies during the last decade. The growing interest towards these investigations can be attributed to several circumstances:

1. It was experimentally established that transitions from the paramagnetic phase to the ordered state in compounds of rare-earth metals, e.g., $RERh_4B_4$, $RERh_6B_6$, $REMo_6S_8$, $REMo_6Se_8$, $RERhSi_2$, and $RERh_2Si_2$ (RE = Y, La, Th, Nd, Sm, Tb, Dy, Ho, Er, Tm, Lu, Yb), are transitions from one steady state to another. In this case, one phase of the system is found to be superconducting while the remaining phases are magnetic. This means that long-range magnetic ordering is possible in the latter type of phases. The superconducting transition temperatures vary in the range $1\ K \lesssim T_{c1}, T_{c2} \lesssim 10\ K$, $T_{c2} < T_{c1}$. In some compounds, a transition from the superconducting state to the magnetic state involves an intermediate phase in which superconductivity and magnetic ordering coexist. Indeed, investigations of the phase diagram [1.1] for the compounds $Ho(Ir_{1-x}Rh_x)_4B_4$ reveal that in the concentration range $0.6 \lesssim x \lesssim 0.8$ of Ir ions the system first undergoes a phase transition to the antiferromagnetic state, followed by a transition to a phase in which antiferromagnetism and superconductivity coexist. In other words, the magnetic transition temperature T_m is higher than the superconducting transition temperature T_c. Superconducting compounds $BaPb_{1-x}Bi_xO_3$, $0 \lesssim x \lesssim 0.3$ with a perovskite structure were also discovered. These compounds were found to have a highest superconducting transition temperature of about 13 K [1.2].

2. In some materials, viz., alloys of transition metals with A-15 structure and rare-earth compounds, the superconducting transition temperature was found to be quite high ($T_{c\ max} = 23.2$ K).

3. A wide range of disordered magnetic systems such as disordered ferro- and antiferromagnets, metal glass, and spin glass was discovered.

Most of these materials were found to possess anomalous electric and magnetic properties.

4. In 1986, **Bednorz** and **Müller** [1.3] discovered a new class of compounds $Ba_xLa_{5-x}Cu_5O_{5(3-y)}$, $x = 0.75$ and 1, $y > 0$, which became superconducting at a temperature of about 30 K. A group of American scientists led by Chu [1.4,5] discovered in early 1987 the compounds $La_{1.85}Ba_{0.15}CuO_4$ and $La_{1.8}Sr_{0.2}CuO_4$ for which the superconducting transition temperature T_c was as high as 40 K. The same group later synthesized ceramics with composition $Y_{1.2}Ba_{0.8}CuO_{4-y}$ having a critical temperature $T_c = 90$ K [1.6]. This heralded the advent of high-temperature superconductivity, i.e., the discovery of superconducting compounds whose critical temperature T_c was higher than the boiling point of liquid nitrogen (77.8 K). So far, high-temperature superconductivity has been discovered in ceramics $YBa_2Cu_3O_{7-y}$ [1.7], as well as in complex systems Tl - Ba - Cu - O [1.8], Bi - Al - Ca - Sr - Cu - O [1.9], and Ba - K - Bi - O.

All these facts stimulated theoretical and experimental studies of condensed media near the phase transition points, and made it possible to describe the behavior of a system in the immediate vicinity of the critical temperature. The modern theory of phase transitions provides a quantitative description of superconducting and magnetic properties of these materials based on the mathematical apparatus of nonlinear equations of the renormalized Wilson-Fisher group ($\tau \lesssim 10^{-3} - 10^{-4}$, $\tau = (T - T_c)/T_c$).

In this book, we shall consider the class of compounds based on rare-earth metals and their alloys (Er-Dy, Er-Tb, Er-Ho, $Er_{1-x}Ho_xRh_4B_4$), as well as ceramics (magnetoelectric materials, viz., oxides with a perovskite structure) having a multicomponent order parameter. The analysis is based on the mechanism of lamination into phases during transitions from the paramagnetic phase to the ordered state. This approach, which was proposed and developed by the authors, takes into consideration the exchange enhancement of relativistic and nonrelativistic interactions in a system followed by a freezing of the individual components of the multicomponent order parameter. The phase lamination effect consists in that phase transitions from the paramagnetic phase to the ordered ground state occur through a number of intermediate phases, the order parameter in each phase having its own topological degeneracy space R_i. The degeneracy spaces are connected through the relation $R_1 \subset R_2 \subset R_3 \subset \dots \subset R_N$, and the phase transition need not occur according to the irreducible representation of the order parameter for which the first term in the expansion of the free energy in the vicinity of the phase transition point tends to zero. The nature of the transition is determined by the interaction between fluctuations of different components of the order parameter, which increase sharply near the phase transition point in the presence of the exchange enhancement effect.

The exchange enhancement effect and its role in phase transitions in magnetically ordered crystals, e.g., ferromagnets with quadrupole interaction, antiferromagnets, and ferroelectro-antiferromagnetic materials in a strong magnetic field, are described in Chap. 2. (Below the word "segnetoantiferromagnet" is equivalent to "ferroelectroantiferromagnet" and is introduced for the sake of brevity.) The exchange enhancement theory was first constructed by **Van Vleck** [1.10] for ferromagnets and **Savchenko** [1.11] for antiferromagnets and was first verified experimentally by **Ozhogin** and **Maximenkov** and **Seavy** [1.12,13] during studies of magnetoelastic waves in a hematite (α-Fe_2O_3) single crystal which is an "easy-plane" weak ferromagnet. This Savchenko effect involves the renormalization of the parameters of relativistic and nonrelativistic interactions (such as the spin-orbit interaction which determines the magnetic anisotropy, magnetostriction, quadrupole exchange interaction, magnetoelectric interaction, and electron-phonon interaction) by the exchange interaction parameter. Such a renormalization may increase the corresponding parameters by an order of magnitude or more. The exchange enhancement considerably increases the critical fluctuation region during phase transitions. Hence the phase transitions in the systems under consideration are not second-order phase transitions as predicted by the self-consistent mean field theory, but rather first-order phase transitions close to second-order transitions.

The interaction mechanism for fluctuations in systems with a multicomponent order parameter is discussed in detail in Chap. 3, where the temperature phase transitions are considered in heavy rare-earth metals and their alloys Er-Dy, Er-Ho, Er-Tb, Er-Tm (space group $P6_3/mmc$ - D^4_{6h}), as well as in ferroelectro-antiferromagnetic compounds. Fluctuations in complex magnetic systems were first considered by **Savchenko** and **Shishkin** [1.14] who studied the effect of magnetization fluctuations on the parametric ferrite amplification mode of electromagnetic type.

The phase diagram of heavy rare-earth metals and their alloys, constructed in the temperature vs. axial magnetic anisotropy parameter

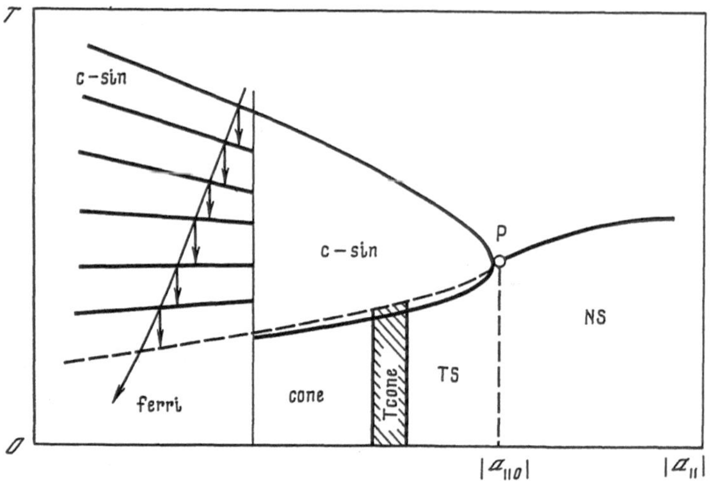

Fig. 1.1. Temperature and axial magnetic anisotropy phase diagram of rare-earth metals and their alloys having a complex magnetic structure in the lowest ordered state; c-sin is the state corresponding to the longitudinal sinusoidal spin-density wave; "ferri" stands for ferrimagnetic antiphase domain, "cone" indicates a conic spiral, "T cone" a tilted conic spiral, NS a normal spiral, TS a tilted spiral, and P the paramagnetic phase

coordinates, is shown in Fig. 1.1. This diagram describes the magnetic phase transitions and contains a large number of phases with a complex magnetic structure. The trajectory with arrows indicates the possibility of multistage phase transitions in such systems. In other words, the system may pass through a large number of magnetic phases before moving from the paramagnetic phase P to the ordered "ferri" ground state (ferrimagnetic antiphase domain), and each of the intermediate magnetic states may have a complex magnetic structure. Solid lines indicate the first-order phase transition, while the dashed lines correspond to the second-order phase transition.

First-order phase transitions occur as a result of the interaction of fluctuations of a strongly fluctuating order parameter with those of a weakly fluctuating one. The correlation length of fluctuations is determined by the coefficient of the first term in the expansion of the free energy in the order parameter in the vicinity of the phase transition point. For a strongly fluctuating order parameter, this coefficient tends to zero, while for a weakly fluctuating order parameter it does not vanish altogether but becomes much smaller than unity due to a sharp renormalization of the strongly fluctuating order parameter by fluctuations. Thus, a sort of "pumping" of fluctuation energy takes place from one component of the multicomponent order parameter to another. This is mathematically reflected in a sharp renormalization of the free energy coefficients of the corresponding powers in the expansion in the weakly fluctuating component. As a consequence, the coefficient of the fourth power in the expansion is no longer positive definite, and this results in a first-order phase transition in irreducible representation corresponding to the weakly fluctuating order parameter.

It was shown on the basis of differential geometry and topological considerations that the following relation is satisfied for degeneracy spaces corresponding to the order parameter in two adjacent phases during phase transitions between states indicated in the phase diagram in Fig. 1.1:

$$R_i \xleftarrow{w_i} R_{i+1} \quad .$$

This relation, called fibration in topology, is the topological definition of the effect of phase lamination. An identical effect of lamination into phases was also observed for magnetoelectrics.

Chapters 4-6 are devoted to the study of magnetic superconductors. Superconducting compounds of rare-earth metals are materials of this type. The fluctuation theory of superconductivity is constructed in Chaps. 4 and 5 for the compounds of rare-earth metals. This theory is based on the exchange enhancement effect and on the idea that magnetic oscillations exist in all phases, viz., paramagnetic, superconducting, and magnetically ordered phases. In order to illustrate this statement, we consider a transition from paramagnetic to superconducting phase in a magnetic superconductor. It follows from Landau's theory of phase transitions that the expansion of the free energy of a system in the order parameter $|\Delta_0|$ near the superconducting transition point has the form

$$F = a(\tau)|\Delta_0|^2 + b|\Delta_0|^4 + c|\nabla\Delta_0|^2 + \dots \qquad (1.1.1)$$

where

$$a(\tau) = \alpha\tau \quad ; \quad \alpha \approx 1 \quad ; \quad \tau = (T - T_c)/T_c \quad ; \quad b = \frac{\pi^2\hbar^3}{2\sqrt{2}\mu_e^{3/2}\varepsilon_F^{3/2}v_0} \quad ;$$

$$c = \frac{\hbar^2\varepsilon_F A v_0^{-2/3}}{6\pi^2\mu_e T_c^2} \quad ; \quad A = \sum_{n=0}^{\infty} \frac{\Gamma(n+3)\zeta(n+3)}{2^n(2n)!} \quad ;$$

μ_e is the electron mass, ε_F the Fermi energy, and v_0 the volume of a unit cell. In order to find the nature of the phase transition to the superconducting state, we must determine the influence of the order parameter fluctuations on the phase transition. This can be done with the help of the Ginzburg-Levanyuk criterion. The critical fluctuation region is then found to be equal to

$$\tau_{Su} = b^2/\alpha c^3 \approx 10^{-14} \quad . \tag{1.1.2}$$

This means that Landau's theory must remain valid right up to the phase transition point and the superconducting fluctuations must not affect the phase transition which is a second-order transition. However, since we are dealing with magnetic superconductors, the system contains localized magnetic moments whose fluctuations must be taken into consideration. The fluctuation region in the vicinity of the phase transition into a magnetically ordered state with localized magnetic moments is equal to [1.15]

$$\tau^{l}_{spin} = 10^{-5} \frac{3[1-(2/e)]^2(2S^2+2S+1)^2}{32\pi^4 S(S+1)} \quad ; \; (e = 2.73 \ldots) \tag{1.1.3}$$

where S is the magnetic ion spin. The quantity τ^{l}_{spin} is found to be of the order of 10^{-5}, which means that localized spin fluctuations may affect the phase transition to the superconducting state only in the vicinity of the point of intersection of the superconducting and magnetic phase transition lines (Fig. 1.2). However, the system also contains spins of delocalized s- and d(f)-electrons whose fluctuations must be taken into account. The fluctuation region of these electrons is found to be equal to

$$\tau^{dl}_{spin} = \frac{1}{36}\left(\frac{<r_c>}{a}\right)^2 \tag{1.1.4}$$

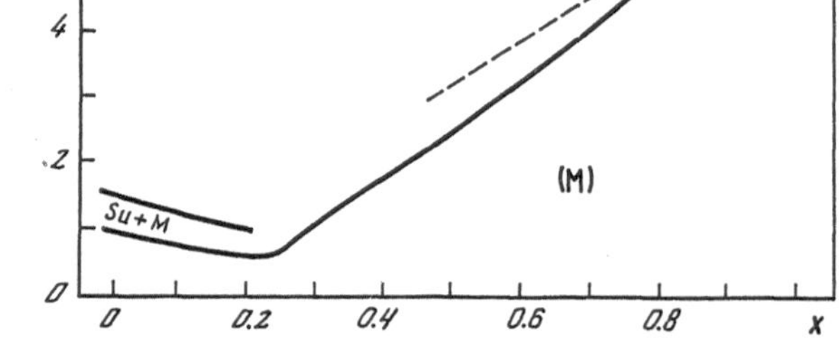

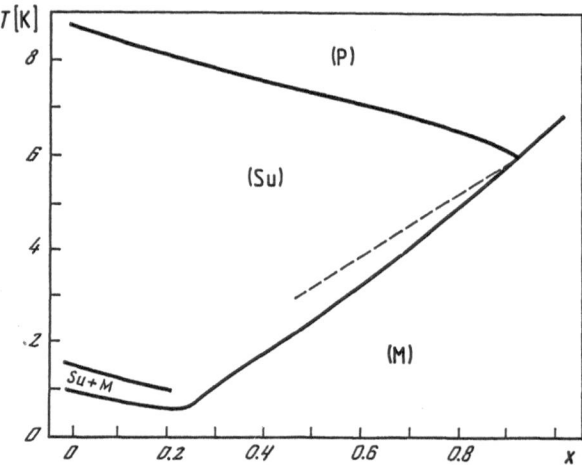

Fig. 1.2. Temperature vs. rare-earth ion concentration phase diagram of pseudo-ternary superconducting compounds of rare-earth metals. P – paramagnetic phase; Su – superconducting; M – magnetic

where 'a' is the lattice constant, $\langle r_c \rangle$ is the exchange correlation radius in the system of collective electron spins

$$\langle r_c \rangle = \left(\int d\mathbf{x} x^2 J(\mathbf{x}) / \int d\mathbf{x} J(\mathbf{x}) \right)^{1/2} \tag{1.1.5}$$

and $J(\mathbf{x})$ is the exchange potential in the delocalized spin system. This parameter was first introduced for a localized spin system by **Vaks**, **Larkin**, and **Pikin** [1.16] while studying the thermodynamic, high-frequency, and kinetic properties of an ideal ferromagnet near the Curie temperature T_C. For delocalized spins, this parameter was first introduced by the authors [1.17]. The quantity τ_{spin}^{dl} was estimated and found to be of the order of 3, which means that spin fluctuations of delocalized spins exist in all phases of magnetic superconductors and considerably broaden the range of superconducting fluctuations. Consequently, Landau's theory is inapplicable in the present case and we must construct a fluctuation theory of magnetic superconductors. This theory is constructed in Chaps. 4 and 5. It is shown that the phase transition to a homogeneous superconducting state is described by the functionals

$$F = \tilde{a}(\tau)|\Delta_0|^2 + \tilde{b}|\Delta_0|^4 + \dots \tag{1.1.6}$$

where $\tilde{a}(\tau)$ and $\tilde{b}$ are the parameters $a(\tau)$ and b but renormalized by magnetic fluctuations: $\tilde{a}(\tau)=\tilde{\alpha}\tau$, $\tilde{\alpha}\approx 1$, $\tilde{b}=b \ln \dfrac{|\Delta_0|^2}{g(\tau)}$, $g(\tau)\sim\tau$.

The appearance of the function $\tilde{b} = b \ln \dfrac{|\Delta_0|^2}{g(\tau)}$ in the expression for the free energy means that the phase transition from the paramagnetic phase to the superconducting state is a first-order phase transition close to a second-order one, and not a second-order phase transition as predicted by the Landau theory. Thus, magnetic fluctuations stimulate a phase transition to the superconducting state. A subsequent decrease in temperature leads to a first-order fluctuation phase transition to the Su+M phase in which superconducting and magnetically ordered phases coexist (Fig. 1.2). In this phase, the fluctuations of the collective electron spins modulate the system of localized magnetic moments and hence a longwave sinusoidal modulated magnetic structure is formed [1.18]. A further decrease in temperature results in the appearance of a ferromagnetic component in the magnetic structure, causing a destruction of superconductivity and a first-order phase transition to the magnetically ordered state.

The microscopic theory of phase transition from paramagnetic to superconducting phase is considered in Chap. 5. This theory enables us to calculate the effective electron-phonon interaction parameter in the case of fluctuations near the phase transition point, and to determine the superconducting transition temperature T_c. It also explains the nature of phase transitions in rare-earth compounds and can be used to predict a new class of magnetic superconducting materials with $T_m > T_c$.

It follows from the self-consistent field theory based on the s-d(f)-electron exchange model that the existence of localized magnetic moments in a system must lead to a suppression of superconductivity, while spin fluctuations (paramagnons) weaken the effective attraction of conduction electrons due to an increase in the parameter of Coulomb repulsion of electrons [1.19]. In some magnetic systems, the fluctuations of the localized magnetic moment do not enhance the Coulomb repulsion of conduction electrons, but rather weaken it near the point of phase transition to the superconducting state (Chap. 5). This can happen in systems with antiferromagnetic ordering. It will be shown in Chap. 5 that antiferromagnetic ordering may play a significant role in the establishment of a high-temperature superconducting phase in rare-earth compounds. We shall also calculate in Chap. 5 the effective electron-phonon interaction parameter in the vicinity of the point of a phase transition to the superconducting state:

$$\lambda_{e-ph} = \ln^{-1}\left(\frac{2\gamma}{\pi}\frac{\langle\omega_D\rangle}{T_N}\right) \tag{1.1.7}$$

where $\langle\omega_D\rangle$ is the mean Debye energy, $\gamma = e^C$, and C is the Euler constant. Equation (1.1.7) expresses the condition for the emergence of superconductivity in magnetic superconductors. This, however, is not a sufficient condition for the existence of the high-temperature superconducting phase. Indeed, antiferromagnets with a low value of Néel temperature ($T_N \simeq 1$ K) also exist. Hence, in order to write a sufficient condition, we must impose a restriction setting a lower bound for the Néel temperature. This restriction is

$$T_N > \varepsilon_F(a/\langle r_c\rangle)^2 \quad . \tag{1.1.8}$$

The necessary and sufficient condition for the emergence of a high-temperature superconducting phase in superconducting compounds of rare-earth metals was first documented in [1.17]. The quantity $(a/\langle r_c\rangle)^2$ can be estimated and is found to be of the order $(a/\langle r_c\rangle) \sim 10^{-2}$. This leads to the estimate $T_N \to T_c \gtrsim 10^2$ K. Thus the microscopic theory makes it possible to obtain the necessary and sufficient condition for the emergence of a high-temperature superconducting phase in rare-earth compounds and in ceramic systems with an antiferromagnetic instability. The predictions of the theory developed by us were supported experimentally by **Bednorz** and **Müller**, as well as by the group led by **Chu** [1.3-6). Among the high-temperature superconducting compounds discovered was $La_{2-x}CuO_{4-y}$ [1.20], which undergoes an antiferromagnetic phase transition at $T_N \approx 200$ K and a superconducting transition at $T_c \approx 30$ K. In other words, this is a high-temperature superconductor with $T_N > T_c$. Moreover, the experimentally discovered high-temperature superconductors RE-Ba-Cu-O (RE stands for the entire range of rare-earth elements except Y, La, Ce, and Pm) [1.21] were found to undergo a low-temperature antiferromagnetic transition ($T_N < T_c$).

The general nonlinear equations for the superconducting phase in this class of compounds are obtained in Chap. 6. It is shown that a transition from the paramagnetic phase to the superconducting state in these compounds may result in the emergence of an unstable topological soliton whose destruction is accompanied by the onset of superconductivity. Moreover, the solution of the general nonlinear equations is used to analyze the possible vortex structures which may appear in the superconducting phase as well as in the phase where superconductivity and magnetic ordering coexist.

At present, disordered magnetic systems are drawing the attention of an ever increasing number of researchers. The central problem in this field is that of a spin glass. The theory of disordered ferro- and antiferromagnets has been developed to a fairly advanced level, while there is no unified theory for spin glasses at present. This is due to the absence of a universal microscopic model which could describe the static properties of spin glasses, i.e., the temperature dependence of the magnetic susceptibility and heat capacity, as well as the spin-wave properties. The static properties of spin glasses can be explained satisfactorily with the help of Heisenberg's microscopic exchange model, while the spin-wave dynamics can be described by using the macroscopic continuum theory of spin waves. This creates certain problems in the interpretation of experimental data. Experimenters have meanwhile switched their attention to the electromagnetic properties of spin glasses, e.g., the Kondo effect.

A new model of spin glasses, based on lamination into phases, is proposed in Chap. 7. This model makes it possible to predict a whole range of new materials. A spin glass is a system in which the spins are randomly distributed in space and have random directions. Such a material is usually produced by introducing magnetic impurities in a magnetic or nonmagnetic crystal. In the case under consideration, disorder in a magnetic system is created by introducing dislocations into the crystal. Dislocations of a certain type, e.g., ring dislocations, do not produce any singularity in the crystal lattice, thus allowing the construction of a space group G_p of the material. In this case, the system can be described by a multicomponent order parameter and possible phase transitions in it can be determined. These transitions are found to be of multiple-step type, i.e., a series of peaks is observed in the temperature dependence of magnetic susceptibility. Further, it is shown that if a dislocation is introduced by radiation, the large number of ensuing phases may also include the superconducting phase although the superconducting transition will occur at a low temperature. At low temperatures, the system is magnetically disordered but contains longwave fluctuations which result in the appearance of new vibrational modes in the spin wave spectrum. Thus, it is possible to construct a unified theory of spin glasses. The universal nature of this model is also confirmed by the fact that it describes the "impurity" model of spin glasses in the case when ring dislocations of small radius are introduced in the material.

The microscopic theory of high-temperature superconductivity in rare-earth compounds and ceramic systems is described in Chaps. 8 and 9. The main problem in the construction of the microscopic theory is to determine the spectrum of quasiparticles exchanged by pair-forming electrons. According to the BCS theory of low-temperature superconductivity, these quasi-particles are phonons. This theory is constructed by using weak electron-phonon interaction [1.22] and it can be shown that the critical temperature for a superconductor obtained with the help of the Bardeen, Cooper, Schrieffer (BCS) theory does not exceed 30-40 K. Generalization of the BCS theory to the case of strong electron-phonon interactions, $\lambda_{e-ph} > 1$ (strong bond theory) [1.23], does not lead to a significant increase in the critical temperature T_c since an increase in the electron-phonon coupling parameter is mainly possible through a decrease in the mean phonon frequency (or mean Debye energy $\langle\omega_D\rangle$), and this checks the increase in the critical temperature. Hence the phonon mechanism cannot be used for explaining high-temperature superconductivity. From our perspective, the high-temperature superconductivity of rare-earth compounds and ceramic systems is due to the enhancement of electron-phonon interaction by spin fluctuations of exchange type. In the class of compounds under consideration, the electron-phonon coupling is enhanced due to the spin fluctuations of the conduction electrons that are bound in pairs and constitute the superconducting Bose condensate. In the general case, the interaction of spin fluctuations with phonons is nonlinear. However, it is sufficient to use the linear approximation of this interaction to determine the spectrum of quasiparticles exchanged by electron pairs. Consequently, a linear relation is found to exist between the phonons and the spin fluctuations corresponding to the longitudinal spin mode whose spectrum has the form

$$\omega_{\parallel s\mathbf{k}} = J_0 s \sqrt{(k/k_c)^2 - 1} \quad . \tag{1.1.9}$$

Here, $J_0 = \int d\mathbf{x}\, J(\mathbf{x})$ and s is the electron spin.
Since there is no long-range magnetic order in a system of electron spins, this mode is found to be real only for $|k| > k_c = 2\pi/\langle r_c\rangle$. For $|k| < 2\pi/\langle r_c\rangle$, this is a diffusive mode. Resonance interaction of this mode with phonons due to a linear coupling leads to a renormalization of the phonon spectrum and to coupled spin-phonon oscillations. Figure 1.3 shows the spectrum of coupled oscillations. It can be seen that on account of the spin-phonon interaction, the phonon branch

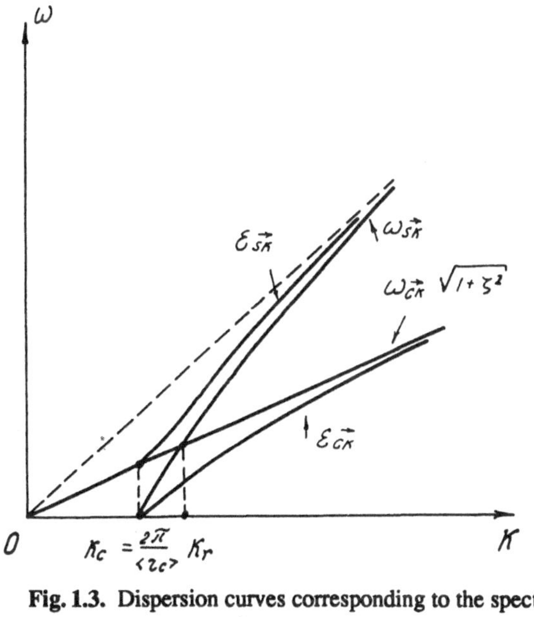

Fig. 1.3. Dispersion curves corresponding to the spectrum of coupled spin-phonon oscillations in high-temperature superconductors

reaches the asymptotic values in the region of large k, corresponding to the spin branch. This results in a sharp increase in the phonon frequency, i.e., to an effective increase in the region of energies near the Fermi surface, in which the electrons are attracted and form a Bose condensate. Consequently, the electron-phonon coupling becomes stronger. This enhancement in the electron-phonon coupling continues to increase with the spin-phonon interaction parameter ζ defined by the expression

$$\zeta = \frac{\sqrt{3}\, g\hbar k_c}{\sqrt{J_0 s M_i}} \quad . \tag{1.1.10}$$

Here $g = U/J_0$, U the electron-ion potential, and M_i the mass of an ion. Thus the pair-forming electrons in high-temperature superconductors exchange quasi-particles with coupled spin-phonon oscillations, i.e., quasi-phonons. In this approximation, the critical temperature is defined by the formula

$$T_c = \frac{2\gamma}{\pi} \langle\omega_D\rangle \exp\left\{-\frac{1}{K_{enh}(\lambda_{e-ph} - \mu^*)}\right\} \tag{1.1.11}$$

where μ^* is the effective Coulomb repulsion potential, and K_{enh} (ζ, λ_{e-ph}, μ^*, $\omega_s/\langle\omega_D\rangle$) is the enhancement coefficient and a monotonically increasing function of parameters ζ and $\omega_s/\langle\omega_D\rangle$ (ω_s is the spin frequency). For ζ, $\omega_s/\langle\omega_D\rangle \simeq 1$-$10$, the coefficient K_{enh} may attain values between 1.3 and 1.9, and this is sufficient for obtaining critical temperatures T_c of about 100 K under the condition that $\langle\omega_D\rangle \simeq 300$-$400$ K, $\lambda_{e-ph} \simeq 0.6$-$0.8$, and $\mu^* \simeq 0.1$-0.2. These values are the experimental data of [1.24]. It should be noted that as $\zeta \rightarrow 0$, $K \rightarrow 1$ and (1.1.11) is transformed into the familiar expression from the BCS theory describing low-temperature superconductors in the absence of strong spin fluctuations.

It should be recalled that (1.1.8) is the lower limit on the Neél temperature (T_N) when high-temperature superconductivity arises in an antiferromagnetic superconductor. Equations (1.1.9-11) were obtained for the case when the antiferromagnetic phase transition in high-temperature superconductors is absent. As in high-temperature superconductors where there is singlet pairing (**S** = 0), the effective electron-phonon interaction of paired electrons will be enhanced by exchange fluctuations of the electron spins, which for s-pairing tend to antiferromagnetic ordering, since at the phase transition point the coherence length ξ_ℓ and exchange correlation length $\langle r_c\rangle$ strongly increase. In order to calculate the superconducting transition temperature, it is necessary to use the linearized equation for the superconducting order parameter $|\Delta_0|$ (Chap. 8). Such linearization is possible if the pair correlation is dominated over the multiparticle correlation (Sects. 8.2,9.3). Nevertheless, in order for the high-temperature superconducting state to be produced it is necessary to drop the spin-phonon interaction (coupling) parameter restriction of $\zeta \lessdot 10$. Then from (1.1.10) the following inequality is obtained [cf. (9.5.27)]

$$J_0 s > \left(\frac{2\pi\sqrt{3}}{10}\right)^2 \left(\frac{\hbar^2 g^2}{\langle r_c \rangle^2 M}\right) , \tag{1.1.12}$$

where $(2\pi\sqrt{3}/10) \simeq 1$. Multiplying and dividing the right-hand side of (1.1.12) by a^2 (or $v_0^{2/3}$) where $\underline{a}$ is the lattice constant and taking into account that $g = U/J_0$, we have:

$$J_0 s > \left(\frac{\hbar^2}{v_0^{2/3} M^*}\right) \left(\frac{v_0^{1/3}}{\langle r_c \rangle}\right)^2 , \tag{1.1.13}$$

where $M^* = (J_0/U)^2 M$ is the effective ion mass and $(J_0/U) \simeq Z(\langle r_c \rangle / r_D)$, with Z the average ion valence and r_D the Debye screening radius. Because $\langle r_c \rangle$ and r_D are both approximately equal to $n_e^{-1/3}$ where n_e is the electron density, (J_0/U) is constant, about 10^{-1} to 10^{-2}, since Z is about 2 to 3 and $(\langle r_c \rangle / r_D)$ is 5 to 10 (Sect. 8.3). Now we consider the value

$$\tilde{\omega}_s = \hbar^2 / v_0^{2/3} M^* . \tag{1.1.14}$$

At $v_0^{1/3} \simeq 10^{-8}$ cm, $M^* \simeq \mu_e^*$, the effective electron mass, $\mu_e^* \simeq 5\mu_e$ (Sect. 8.3). The value of $\tilde{\omega}_s$ is in the range of 0.1-1.0 eV, i.e., $\tilde{\omega}_s$ is close to the order of ε_F. If in the collective electron system the antiferromagnetic phase transition were possible, $J_0 s \simeq T_N$ and (1.1.13) assumes the form [cf. (9.5.17)]

$$T_N > \varepsilon_F \left(\frac{v_0^{1/3}}{\langle r_c \rangle}\right)^2 .$$

The restriction for the value of exchange interaction between the magnetic moments follows from (1.1.8,13). If the antiferromagnetic transition point is close to T_c, then (1.1.8) is used and if the antiferromagnetic phase transition is absent, then (1.1.13), in order for the high-temperature superconducting state to be produced.

In order to construct a rigorous quantum theory of high-temperature superconductivity and to find means of increasing the critical temperature to values close to room temperature, we must consider nonlinear interaction of phonons with spin fluctuations. The general microscopic theory of high-temperature superconductivity is constructed in Chap. 9, and its connection with the phenomenological theory described in Chaps. 4 and 6 is established. The phenomenological theory is used in Chap. 8 for calculating the temperature dependence of the lower and upper critical fields H_{c1} and H_{c2} of new high-temperature superconductors, and for comparing the obtained results with the experimental data. The temperature dependence of electrical resistivity $\rho(T)$ in the normal phase and magnetoresistance as a function of the external magnetic field H are also calculated for the given class of materials. The high-temperature superconductivity mechanism developed by us is used to propose ways of increasing the critical temperature of new high-temperature superconductors.

The microscopic theory of high temperature superconductivity is used in Chap. 10 for studying the possibility of high-temperature superconductivity in polymer systems, including polyacetylene $(-CH-)_x$. Organic and polymer systems offer an alternative class of materials that can be used for synthesizing high-temperature superconductors. Indeed, a careful study of (1.1.10) reveals that the lower the mass of the ion and the smaller the exchange correlation length $\langle r_c \rangle$, the higher the spin-phonon interaction parameter. Polymer systems are based on the -CH- group formed by comparatively light atoms. The atomic radius of carbon is close to that of oxygen, which is among the smallest for the elements of the Periodic Table. If the carbon atoms participate in the formation of a covalent bond, it can be expected that the exchange forces between electrons participating in the formation of such a bond will be short-range forces. However, a particular feature of polymer systems is that they are generally semiconductors or insulators with a forbidden band of width $\Delta E \simeq$ 1-2 eV. The conductivity of polymer systems can be increased by the introduction of paramagnetic impurities, or on account of the topological defects, viz. solitons, which form energy levels near the bottom of the conduction band and increase the number of "free" electrons. If the concentration of such defects is quite high, the forbidden gap would be considerably suppressed and the polymer (e.g., polyacetylene) would possess metallic conductivity. Solitons may have a spin equal to that of electrons, and this opens the possibility of enhancement of electron-phonon interaction by spin fluctuations, and consequently of a phase transition to the superconducting state. The necessary and sufficient conditions for the emergence of a superconducting phase in polyacetylene are worked out and high-frequency properties of this material in the nonsuperconducting state are described.

2. Exchange Enhancement Effect During Phase Transitions in Magnetically Ordered Crystals

2.1 Phase Transition in a Ferromagnet with Quadrupole Interaction

According to Landau's theory of phase transitions, if the symmetry of a system "allows" a second-order phase transition, such a transition does take place in most cases near the Curie temperature T_C for ferromagnets and near the Néel temperature T_N for antiferromagnets. This is the self-consistent (mean) field theory and is based on the hypothesis that the order parameter experiences small fluctuations in the vicinity of the phase transition point.

However, recent experimental results [2.1] on magnetization and specific heat are not in agreement with the results obtained from the self-consistent field theory. This is possibly due to an increase in the order parameter fluctuations near the phase transition point, and to a strong influence of these fluctuations on the thermodynamics of the system.

It was mentioned in Chap. 1 that the thermodynamic, high-frequency, and kinetic properties of ferromagnets in the iron group (Co, Ni, Fe) were explained by Van Vleck [2.2] who introduced into the normal Heisenberg exchange Hamiltonian a fourth-order invariant called the biquadratic exchange. Van Vleck predicted that the biquadratic exchange plays a decisive role in the formation of magnetic anisotropy.

In this section, we shall show that a consideration of the biquadratic exchange radically changes the critical thermodynamics of an axial ferromagnet. As a result of the exchange enhancement effect considered by us in Chap. 1, the system experiences a first order phase transition instead of a second-order transition, accompanied by a jump in magnetization at the transition point even for an arbitrarily small quadrupole interaction constant ($K_0/J_0 \ll 1$, where J_0 and K_0 are zero-order Fourier components of the exchange and quadrupole exchange integrals).

Let us consider the Hamiltonian of a ferromagnet in the form

$$\mathbb{H} = -\frac{1}{2}\sum_{\mathbf{x},\mathbf{x}'} J(\mathbf{x}-\mathbf{x}')S_{\mathbf{x}}S_{\mathbf{x}'} - \frac{1}{2}\sum_{\mathbf{x},\mathbf{x}'} K(\mathbf{x}-\mathbf{x}')Q_{\mathbf{x}}Q_{\mathbf{x}'} . \quad (2.1.1)$$

Here Q_x is the quadrupole moment operator, $Q^z_{\mathbf{x}} = S^{z^2}_{\mathbf{x}} - (1/3)S(S+1)$; and S is the spin of an atom. The first term in (2.1.1) corresponds to the exchange interaction between the electron spins of the atoms, and the second to the biquadratic exchange.

Let us study the behavior of the system near the phase transition point in the paramagnetic region. In this case, the statistical sum can be presented as a functional integral over the fluctuating fields [2.3]:

$$Z = \prod_{\mathbf{x}} \int D\varphi_{1\mathbf{x}} D\varphi_{2\mathbf{x}} \exp[-\mathbb{H}(\varphi_1, \varphi_2)], \quad (2.1.2)$$

$$\mathbb{H} = \frac{1}{2}\sum_{\mathbf{x},\mathbf{x}'} [\beta J(\mathbf{x}-\mathbf{x}')]^{-1}\varphi_{1\mathbf{x}}\varphi_{1\mathbf{x}'} + \frac{1}{2}\sum_{\mathbf{x},\mathbf{x}'} [\beta K(\mathbf{x}-\mathbf{x}')]^{-1}$$

$$\cdot\varphi_{2\mathbf{x}}\varphi_{2\mathbf{x}'} - \sum_{\mathbf{x}} \ln \mathrm{Tr}\exp[f_1(\varphi_{1\mathbf{x}})S^z_{\mathbf{x}} + f_2(\varphi_{2\mathbf{x}})Q^z_{\mathbf{x}}], \quad (2.1.3)$$

and $\varphi_{1\mathbf{x}}$, $\varphi_{2\mathbf{x}}$ are the fluctuating fields corresponding to the order parameters $\langle S^z_{\mathbf{x}}\rangle$ and $\langle Q^z_{\mathbf{x}}\rangle$, respectively. Expanding the Hamiltonian (2.1.3) into a series in fluctuating fields $\varphi_{1\mathbf{x}}$ and $\varphi_{2\mathbf{x}}$ and confining ourselves to the third- and fourth-order terms, we obtain

$$\begin{aligned}\mathbb{H}(\varphi_1,\varphi_2) = &-\frac{1}{2}(\beta J_0)^{-1}\sum_{\mathbf{x}}[\varphi^2_{1\mathbf{x}} + \alpha_1(\nabla\varphi_{1\mathbf{x}})^2] + \frac{1}{2}\langle S^{z^2}\rangle_{\rho_0}\sum_{\mathbf{x}}\varphi^2_{1\mathbf{x}} - \\ &\frac{1}{2}(\beta K_0)^{-1}\sum_{\mathbf{x}}[\varphi_{2\mathbf{x}} + \alpha_2(\nabla\varphi_{2\mathbf{x}})^2] + \frac{1}{2}\langle Q^{z^2}\rangle_{\rho_0}\sum_{\mathbf{x}}\varphi^2_{2\mathbf{x}} + \\ &\frac{1}{2}\langle S^{z^2}Q^z\rangle_{\rho_0}\sum_{\mathbf{x}}\varphi^2_{1\mathbf{x}}\varphi_{2\mathbf{x}} + \frac{1}{6}\langle Q^{z^3}\rangle_{\rho_0}\sum_{\mathbf{x}}\varphi^3_{2\mathbf{x}} - \\ &\frac{1}{8}\left(\langle S^{z^2}\rangle^2_{\rho_0} - \frac{1}{3}\langle S^{z^4}\rangle_{\rho_0}\right)\sum_{\mathbf{x}}\varphi^4_{1\mathbf{x}} - \frac{1}{4}\left(\langle S^{z^2}\rangle_{\rho_0}\langle Q^{z^2}\rangle_{\rho_0} - \right. \\ &\left.\langle S^{z^2}Q^{z^2}\rangle_{\rho_0}\right)\sum_{\mathbf{x}}\varphi^2_{1\mathbf{x}}\varphi^2_{2\mathbf{x}} - \frac{1}{8}\left(\langle Q^{z^2}\rangle^2_{\rho_0} - \frac{1}{3}\langle Q^{z^4}\rangle_{\rho_0}\right)\sum_{\mathbf{x}}\varphi^4_{2\mathbf{x}} + \ldots\end{aligned} \quad (2.1.4)$$

In the expansion (2.1.4),

$$\alpha_1 = \frac{1}{3}S(S+1)(\beta J_0)^{-1}\sum_{\mathbf{x}}\beta J(\mathbf{x})\mathbf{x}^2, \quad \varphi_{1\mathbf{x}} = \beta J_0\langle S^z_{\mathbf{x}}\rangle_{\rho_0},$$

$$\alpha_2 = \frac{1}{45}S(S+1)(4S^2+4S-3)(\beta K_0)^{-1}\sum_{\mathbf{x}}\beta K(\mathbf{x})\mathbf{x}^2,$$

$$\varphi_{2\mathbf{x}} = \beta K_0\langle Q^z_{\mathbf{x}}\rangle_{\rho_0}, \quad \beta = 1/T, \quad T \to T^0_C + \delta T, \quad T^0_C = \frac{1}{3}J_0 S(S+1);$$

and T is the temperature in energy units ($T \to kT$, where k is the Boltzmann constant). In order to obtain the effective Hamiltonian of a ferromagnet near the phase transition point, we must calculate averages over the paramagnetic phase density matrix ρ_0. The computation of these mean values involves the evaluation of sums of the type

$$(2S+1)^{-1} \sum_{n=-S}^{S} n^{\ell}, \quad \ell = 2k, \quad k = 0,1,2, \ldots \tag{2.1.5}$$

Such sums can be easily obtained by introducing the function

$$f(y,S) = \sum_{n=-S}^{S} e^{ny} . \tag{2.1.6}$$

This gives

$$\sum_{n=-S}^{S} n^{2k} = \lim_{y\to 0} \frac{d^{2k}}{(dy)^{2k}} f(y,S) . \tag{2.1.7}$$

Hence, averaging over ρ_0 in (2.1.4), we obtain

$$\langle S^{z^2}\rangle_{\rho_0} = (1/3)S(S+1), \qquad \langle Q^{z^2}\rangle_{\rho_0} = (1/45)S(S+1)(4S^2+4S-3),$$

$$\langle S^{z^2}Q^z\rangle_{\rho_0} = \langle Q^{z^2}\rangle_{\rho_0} ,$$

$$\langle Q^{z^3}\rangle_{\rho_0} = (1/945)S(S+1)(16S^4 + 32S^3 - 56S^2 - 72S + 45) ,$$

$$\langle S^{z^2}\rangle^2_{\rho_0} - (1/3)\langle S^{z^4}\rangle_{\rho_0} = (1/45)S(S+1)(2S^2 + 2S + 1) ,$$

$$\langle Q^{z^2}\rangle^2_{\rho_0} - (1/3)\langle Q^{z^4}\rangle_{\rho_0} = (1/14175)S(S+1)(32S^6 + 96S^5 + 280S^4 + 416S^3 - 390S^2 - 582S + 315). \tag{2.1.8}$$

Then the effective Hamiltonian of a ferromagnet near the phase transition point as a functional of the fluctuating fields is

$$\mathbb{H}^{eff} = \frac{1}{2}\sum_{\mathbf{x}}[\tau\varphi^2_{1\mathbf{x}} + (\nabla\varphi^2_{1\mathbf{x}})^2] + \frac{1}{2}\sum_{\mathbf{x}}[a_0\varphi^2_{2\mathbf{x}} + \alpha(\nabla\varphi_{2\mathbf{x}})^2] + \Gamma_{10}\sum_{\mathbf{x}}\varphi^2_{1\mathbf{x}}\varphi_{2\mathbf{x}} + \Gamma_{20}\sum_{\mathbf{x}}\varphi^4_{1\mathbf{x}} ,$$

$$\tau = (1/3)S(S+1)(T-T_C^0)/T_C^0, \quad a_0 = (1/3)S(S+1)(K_0/J_0)^{-1},$$

$$\alpha = (1/15)(4S^2+4S-3)(K_0/J_0)^{-1}, \quad \mathbf{x}'_\nu = \mathbf{x}_\nu(3/S(S+1)\sum_{\mathbf{x}} g(\mathbf{x})x^2)^{\frac{1}{2}},$$

$$g(\mathbf{x}) = (1/K_0)K(\mathbf{x}) = (1/J_0)J(\mathbf{x}),$$

$$\Gamma_{10} = (-1/2)\langle S^{z^2}Q^z\rangle_{\rho_0} = (-1/90)S(S+1)(4S^2+4S-1),$$

$$\Gamma_{20} = (1/8)[\langle S^{z^2}\rangle^2_{\rho_0} - (1/3)\langle S^{z^4}\rangle_{\rho_0}] = (1/360)S(S+1)(2S^2+2S+1). \quad (2.1.9)$$

In this formula, we have omitted terms φ_2^3, $\varphi_1^2\varphi_2^2$, and φ_2^4, since the fluctuating field φ_2 is of the order K_0/J_0 and the quantity $K_0/J_0 \ll 1$ in ferromagnets of the iron group. Let us now study the behavior of a system near the phase transition point. For this purpose, we must determine its fluctuational free energy. We shall compute the free energy of a system by using the renormalization group technique [2.4-10]. Going over to Fourier transforms in (2.1.9), we obtain

$$\mathbb{H}^{eff} = \frac{1}{2V}\sum_{\mathbf{k}}(\tau+k^2)\varphi_{1\mathbf{k}}\varphi_{1-\mathbf{k}} + \frac{1}{2V}\sum_{\mathbf{k}}(a_0+\alpha k^2)\varphi_{2\mathbf{k}}\varphi_{2-\mathbf{k}} +$$

$$\frac{1}{V^2}\Gamma_{10}\sum_{\mathbf{k}_1+\mathbf{k}_2+\mathbf{k}_3=0}\varphi_{1\mathbf{k}_1}\varphi_{1\mathbf{k}_2}\varphi_{2\mathbf{k}_3} + \frac{1}{V^3}\Gamma_{20}$$

$$\sum_{\mathbf{k}_1+\mathbf{k}_2+\mathbf{k}_3+\mathbf{k}_4=0}\varphi_{1\mathbf{k}_1}\varphi_{1\mathbf{k}_2}\varphi_{1\mathbf{k}_3}\varphi_{1\mathbf{k}_4}. \quad (2.1.10)$$

The free energy of the system is defined as

$$F = -\ln\prod_{\mathbf{k}}\int D\varphi_{1\mathbf{k}}\, D\phi^{(0)}_{2\mathbf{k}}\, D\phi^{(1)}_{2\mathbf{k}}\, \exp[-\mathbb{H}^{eff}(\varphi_{1\mathbf{k}}, \varphi_{2\mathbf{k}})]. \quad (2.1.11)$$

The variable $\varphi^{(0)}_{2\mathbf{k}}$ in this relation corresponds to the fluctuations of the field φ_2, which are rigidly coupled to the fluctuations of the field φ_1 as a result of the exchange enhancement effect. The variable $\varphi_{2\mathbf{k}}$ corresponds to free thermal fluctuations of the field φ_2.

Next we introduce Green function of noninteracting fields:

$$G_1(\mathbf{k}) = \langle\varphi_{1\mathbf{k}}\varphi_{1-\mathbf{k}}\rangle_0 = 1/(\tau+k^2),$$

$$G_2(\mathbf{k}) = \langle\varphi_{2\mathbf{k}}\varphi_{2-\mathbf{k}}\rangle_0 = 1/(a_0+\alpha k^2). \quad (2.1.12)$$

The complete Green functions are defined as follows:

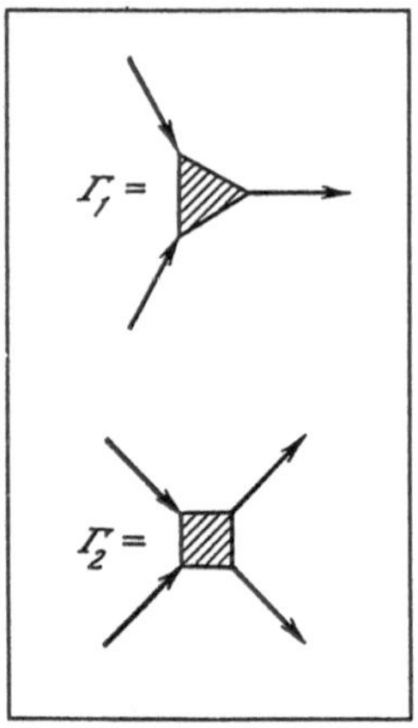

Fig. 2.1.

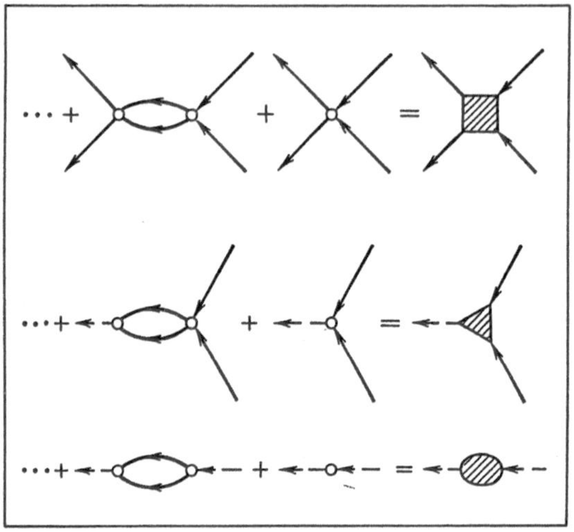

Fig. 2.2.

Fig. 2.1. Schematic representation of interaction amplitudes Γ_1 and Γ_2

Fig. 2.2. Quadrupole interaction parameter a and amplitudes Γ_1 and Γ_2

$$\tilde{G}_1(\mathbf{k}) = \langle \varphi_{1\mathbf{k}} \varphi_{1-\mathbf{k}} \rangle = [\tau + k^2 + \Sigma_1(\mathbf{k},\tau,a_0)]^{-1} ,$$

$$\tilde{G}_2(\mathbf{k}) = \langle \tilde{\varphi}_{2\mathbf{k}} \varphi_{2-\mathbf{k}} \rangle = [a_0 + \alpha k^2 + \Sigma_2(\mathbf{k},\tau,a_0)]^{-1} . \qquad (2.1.13)$$

Here, $\Sigma_1(\mathbf{k},\tau,a_0)$ and $\Sigma_2(\mathbf{k},\tau,a_0)$ represent the sums of all nonequivalent diagrams. Let us now juxtapose the third and fourth terms in (2.1.10) with the diagrams shown in Figs. 2.1 and 2.2. The hatched parts of the diagrams represent the complete vertices Γ_1 and Γ_2. The diagrams in Fig. 2.1 can be presented in the same form as the diagrams in Fig. 2.2. The functional integral with respect to the field $\varphi^{(1)}_{2\mathbf{k}}$ in formula (2.1.11) is Gaussian and can therefore be evaluated easily. Hence the points on the right-hand side of Fig. 2.2 correspond to Γ_{10} and $\tilde{\Gamma}_{20} = \Gamma_{20} - (\Gamma^2_{10}/a_0)$. Each line with momentum $\mathbf{q}_j$ is juxtaposed to a Green function $G_i(\mathbf{q}_j)$ from (2.1.12), and the summation is carried out over all intrinsic momenta $\mathbf{q}_j$: $G_i(\mathbf{q}_j) = \langle \varphi_{i\mathbf{q}_j} \varphi_{i-\mathbf{q}_j} \rangle$. All internal lines in Fig. 2.2 correspond to the Green function $G_1(\mathbf{q})$. The most divergent diagrams must be considered near the phase transition region, since the long-wave correlations of the fluctuating fields play an important role in this region. Hence we put the momenta of the outer lines equal to zero. Since the axial ferromagnets always have a dipole-dipole interaction, the summation of the diagrams will be carried out for the case when the space has a dimensionality $d = 4$. Summing the diagrams for $d = 4$

in the "parquette" (leading logarithmic) approximation, we obtain an equation for the vertices Γ_1 and Γ_2, which are identical to the Gell-Mann and Low-renormalization group equations [2.4]:

$$-a' = (1/2)\Gamma_1^2 \; , -\Gamma_1' = 3\Gamma_1\Gamma_2 \; , \quad -\Gamma_2' = 9\Gamma_2^2 \; , \qquad (2.1.14)$$

where

$$x = \frac{1}{(\varepsilon/2)(\Lambda^2)^{\varepsilon/2}} \left\{ \left(\frac{\Lambda^2}{\max[\lambda^2(\tau), s_0^2]} \right)^{\varepsilon/2} - 1 \right\} , \varepsilon \to 0 \; ,$$

Λ is the cut-off momentum and $\lambda^2(\tau)$ is the scaling parameter as a function of τ. Introducing new variables $y = \Gamma_1/\Gamma_2$ and $z = -\ln \Gamma_2$, we obtain

$$dy/dz = (2/3)y \; . \qquad (2.1.15)$$

Equation (2.1.15) has a single stationary point $y = 0$ which turns out to be unstable (Fig. 2.3). Since $y_0 < 0$, the system will move towards decreasing y until it reaches the stability boundary of the paramagnetic phase, when a first-order phase transition takes place in the system. Since $a_0 >> 1$, the system will take a very long time x before reaching the first-order phase transition point.

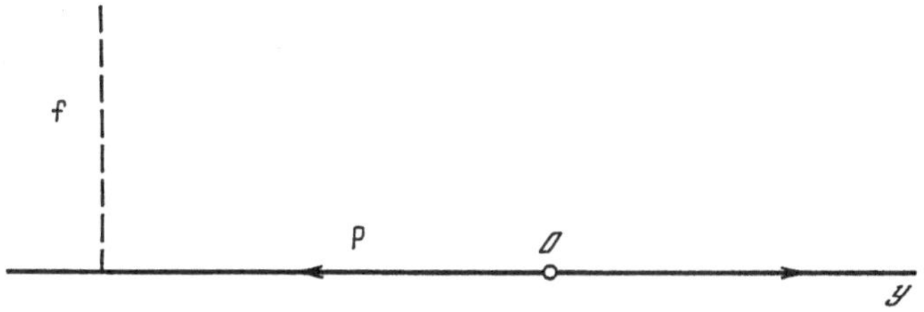

Fig. 2.3. Phase trajectory of a ferromagnet with quadrupole interaction. The *dashed line* corresponds to the stability boundary of the paramagnetic phase

Let us turn to the analysis of the free energy of the system. Integrating (2.1.14), we obtain the exact dependences $a(x)$, $\Gamma_1(x)$, $\Gamma_2(x)$:

$$a(x) = a_0 - (\Gamma_{10}^2/6\tilde{\Gamma}_{20})[(1 + 9\tilde{\Gamma}_{20}x)^{1/3} - 1] \; ,$$

$$\Gamma_1(x) = \Gamma_{10}/(1 + 9\tilde{\Gamma}_{20}x)^{1/3} \; ,$$

$$\Gamma_2(x) = \tilde{\Gamma}_{20}/(1 + 9\tilde{\Gamma}_{20}x) \; , \quad \tilde{\Gamma}_{20} \to \Gamma_{20} \; . \qquad (2.1.16)$$

The fluctuation free energy per lattice site of a crystal has the form

$$F = (1/2)\tau\varphi_{10}^2 + (1/2)a\varphi_{20}^2 + \Gamma_1\varphi_{10}^2\varphi_{20} + \Gamma_2\varphi_{10}^4. \tag{2.1.17}$$

Minimizing (2.1.17) with respect to φ_{20}, we obtain

$$\varphi_{20} = -(\Gamma_1/a)\varphi_{10}^2 . \tag{2.1.18}$$

Substituting this expression into (2.1.17), we obtain

$$F = (1/2)\tau\varphi_{10}^2 + \Gamma\varphi_{10}^4 , \tag{2.1.19}$$

where

$$\Gamma = \Gamma_2 - (1/2a)\Gamma_1^2 .$$

From (2.1.16) and (2.1.19), we can obtain the stability boundary of the free energy (2.1.19). This boundary is determined from the condition $\Gamma = 0$ and is found to be equal to

$$x^* = \frac{1}{9\Gamma_{20}}\left[\left(\frac{3a_0\Gamma_{20}}{2\Gamma_{10}^2}\right)^3 - 1\right]. \tag{2.1.20}$$

It follows from (2.1.20) that as the quadrupole exchange interaction constant $K_0 \to 0$, $a_0 \to \infty$, $x^* \to \infty$, and $y \to -\infty$. In other words, a second-order phase transition takes place in the system. We expand the quantity Γ into a series in powers of $x - x^*$, confine ourselves to the first term in the expansion, and substitute it into (2.1.19). This gives

$$F = (1/2)\tau\varphi_{10}^2 + b(x^*)(x^* - x)\varphi_{10}^4 ,$$

$$b(x^*) = 3\Gamma_2^2(x^*) + \frac{\Gamma_1^4(x^*)}{4a^2(x^*)} \to 3\Gamma_2^2(x^*) . \tag{2.1.21}$$

Changing variables from φ_{10} in (2.1.19) to the order parameter s_0, we obtain the following expression for the free energy of the system:

$$F = \frac{1}{2}\tau s_0^2 + b(x^*)s_0^4 \ln\frac{s_0^2}{\lambda^{2*}(\tau)} . \tag{2.1.22}$$

This expression can also be obtained by summation of the looped graphs in the condensed phase in all orders of the perturbation theory (Fig. 2.4). The jump in the order parameter s_0 at the phase transition point and the transition temperature τ_0 are determined from the conditions $F = 0$, $\partial F/\partial s_0^2 = 0$:

$$\tau_0 = 2b(x^*)\left(\frac{1 - \varepsilon/2}{1 + x^*\varepsilon/2}\right)^{2/\varepsilon}, \quad \varepsilon \to 0 , \tag{2.1.23}$$

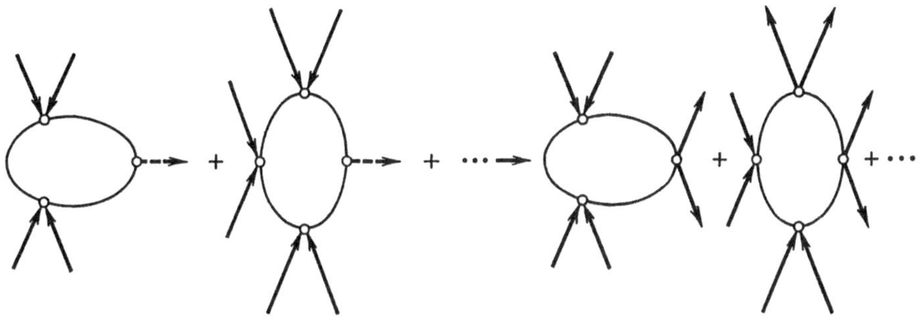

Fig. 2.4. Loop graphs whose summation allows the determination of the conditions for stabilization of the thermodynamic potential near the point of a first-order phase transition located close to a second-order one

$$s_0^2 = \left(\frac{1 - \varepsilon/2}{1 + x^* \varepsilon/2} \right)^{2/\varepsilon} , \quad \varepsilon \to 0 . \tag{2.1.24}$$

When $x^* \to \infty$, $\tau_0 \to 0$ and $s_0^2 \to 0$, which means that a second-order phase transition occurs in the system.

It follows from what has been stated above that the exchange enhancement effect radically changes the thermodynamics of an axial ferromagnet: instead of a second-order phase transition the system experiences a first-order phase transition close to a second-order one. This follows from the fact that the specific heat and magnetic susceptibility of the system remain finite at the phase transition point. Moreover, the exchange enhancement effect is associated with an increase in the fluctuations of one fluctuating subsystem on account of the other. As a result of this mechanism, a first-order phase transition is accompanied by the emergence of a quadrupole moment in addition to the magnetic moment.

First-order phase transitions close to the second were studied in detail in [2.11], where it was shown that the instability causing a first-order phase transition close to second-order one in a ferromagnet may be caused by exchange striction and attraction resulting from an exchange of acoustic phonons.

In most magnets, the symmetry allows the existence of invariants corresponding to the quadrupole exchange interaction [2.12]. This means that the mechanism of magnetization breakdown considered by us in the neighbourhood of the phase transition point is universal, since the exchange enhancement effect and the quadrupole interaction occur simultaneously. Moreover, substances in which a weak striction is observed upon a phase transition to the paramagnetic state, e.g., in rare-earth ferro- and antiferromagnets, the instability responsible for a first-order phase transition close to the second may be caused by quadrupole interaction.

2.2 Phase Transition in an Antiferromagnet in a Strong Magnetic Field

It is well known that as the strength of a magnetic field applied to an antiferromagnet is increased, a gradual "flipping" of the magnetic moments of the sublattices is observed (Fig. 2.5). When the external magnetic field H attains a value H_c [2.13], the magnetic moments of the sublattice are oriented along the magnetic field and the equilibrium angle between the direction of the external magnetic field and the magnetic moment of the sublattice tends to zero.

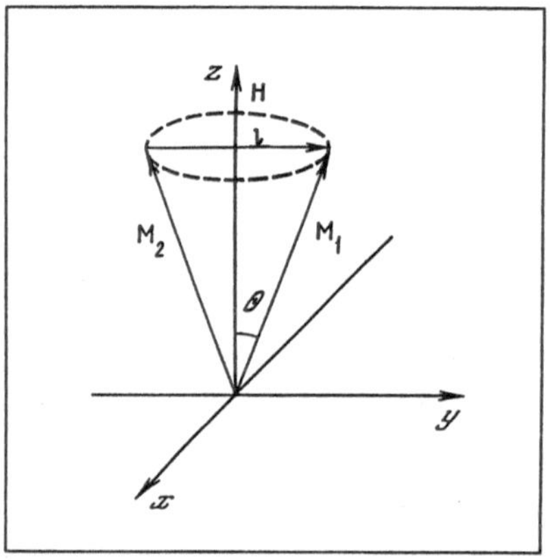

Fig. 2.5. Flipping of magnetic moments of sublattices in an antiferromagnet in a strong magnetic field

According to Landau's theory of phase transitions, a second-order phase transition takes place in this case. Recent experimental results obtained from an investigation of specific heat and magnetization in antiferromagnets in strong magnetic fields [2.14] are in disagreement with the results obtained from the self-consistent field theory [2.15]. This can be explained by the fact that the order parameter fluctuations, which are not taken into consideration in the self-consistent field theory, increase considerably near the phase transition point.

In this section, we shall use the renormalization group (RG) method of ε-expansion [2.6-8] to study the phase transition in an antiferromagnet in a strong magnetic field upon a flipping of the magnetic moments of sublattices at low temperatures $T/T_N \ll 1$. The exchange enhancement effect in an antiferromagnet upon a flipping of the magnetic moments of sublattices is accompanied by a bulk compression of the crystal due to a strong magnetoelastic bond in the spin-flip phase (bulk striction) [2.13,2.16]. In this case, the values of the uniform deformation u^0_{ik} are determined from the free-energy minimum $(\partial F/\partial u^0_{ik}) = 0$ and are functions of the equilibrium value of the antiferromagnetism vector: $u^0_{ik}(L_s)$, where $L_s = M_1 - M_2$, $\mathbf{M}_1$ and $\mathbf{M}_2$ being the magnetizations of the sublattices (Fig. 2.5). Hence the Hamiltonian of an antiferromagnet taking into account the interaction of the magnetic subsystem with a crystal lattice can be written in the form

$$\mathbb{H} = \mathbb{H}_m + \mathbb{H}_u + \mathbb{H}_{mu} ,$$

where

$$\mathbb{H} = \int dx \left[\delta(\mathbf{M}_1\mathbf{M}_2) - \alpha_{ik} \frac{\partial \mathbf{M}_1}{\partial x_i} \frac{\partial \mathbf{M}_2}{\partial x_k} + \frac{1}{2} \alpha'_{ik} \left(\frac{\partial \mathbf{M}_1}{\partial x_i} \frac{\partial \mathbf{M}_1}{\partial x_k} + \frac{\partial \mathbf{M}_2}{\partial x_i} \frac{\partial \mathbf{M}_2}{\partial x_k} \right) - \right.$$

$$\left. \frac{1}{2} \beta(M^2_{1z} + M^2_{2z}) + \beta' M_{1z} M_{2z} - (\mathbf{H}, \mathbf{M}_1 + \mathbf{M}_2) \right] ,$$

$$\mathbb{H}_u = \frac{1}{2} \int dx \ \lambda_{iklm} u_{ik} u_{lm} ,$$

$$\mathbb{H}_{mu} = \int dx \ \gamma^{ik}_{10} \ u_{ik}(\mathbf{M}_1\mathbf{M}_2) . \qquad (2.2.1)$$

The elastic deformation tensor can be presented in the following form:

$$u_{ik} = u^{(0)}_{ik} + \frac{1}{2} \left(\frac{\partial u_i}{\partial x_k} + \frac{\partial u_k}{\partial x_i} \right) . \qquad (2.2.2)$$

Here $u^{(0)}_{ik}$ is the longwave bulk deformation tensor whose uniform part is determined by the value of the external magnetic field (Fig. 2.6). Introducing new variables

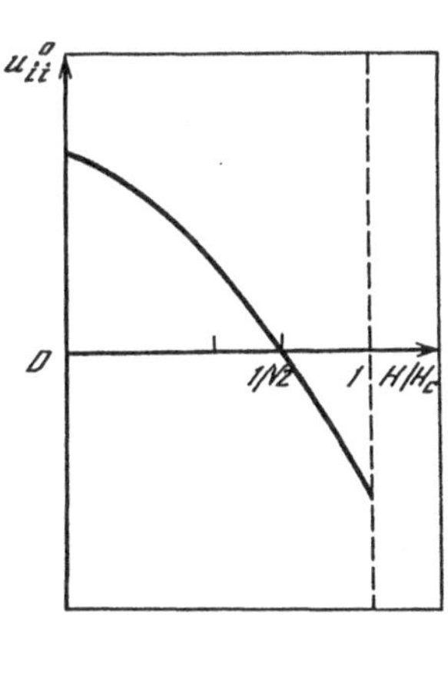

Fig. 2.6. Dependence of the diagonal components of the strain tensor on magnetic field in the spin-flip phase

$$\mathbf{m} = \frac{1}{\sqrt{2}\, M_0}(\mathbf{M}_1 + \mathbf{M}_2) , \quad \mathbf{l} = \frac{1}{\sqrt{2}\, M_0}(\mathbf{M}_1 - \mathbf{M}_2) \tag{2.2.3}$$

(M_0 is the rated magnetization of a lattice site), we can obtain the effective Hamiltonian in the vicinity of the phase transition point in a strong magnetic field, normalized to HM_0 [2.16]:

$$\mathbb{H}^{eff} = \int dx \left[\frac{1}{2}\tau \ell^2 + \frac{1}{2}\alpha_{ik}\left(\frac{\partial \boldsymbol{\ell}}{\partial x_i}\frac{\partial \boldsymbol{\ell}}{\partial x_k}\right) + \frac{1}{2}\lambda^{(0)}_{ik\ell m} u_{ik} u_{\ell m} + \Gamma^{ik}_{10} u_{ik}\ell^2 + \Gamma_{20}\ell^4 \right] , \tag{2.2.4}$$

where $\tau = 1 - H^0_c/H$; $H^0_c = (2\delta - \beta + \beta')M_0 + 2\gamma^{ik}_{10} u^{(00)}_{ik} M_0$; $u^{(00)}_{ik}$ are the uniform deformations that are independent of the quantity $|\ell|$, and

$$\lambda^{(0)}_{ik\ell m} = \lambda_{ik\ell m}/HM_0 \; ; \quad \Gamma^{ik}_{10} = -\gamma^{ik}_{10} M_0/H \; ; \quad \Gamma_{20} = 1/16 \; .$$

Changing to Fourier transforms

$$\mathbf{l}(\mathbf{x}) = \frac{1}{V}\sum_{\mathbf{q}} \mathbf{l}(\mathbf{q})e^{i\mathbf{qx}} , \quad u_{ik}(\mathbf{x}) = \frac{1}{V}\sum_{\mathbf{q}} u_{ik}(\mathbf{q})e^{i\mathbf{qx}} ,$$

we can write (2.2.4) in the form

$$\mathbb{H}^{eff} = \frac{1}{V}\sum_{\mathbf{q}} \frac{1}{2}(\tau + \tilde{\alpha}_{ik}q_i q_k)\mathbf{l}(\mathbf{q})\mathbf{l}(-\mathbf{q}) + \frac{1}{V}\sum_{\mathbf{q}} \frac{1}{2}\lambda^{(0)}_{ik\ell m} u_{ik}(\mathbf{q})u_{\ell m}(-\mathbf{q})$$
$$+ \frac{1}{V^2}\sum_{\mathbf{q}_1+\mathbf{q}_2+\mathbf{q}_3=0} \Gamma^{ik}_{10}\, u_{ik}(\mathbf{q}_1)\mathbf{l}(\mathbf{q}_2)\mathbf{l}(\mathbf{q}_3)$$
$$+ \frac{1}{V^3}\sum_{\mathbf{q}_1+\mathbf{q}_2+\mathbf{q}_3+\mathbf{q}_4=0} \Gamma_{20}\ell_p(\mathbf{q}_1)\ell_p(\mathbf{q}_2)\ell_s(\mathbf{q}_3)\ell_s(\mathbf{q}_4) \; . \tag{2.2.5}$$

In this case, the free energy of the system can be written in the form

$$-F = \ln \prod_{\mathbf{q}} \int D\ell_p(\mathbf{q})Du^{(0)}_{ik}(\mathbf{q})Du_\ell(\mathbf{q})\exp(-\mathbb{H}^{eff}) \; . \tag{2.2.6}$$

The number of components of the field $\ell(\mathbf{q})$ may differ according to the type and magnitude of the crystallographic anisotropy in the direction of the external magnetic field (n = 1,2). The Green function of noninteracting fields $\ell_p(\mathbf{q})$ for $\tau \ll q^2$ is

$$G_{ps}(\mathbf{q}) = \langle \ell_p(\mathbf{q})\ell_s(-\mathbf{q})\rangle_0 = \frac{\delta_{ps}}{\tilde{\alpha}_{ik}q_i q_k} = \delta_{ps}\frac{f(\Theta_k)}{q^2} , \tag{2.2.7}$$

where $f(\Theta_k) = 1/\tilde{\alpha}_{ik}q_{oi}q_{ok}$; $q_{oi} = q_i/q$ are the direction cosines of the angles between the direction of the wave vector and the crystallographic axes. Thus, the RG equations can be written in the form

$$\lambda'_{ik\ell m} = -(1/2)n\Gamma_1^{ik}\Gamma_1^{\ell m} \langle f^2\rangle_{\rho(\Theta_k)} , \quad \varepsilon \to +0 , \tag{2.2.8}$$

$$\Gamma_1^{ik'} = (\varepsilon/2)\Gamma_1^{ik} - n\Gamma_1^{ik}\Gamma_2 \langle f^2\rangle_{\rho(\Theta_k)} - 2\,\Gamma_1^{ik} \langle f^2\Gamma_2\rangle_{\rho(\Theta_k)} , \quad \varepsilon \to +0 , \tag{2.2.9}$$

$$\Gamma'_2 = \varepsilon\Gamma_2 - n\,\Gamma_2^2 \langle f^2\rangle_{\rho(\Theta_k)} - 4\Gamma_2 \langle f^2\Gamma_2\rangle_{\rho(\Theta_k)} - 4\,\langle f^2\Gamma_2^2\rangle_{\rho(\Theta_k)} , \quad \varepsilon \to +0 . \tag{2.2.10}$$

Here $\Gamma_1^{ik}(x,\Theta_k)$ is the amplitude of scattering of the order parameter fluctuations on longwave bulk deformations $u_{ik}^{(0)}(q)$, $\Gamma_2(x,\Theta_k)$ is an effective four-particle vertex whose initial value is equal to $\tilde{\Gamma}_{20} = \Gamma_{20} - b(\Theta_k)$, $b(\Theta_k)$ being the amplitude of order parameter scattering by thermal phonons,

$$x = \frac{2}{\varepsilon(\Lambda^2)^{\varepsilon/2}} \left\{ \left(\frac{\Lambda^2}{\max[\lambda^2(\tau),\ell_o^2]} \right)^{\varepsilon/2} - 1 \right\}, \quad \varepsilon \to +0 ; \tag{2.2.11}$$

and $\rho(\Theta_k)$ is the angular distribution function. The diagrams corresponding to λ_{iklm} , Γ_1^{ik} and Γ_2 are shown in Fig. 2.7.

Let us now consider the critical thermodynamics of an antiferromagnet in a strong constant magnetic field. We consider an antiferromagnet with isotropic magnetic and elastic properties. The effective Hamiltonian of the system in this case has the form

$$\mathbb{H}^{eff} = \frac{1}{V}\sum_{\mathbf{q}} \frac{1}{2}(\tau + q^2)\ell_p(\mathbf{q})\ell_p(-\mathbf{q}) + \frac{1}{V}\sum_{\mathbf{q}} \frac{1}{2}[\lambda_0 u_{ii}(\mathbf{q})u_{ii}(-\mathbf{q}) + 2\varkappa u_{ik}(\mathbf{q})u_{ik}(-\mathbf{q})] + \frac{1}{V^2}\Gamma_{10} \sum_{\mathbf{q}_1+\mathbf{q}_2+\mathbf{q}_3=0} u_{ii}(\mathbf{q}_1)\ell_p(\mathbf{q}_2)\ell_p(\mathbf{q}_3) + \frac{1}{V^3}\Gamma_{20} \sum_{\mathbf{q}_1+\mathbf{q}_2+\mathbf{q}_3+\mathbf{q}_4=0} \ell_p(\mathbf{q}_1)\ell_p(\mathbf{q}_2)\ell_s(\mathbf{q}_3)\ell_s(\mathbf{q}_4) , \tag{2.2.12}$$

where λ_0 and $\varkappa$ are relative moduli of uniform compression and shear; and

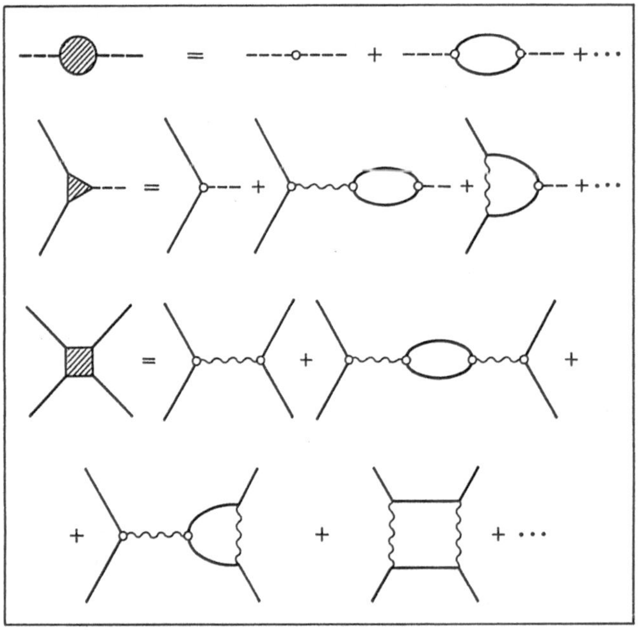

Fig. 2.7. Diagrams corresponding to the bulk modulus and to the interaction amplitudes Γ_1 and Γ_2 in an antiferromagnet in a strong magnetic field

$$H_c^0 = (2\delta - \beta + \beta')M_0 + 2\delta u_{ii}^{(00)} M_0 \ ; \ \Gamma_{10} = -\delta M_0/H \ , \ \Gamma_{20} = 1/16 .$$

The RG equations then assume the form

$$\lambda' = -(3/2)n\Gamma_1^2 \ , \quad \varepsilon \to +0 \ , \tag{2.2.13}$$

$$\Gamma_1' = (\varepsilon/2)\Gamma_1 - (n+2)\Gamma_1\Gamma_2 \ , \ \varepsilon \to +0 \ , \tag{2.2.14}$$

$$\Gamma_2' = \varepsilon\Gamma_2 - (n+8)\Gamma_2^2 \ , \ \varepsilon \to +0 \ , \tag{2.2.15}$$

and the variable x is defined by (2.2.11). The equations for the amplitudes Γ_1 and Γ_2 have two unstable stationary points $A_1(0,0)$ and $A_2[0, \varepsilon/(n+8)]$ (Fig. 2.8), and the point A_1 is absolutely unstable in view of the fact that the point $\Gamma_2 = \varepsilon/(n+8)$ is a stationary point of the Hamiltonian (2.2.10). This means that the system can undergo only a first-order phase transition.

Let us now analyze the fluctuation free energy of the system. Integrating (2.2.13-15), we obtain the exact dependences $\lambda(x)$, $\Gamma_1(x)$, $\Gamma_2(x)$:

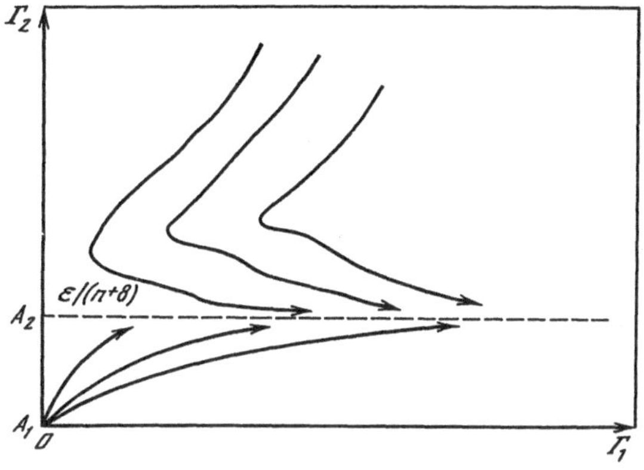

Fig. 2.8. Phase trajectories of an antiferromagnet in a strong magnetic field during a spin-flip transition

$$\lambda(x) = \lambda_0 - \frac{3n\,\Gamma_{10}^2}{2(4-n)\tilde{\Gamma}_{20}\;\varepsilon^{(4-n)/(n+8)}} \cdot \left\{[\varepsilon - (n+8)\tilde{\Gamma}_{20}\,(1 - e^{\varepsilon x})] - \varepsilon\right\}^{(4-n)/(n+8)}, \quad \varepsilon \to +0, \quad (2.2.16)$$

$$\Gamma_1(x) = \Gamma_{10}\exp\left[\frac{4-n}{n+8}\,\varepsilon x\right]\left[\frac{\varepsilon}{\varepsilon e^{-\varepsilon x} + (n+8)\tilde{\Gamma}_{20}(1 - e^{-\varepsilon x})}\right]^{\frac{n+2}{n+8}}, \quad \varepsilon \to +0 \quad (2.2.17)$$

$$\Gamma_2(x) = \varepsilon\tilde{\Gamma}_{20}/[\varepsilon e^{-\varepsilon x} + (n+8)\tilde{\Gamma}_{20}(1 - e^{-\varepsilon x})], \quad \varepsilon \to +0,$$

$$\tilde{\Gamma}_{20} = \Gamma_{20} - 3\Gamma_{10}^2/(\lambda_0 + 2\varkappa). \quad (2.2.18)$$

Evaluating the functional integral (2.2.6) with respect to the slow variables ℓ_0 and $u_{ii}^{(0)}$ by the method of steepest descent, we can obtain the fluctuation free energy of the system per unit volume:

$$F = (1/2)\tau^{\gamma(\varepsilon)}\ell_0^2 + (1/2)\,\lambda u_{ii}^{(0)^2} + \Gamma_1 u_{ii}^{(0)}\,\ell_0^2 + \Gamma_2\ell_0^4\,. \quad (2.2.19)$$

Here $\gamma(\varepsilon) = 1 + \frac{n+2}{2(n+8)}\varepsilon + \dots$ is the critical index for the magnetic field, which is identical to the temperature index [2.17]. Minimizing (2.2.19) with respect to $u_{ii}^{(0)}$, we obtain

$$F = (1/2)\tau^{\gamma(\varepsilon)}\ell_0^2 + \Gamma\ell_0^4\,, \quad (2.2.20)$$

where $\Gamma = \Gamma_2 - (3/2)\Gamma_1^2/\lambda$.

Thus, the existence of an elastic subsystem results in the renormalization of the magnetic free energy parameters. Formulas

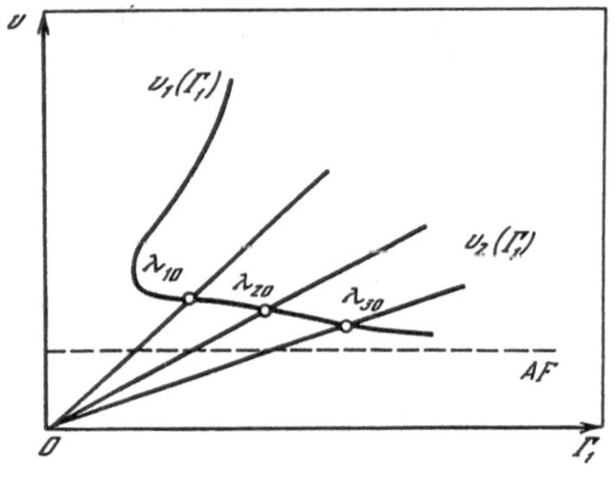

Fig. 2.9. Phase diagram of an antiferromagnet in a strong magnetic field constructed in variables v, $\Gamma_1[v_1(\Gamma_1) = \Gamma_2/\Gamma_1; v_2(\Gamma_1) = \Gamma_1/2\lambda]$. The points of intersection of trajectories $v_1(\Gamma_1)$ and $v_2(\Gamma_1)$ correspond to a first-order phase transition close to second-order ones in the spin-flip phase (AF)

(2.2.16-18,20) can be used to determine the stability boundary of the free energy. The stability boundary is determined in the same way as in Sect. 2.1 from the condition $\Gamma = 0$. It can also be obtained from the phase diagram of the system when $v_1(\Gamma_1) = v_2(\Gamma_1)$ (Fig. 2.9). With the help of either of these methods, we obtain the following expression:

$$x^* = \frac{1}{\varepsilon} \ln \left\{ 1 + \frac{\varepsilon}{(n+8)\tilde{\Gamma}_{20}} \left\{ \left[\frac{(4-n)\tilde{\Gamma}_{20}[\lambda_0 + 3n\Gamma_{10}^2/(2(4-n)\tilde{\Gamma}_{20})]}{6\Gamma_{10}^2} \right]^{\frac{n+8}{4-n}} - 1 \right\} \right\}, \quad \varepsilon \to +0. \tag{2.2.21}$$

For $x < x^*$, the free energy is positive definite and has a minimum for $\ell_0^2 = 0$. For $x > x^*$, there exists a point τ_0 in the interval $0 < \tau < \tau^*$, at which the free energy has its minimum value for $\ell_0^2 \neq 0$. This corresponds to a first-order phase transition. The quantities ℓ_0^2 and τ_0 are given by the expressions

$$\ell_0^2 = [(1 - \varepsilon/2)/(1 + x^*\varepsilon/2)]^{2/\varepsilon}, \quad \varepsilon \to +0, \tag{2.2.22}$$

$$\tau_0 = [2b(x^*)]^{1/\gamma(\varepsilon)} \left(\frac{1 - \varepsilon/2}{1 + x^*\varepsilon/2} \right)^{2(1-\varepsilon/2)/\varepsilon\gamma(\varepsilon)}, \quad \varepsilon \to +0. \tag{2.2.23}$$

It follows from (2.2.21-23) that as $\lambda_0 \to \infty$, $x^* \to \infty$, $\tau_0 \to 0$, and $\ell_0^2 \to 0$, i.e., the system undergoes a second-order phase transition. In this connection, the following circumstance is worth noting. The free energy can be presented in the form

$$F = F_{reg} + F_{sing}, \tag{2.2.24}$$

where F_{reg} is an analytical function of τ and $F_{sing} = B\,\tau^{2-\alpha}$. The index α is the magnetic susceptibility index and not the specific heat index as in the case of a temperature-induced phase transition. Its value is determined by using the standard procedure of Wilson's ε-expansion [2.8]:

$$\alpha = \frac{4 - n}{2(n + 8)}\varepsilon - \ldots \tag{2.2.25}$$

As $\varepsilon \to 0$, the quantity χ^{zz} is found to be proportional to

$$\int_0^x dy\, P_\chi(y) \; . \tag{2.2.26}$$

The vertex $P_\chi(x)$ is determined from Ward's identity

$$\frac{d\tilde{G}^{-1}(0,\tau)}{d\tau^{(0)}} = P_\chi(x) \; , \tag{2.2.27}$$

[here, $\tilde{G}(q,\tau)$ is the complete Green function of the system, and $\tau^{(0)}$ is the initial value of τ], and satisfies the equation

$$-P'_\chi = (n + 2)P_\chi\Gamma_2 . \tag{2.2.28}$$

The amplitude Γ_2 satisfies (2.2.15) for $\varepsilon \to 0$. Integrating (2.2.28) and substituting its solution into (2.2.26), we obtain

$$\chi^{zz} \sim x^{(4-n)/(n+8)} \; , \tag{2.2.29}$$

which is identical to the characteristic specific heat dependence for a temperature-induced phase transition. The remaining critical indices γ, ν, and β are the same as the temperature indices and the specific heat of the system in the vicinity of the phase transition point is determined by Bloch spin waves while its dependences on temperature and magnetic field for $T/T_N \ll 1$ can be described quite well with the help of the self-consistent field theory.

However, λ_0 is finite in real systems and we obtain a first-order phase transition close to the second. The specific heat and the magnetic susceptibility of the system remain finite at the phase transition point, which corresponds to a first-order phase transition close to the second. It follows from (2.2.16-23) that the bulk modulus suffers a discontinuity at the phase transition point [2.18]. The dependence of λ on the magnetic field is shown in fig. 2.10. From the above analysis, it can be concluded that as the dimensionality d of the space tends to four from above, the discontinuity in

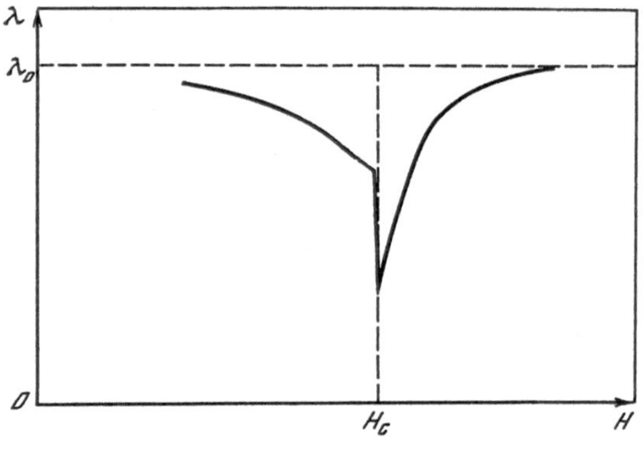

Fig. 2.10. Dependence of the bulk modulus of an antiferromagnet on the magnitude of the external magnetic field during a spin-flip transition

the order parameter is found to be exponentially small, since $\lambda_0 >> 1$ for real crystals. As ε increases, the discontinuity in the order parameter ℓ_0 increases, thus indicating an enhancement in the effect of fluctuations on the phase transition with decreasing dimensionality of the space. The discontinuity in the bulk modulus λ also increases. In spite of all this, the nature of the phase transition remains the same, since the correlation length at the transition point decreases and the specific heat and magnetic susceptibility remain finite at the phase transition point although the jumps in their values increase.

Thus, we have studied the thermodynamics of an antiferromagnet in the vicinity of the critical point for a field-induced phase transition in the region of highly developed fluctuations. Let us now consider the thermodynamic properties of an antiferromagnet in the paramagnetic phase away from the phase transition point in the region of applicability of the self-consistent field theory. The expression for the free energy of an antiferromagnet in the paramagnetic phase can be obtained with the help of the graph technique developed in [2.13,2.16], and has the form

$$-\beta F = \beta V_0 \langle S^z\rangle^2 + 2\ln\frac{\sinh(S+\frac{1}{2})y}{\sinh(\frac{1}{2})y} - \frac{1}{2N}\sum_{j=1}^{2}\sum_{\mathbf{q}}\ln\left\{1 - \beta J_{\mathbf{q}} b' \cdot \frac{\beta J_{\mathbf{q}} b(1+\Delta J_{\mathbf{q}}/J_{\mathbf{q}}) + (-1)^j y(1-\Delta J_{\mathbf{q}}/J_{\mathbf{q}})}{y + (-1)^j \beta J_{\mathbf{q}} b}\right\}$$

$$-\frac{1}{N}\sum_{j=1}^{2}\sum_{\mathbf{q},n}\ln\left(\frac{\beta\omega_{j\mathbf{q}} - i\beta\omega_n}{y - i\beta\omega_n}\right), \quad \beta\omega_{j\mathbf{q}} = y + (-1)^{j+1}\beta J_{\mathbf{q}} b, \tag{2.2.30}$$

where $y = \beta(\mu H - V\langle S^z\rangle)$; $J_{\mathbf{q}} = V_{\mathbf{q}} - J_{\mathbf{q}}$ is the exchange interaction anisotropy, $V_{\mathbf{q}}$ and $J_{\mathbf{q}}$ are the Fourier components of the exchange integrals, b(y) is the Brillouin function, b' = db/dy, and $\omega_n=2\pi n/\beta$ are frequencies in the finite temperature technique. At low temperatures, (2.2.30) is simplified and assumes the form

$$-\beta F = -\beta V_0 \langle S^z \rangle (2S - \langle S^z \rangle) + 2\beta\mu HS - (1/N) \sum_{j=1}^{2} \sum_{\mathbf{q},n} \ln[(\beta\omega_{j\mathbf{q}} - i\beta\omega_n)/(y - i\beta\omega_n)] \ . \qquad (2.2.31)$$

This relation can be used to determine the thermodynamic parameters M^z, χ^{zz}, and C_H of the system as functions of temperature and magnetic field, which are defined by

$$M^z = \frac{2\mu}{v_0} \langle S^z \rangle = \frac{\mu}{v_0} \frac{\partial}{\partial y}(-\beta F) \ , \qquad (2.2.32)$$

$$\chi^{zz} = \partial M^z / \partial H \ , \qquad (2.2.33)$$

$$C_H = \beta^2 \frac{\partial^2}{\partial \beta^2}(-\beta F) \ . \qquad (2.2.34)$$

Evaluating the derivatives (2.2.32) for $\tau \gtrsim 10^{-3}$, $T \gtrsim 1$ K, and $\mu(H - H_c)/T \gtrsim 10^{-1} - 10^{-2}$, we obtain

$$M^z = M_s - \frac{\mu}{4\pi^2 v_0} \left(\frac{v_0}{R_0^3}\right) \left(\frac{S+1}{3}\right)^{3/2} \left(\frac{T}{T_N}\right)^{3/2} Z_1(\xi) \ , \qquad (2.2.35)$$

$$\chi^{zz} = -\frac{\mu^2}{4\pi^2 v_0 J_0 S} \left(\frac{v_0}{R_0^3}\right) \left(\frac{S+1}{3}\right)^{1/2} \left(\frac{T}{T_N}\right)^{1/2} Z_1'(\xi) \ , \qquad (2.2.36)$$

$$C_H = \frac{1}{4\pi^2} \left(\frac{v_0}{R_0^3}\right) \left(\frac{S+1}{3}\right)^{3/2} \left(\frac{T}{T_N}\right)^{3/2} [\tfrac{5}{2} Z_2(\xi) - \xi Z_2'(\xi)] \ ,$$

$$M_s \equiv 2M_0 \ , \quad R_0 = \left[\frac{1}{3} \frac{\sum_{\mathbf{x}} x^2 J(\mathbf{x})}{\sum_{\mathbf{x}} J(\mathbf{x})}\right]^{1/2} , \qquad (2.2.37)$$

where the functions $Z_1(\xi)$ and $Z_2(\xi)$ are defined as follows:

$$Z_1(\xi) = \int_0^\infty \frac{x^{\frac{1}{2}}\, dx}{e^{x+\xi} - 1} \ , \qquad Z_2(\xi) = \int_0^\infty \frac{(x + \xi) x^{\frac{1}{2}}\, dx}{e^{x+\xi} - 1} \ ,$$

$$\xi = \mu(H - H_c)/T ,$$

$$Z_1(0) = (1/2)\pi^{\frac{1}{2}}\zeta(3/2) , \quad Z_1'(0) = -(1/2)\pi^{\frac{1}{2}}\zeta(1/2) ,$$

$$Z_2(0) = (3/4)\pi^{\frac{1}{2}}\zeta(5/2) , \quad Z_2'(0) = -(1/4)\pi^{\frac{1}{2}}\zeta(3/2) .$$

For $\mu(H - H_c)/T \gg 1$, we obtain

$$M^z = M_s - \frac{\mu}{8\pi^{3/2}v_0}\left(\frac{v_0}{R_0^3}\right)\left(\frac{S+1}{3}\right)^{3/2}\left(\frac{T}{T_N}\right)^{3/2} \exp[-\frac{\mu(H-H_c)}{T}], \quad (2.2.38)$$

$$\chi^{zz} = \frac{\mu^2}{8\pi^{3/2}v_0 J_0 S}\left(\frac{v_0}{R_0^3}\right)\left(\frac{S+1}{3}\right)^{1/2}\left(\frac{T}{T_N}\right)^{1/2} \exp[-\frac{\mu(H-H_c)}{T}], \quad (2.2.39)$$

$$C_H = \frac{1}{8\pi^{3/2}}\left(\frac{v_0}{R_0^3}\right)\left(\frac{S+1}{3}\right)^{3/2}\left[\frac{\mu(H-H_c)}{T}\right]^2\left(\frac{T}{T_N}\right)^{3/2} \cdot \exp\left[-\frac{\mu(H-H_c)}{T}\right] . \quad (2.2.40)$$

In the vicinity of the phase transition point, the magnetization of the system may be determined exactly, and is found to have the following temperature dependence:

$$M^z = M_s - \frac{\mu}{8\pi^{3/2}v_0}\left(\frac{v_0}{R_0^3}\right)\left(\frac{S+1}{3}\right)^{3/2}\zeta(\tfrac{3}{2})\left(\frac{T}{T_N}\right)^{3/2} , \quad (2.2.41)$$

which is the same as the normal temperature dependence of a ferromagnet in a weak magnetic field. It is not possible to write the corresponding asymptotic forms for the quantities χ^{zz} and C_H in the vicinity of the phase transition point in the region of applicability of the self-consistent field theory, since the functions $Z_1(\xi)$ and $Z_2(\xi)$ are not analytic at the point $\xi = 0$. In the scaling region, the specific heat C_H and the susceptibility χ^{zz} are porportional:

$$C_H \sim \tau^{-\alpha}\left(\frac{T}{T_N}\right)^2 , \quad \chi^{zz} \sim \tau^{-\alpha} , \quad (2.2.42)$$

where $\alpha = \frac{4-n}{2(n+8)}\varepsilon - \ldots$, and for $\varepsilon \to 0$

$$C_H \sim x^{[(4-n)/(n+8)]}\left(\frac{T}{T_N}\right)^2 , \quad \chi^{zz} \sim x^{[(4-n)/(n+8)]} . \quad (2.2.43)$$

The dependences χ^{zz} and $C_H(\xi)$ are shown in Fig. 2.11.

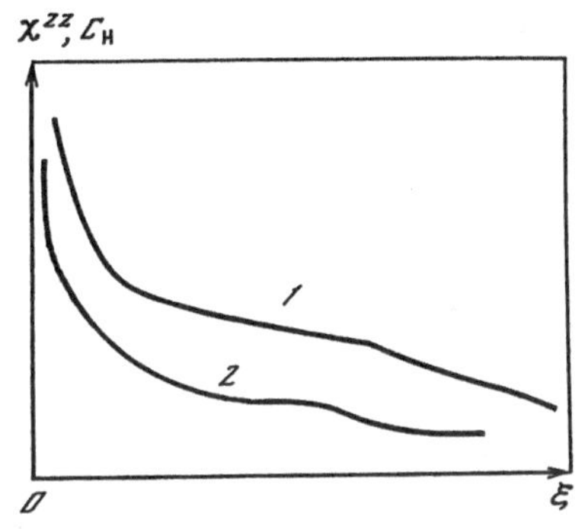

Fig. 2.11. Dependence of the magnetic susceptibility χ^{zz} (*curve 1*) and heat capacity C_H (*curve 2*) on the quantity $\xi = \mu(H - H_c)/T$ in the paramagnetic region for $H > H_c$

2.3 Fluctuations and Spin Waves in Ferroelectro-antiferromagnets in a Strong Magnetic Field

The system considered in this section differs from the normal antiferromagnet considered in Sect. 2.2 in that it contains a spontaneous electric polarization vector. The "flipping" of the magnetic moments of antiferromagnetic sublattices is accompanied by a change in the spontaneous polarization due to the interaction of the magnetic moment with the spontaneous electric polarization, which is enhanced by the exchange interaction parameter (Fig. 2.12). In [2.19], where the exchange enhancement effect is not taken into consideration, it is stated that for certain values of the dielectric permeability, the phase transition in a ferroelectro-antiferromagnet upon a flipping of the magnetic moments of the sublattices in a strong magnetic field may turn out to be a second-order phase transition. We shall now show that on account of the exchange enhancement effect, the phase transition in a ferroelectro-antiferromagnet in a strong magnetic field is a first-order transition close to the second for a nonzero magnetoelectric coupling constant and an arbitrary dielectric constant.

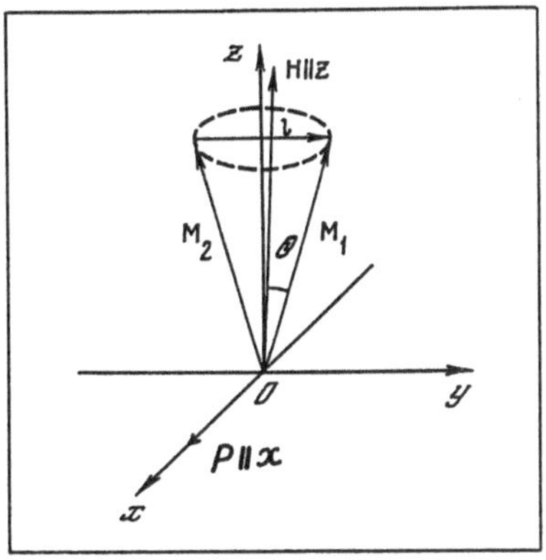

Fig. 2.12. Flipping of magnetic moments of sublattices in a segnetoantiferromagnet in a strong magnetic field (*P* is the spontaneous polarization vector)

The Hamiltonian of a ferroelectro-antiferromagnet in a magnetic field can be written in the form

$$\mathbb{H} = \int d\mathbf{x} \left\{ \delta(\mathbf{M}_1\mathbf{M}_2) + \alpha \frac{\partial \mathbf{M}_1}{\partial x_i} \frac{\partial \mathbf{M}_2}{\partial x_i} + \frac{1}{2} \alpha' \left[\left(\frac{\partial \mathbf{M}_1}{\partial x_i} \right)^2 + \left(\frac{\partial \mathbf{M}_2}{\partial x_i} \right)^2 \right] \right.$$

$$+ \beta M_{1z} M_{2z} + \beta'(M_{1z}^2 + M_{2z}^2) - (\mathbf{H}, \mathbf{M}_1 + \mathbf{M}_2) + \frac{1}{2} \varkappa P^2$$

$$\left. + \frac{1}{2} s \left(\frac{\partial \mathbf{P}}{\partial x_i} \right)^2 - \frac{1}{4} \sigma P^4 + \gamma P^2 (\mathbf{M}_1 \mathbf{M}_2) \right\} . \qquad (2.3.1)$$

As in Sect. 2.2, we introduce the variables

$$\mathbf{m} = (1/\sqrt{2}\, \mathbf{M}_0)(\mathbf{M}_1 + \mathbf{M}_2) , \qquad \mathrm{I} = (1/\sqrt{2}\, \mathbf{M}_0)(\mathbf{M}_1 - \mathbf{M}_2) ,$$

$$\mathbf{P} = \mathbf{n} P_0 (1 + p) ,$$

where $\mathbf{P}$ is the change in the polarization of the system due to the presence of magnetic subsystem and the external magnetic field. This gives the following expression for the effective Hamiltonian of a ferroelectro-antiferromagnet in the vicinity of the phase transition point [2.20]:

$$\mathbb{H}^{eff} = \frac{1}{2V} \sum_{\mathbf{q}} (\tau^{(0)} + q^2) \ell_i(\mathbf{q}) \ell_i(-\mathbf{q}) + \frac{1}{2V} \sum_{\mathbf{q}} \omega_0 p(\mathbf{q}) p(-\mathbf{q})$$

$$+ \frac{1}{V^2} \sum_{\mathbf{q}_1+\mathbf{q}_2+\mathbf{q}_3=0} \Gamma_{10} \ell_i(\mathbf{q}_1) \ell_i(\mathbf{q}_2) p(\mathbf{q}_3)$$

$$+ \frac{1}{V^3} \sum_{\mathbf{q}_1+\mathbf{q}_2+\mathbf{q}_3+\mathbf{q}_4=0} \Gamma_{20} \ell_i(\mathbf{q}_1) \ell_i(\mathbf{q}_2) \ell_k(\mathbf{q}_3) \ell_k(\mathbf{q}_4) , \qquad (2.3.2)$$

where

$$\tau^{(0)} = 1 - H_c^0/H , \quad \omega_0 = -2\sigma P_0^4/HM_0 ,$$

$$\Gamma_{10} = -2\gamma P_0^2 M_0/H , \quad \Gamma_{20} = 1/16 , \quad H_c^0 = (2\delta + \beta + \beta' + 2\gamma P_0^2) M_0 .$$

The thermodynamic potential of the system is defined by an expression similar to (2.1.11):

$$F = -\ln \prod_{\mathbf{q}} \int D\ell_i(\mathbf{q}) Dp^{(0)}(\mathbf{q}) Dp^{(1)}(\mathbf{q}) \exp(-\mathbb{H}^{eff}) . \qquad (2.3.3)$$

Evaluating the functional integrals with respect to the fluctuating fields $\ell_i(q)$, $p^{(0)}(q)$, and $p^{(1)}(q)$, we obtain

$$F = (1/2)\tau \ell_0^2 + (1/2)\, \omega p_0^2 + \Gamma_1 \ell_0^2 p_0 + \Gamma_2 \ell_0^4 , \qquad (2.3.4)$$

where p_0 and ℓ_0 are the equilibrium values of polarization and the magnitude of the antiferromagnetism vector. The quantities ω, Γ_1, and Γ_2 satisfy the RG equations

$$\omega' = -(1/2)n\Gamma_1^2 , \quad \varepsilon \to +0 , \tag{2.3.5}$$

$$\Gamma_1' = (\varepsilon/2)\Gamma_1 - (n+2)\Gamma_1\Gamma_2 , \quad \varepsilon \to +0 , \tag{2.3.6}$$

$$\Gamma_2' = \varepsilon\Gamma_2 - (n+8)\Gamma_2^2 , \quad \varepsilon \to +0 \tag{2.3.7}$$

with initial conditions $\omega(0) = \omega_0$, $\Gamma_2(0) = \Gamma_{20} - \Gamma_{10}^2/\omega_0$, and $\Gamma_1(0) = \Gamma_{10}$. The variable in the ε-expansion method is defined by (2.2.11); n is the dimension of the order parameter. The solutions of (2.3.5-7) have the form

$$\omega(x) = \omega_0 - \frac{n\Gamma_{10}^2}{2(4-n)\tilde{\Gamma}_{20}\,\varepsilon^{[(4-n)/(n+8)]}}\left\{[\varepsilon - (n+8)\tilde{\Gamma}_{20}(1-e^{\varepsilon x})]^{(4-n)/n+8} - \varepsilon^{(4-n)/(n+8)}\right\} , \quad \varepsilon \to +0 , \tag{2.3.8}$$

$$\Gamma_1(x) = \Gamma_{10} \exp\left[\frac{\varepsilon(4-n)}{2(n+8)}x\right]\left[\frac{\varepsilon}{\varepsilon e^{-\varepsilon x}+(n+8)\tilde{\Gamma}_{20}(1-e^{-\varepsilon x})}\right]^{(n+2)/(n+8)}$$

$$\varepsilon \to +0 , \tag{2.3.9}$$

$$\Gamma_2(x) = \frac{\varepsilon\tilde{\Gamma}_{20}}{\varepsilon e^{-\varepsilon x} + (n+8)\tilde{\Gamma}_{20}(1 - e^{-\varepsilon x})} , \quad \varepsilon \to +0 . \tag{2.3.10}$$

Minimizing the free energy (2.3.4) in p_0, we obtain

$$F = (1/2)\tau\ell_0^2 + \Gamma\ell_0^4 , \quad \Gamma = \Gamma_2 - \Gamma_1^2/2\omega. \tag{2.3.11}$$

It follows from (2.3.8-11) that for x equal to

$$x^* = \frac{1}{\varepsilon}\ln\left\{1 + \frac{\varepsilon}{(n+8)\tilde{\Gamma}_{20}}\left[\left(\frac{(4-n)\tilde{\Gamma}_{20}[\omega_0+n\Gamma_{10}^2/2(4-n)\tilde{\Gamma}_{20}]}{2\tilde{\Gamma}_{10}^2}\right)^{\frac{(n+8)}{(4-n)}} - 1\right]\right\}$$

$$\varepsilon \to +0 , \tag{2.3.12}$$

the quantity Γ vanishes, i.e., the system becomes thermodynamically unstable and undergoes a first-order phase transition. The jump in the order parameter and the shift in the critical field are respectively given by

$$\ell_0^2 = [(1 - \varepsilon/2)/(1 + x^*\varepsilon/2)]^{2/\varepsilon} , \quad \varepsilon \to +0 ,$$

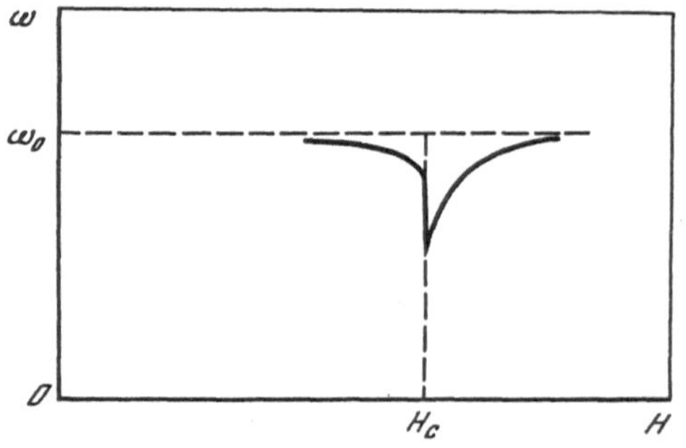

Fig. 2.13. Dependence of the soft ferroelectric mode on magnetic field during a phase transition to a segnetoantiferromagnet in a strong magnetic field

$$\tau_0 = 2b(x^*)[(1 - \varepsilon/2)/(1 + x^*\varepsilon/2)]^{2/\varepsilon}, \ \varepsilon \to +0 ,$$

$$b(x^*) = (4 - n)\Gamma_2^2(x^*) + n\Gamma_1^4(x^*)/4\omega^2(x^*) . \qquad (2.3.13)$$

It is also worthwhile to note that the frequency of the soft ferroelectric mode, as well as the bulk modulus , suffers a discontinuity at the phase transition point. The behavior of $\omega(H)$ is presented in Fig. 2.13. In this section, we have not considered the magnetoelastic interaction [2.18], since for $\gamma \simeq 10^{-4}\ cm^2N^{-1}$ (when the magnetoelectric coupling is relativistic in nature) and $\sigma \simeq 10^{-6} - 10^{-7}\ cm^2N^{-1}$ [2.21], the magnetoelectric interaction plays a decisive role in converting a second-order phase transition into a first-order transition. As $\varepsilon \to +0$, the discontinuity in the order parameter becomes exponentially small, i.e., the phase transition occurs in the scaling region. In this case, the specific heat C_H and the magnetic susceptibility χ^{zz} are defined by (2.2.43), which means that the specific heat of the system for $(T/T_N) \ll 1$ will also be determined by the Bloch spin waves.

2.4 Phase Transition in a Crystal with Dislocations in a Strong Magnetic Field

Let us consider the phase transition in a nonideal antiferromagnet containing crystal defects. It is well known that the defects of the crystal lattice considerably affect the thermodynamic and kinetic properties of crystals. Moreover, the physical properties of crystals depend on the defect size. In this section, we shall consider the influence of one-dimensional defects on a phase transition. Dislocations are a classical type of such defects. The Hamiltonian of interaction of a magnetic subsystem with dislocations has the form

$$\mathbb{H}_{md} = \sum_{\nu} \int d\mathbf{x}\ \gamma_{10}^{ik}\ u_{ik}^{\nu}\ (\mathbf{M}_1\mathbf{M}_2) , \qquad (2.4.1)$$

where u^{ν}_{ik} is the symmetric part of the elastic deformation tensor associated with ν-dislocations:

$$u^{\nu}_{ik}(\mathbf{x}) = \frac{1}{4\pi}\left\{(\eta^2 - 1)\int_{(D^{\nu})} (\delta_{ik} - 3n_i n_k)(b^{\nu}[\mathbf{n}\boldsymbol{\tau}^{\nu}])\frac{d\ell}{|\mathbf{x}|^2}\right.$$

$$\left. + \frac{1}{2}\int_{(D^{\nu})} (b^{\nu}_i[\mathbf{n}\boldsymbol{\tau}^{\nu}]_k + b^{\nu}_k[\mathbf{n}\boldsymbol{\tau}^{\nu}]_i)\frac{d\ell}{|\mathbf{x}|^2}\right\},$$

$$\eta^2 = \varkappa/(\lambda_0 + 2\varkappa), \quad \mathbf{n} = \mathbf{x}/|\mathbf{x}|, \tag{2.4.2}$$

b^{ν} is the Burgers vector for ν-dislocation, τ^{ν} is a vector tangential to the line of ν-dislocation, and $d\ell$ is an element of the dislocation contour. For further analysis, we shall require the values of the Fourier component of the elastic deformation tensor $u_{ik}(x)$ [2.13]:

$$u^{\nu}_{ik}(\mathbf{q}) = -(i/\mathbf{q})\varphi_{ik\ell m}(\mathbf{q}_0)b^{\nu}_{\ell}T^{\nu}_n(\mathbf{q})\exp(i\mathbf{q}\boldsymbol{\rho}_{\nu}), \tag{2.4.3}$$

where $\varphi_{ik\ell m}$ and the vector $T^{\ell}_n(\mathbf{q})$ are defined by the relations

$$\varphi_{ik\ell m}(\mathbf{q}_0) = 2(\eta^2 - 1)\mathbf{q}^i_0\mathbf{q}^k_0\mathbf{q}^m_0 e_{\ell mn} + (1/2)(e_{in\ell}\mathbf{q}^k_0 + e_{kn\ell}\mathbf{q}^i_0$$

$$+ e_{imn}\delta_{k\ell}\mathbf{q}^m_0 + e_{kmn}\delta_{i\ell}\mathbf{q}^m_0), \tag{2.4.4}$$

$$\mathbf{T}^{\nu}(\mathbf{q}) = \int_{(D^{\nu})} d\ell\,\boldsymbol{\tau}^{\nu}\exp(i\mathbf{q}\ell) \tag{2.4.5}$$

$\mathbf{q}_0 = \mathbf{q}/|\mathbf{q}|$ is the vector pointing towards the ν-dislocation. In this case, the effective Hamiltonian of an isotropic antiferromagnet has the form [2.16]

$$\mathbb{H}^{eff} = (1/V)\sum_{\mathbf{q}}(1/2)(\tau + q^2)\ell_p(\mathbf{q})\ell_p(-\mathbf{q}) + (1/V)\sum_{\mathbf{q}}(1/2)[\lambda_0 u_{ii}(\mathbf{q})u_{ii}(-\mathbf{q})$$

$$+ 2\varkappa u_{ik}(\mathbf{q})u_{ik}(-\mathbf{q})] + (1/V^2)\Gamma_{10}\sum_{\mathbf{q}_1+\mathbf{q}_2+\mathbf{q}_3=0} u_{ii}(\mathbf{q}_1)\ell_p(\mathbf{q}_2)\ell_p(\mathbf{q}_3)$$

$$+ (1/V^3)\Gamma_{20}\sum_{\mathbf{q}_1+\mathbf{q}_2+\mathbf{q}_3+\mathbf{q}_4=0}\ell_p(\mathbf{q}_1)\ell_p(\mathbf{q}_2)\ell_s(\mathbf{q}_3)\ell_s(\mathbf{q}_4)$$

$$+ (1/V^2)\sum_{\mathbf{q}_1+\mathbf{q}_2+\mathbf{q}_3=0}\Gamma_{30}(\mathbf{q})\ell_p(\mathbf{q}_1)\ell_p(\mathbf{q}_2), \tag{2.4.6}$$

where

$$\Gamma_{30}(\mathbf{q}) = -(\delta M_0/H)\sum_{\nu}\mathrm{Tr}[u^{\nu}_{ik}(\mathbf{q})].$$

The corresponding RG equations for λ, Γ_1, and Γ_2, as well as the amplitude U of scattering of fluctuations by dislocations can be written in the form

$$\lambda' = -(3/2)n\Gamma_1^2(x,y(x),\Theta_k)\Big|_{y(x)\to 0}, \quad \varepsilon \to +0, \tag{2.4.7}$$

$$\begin{aligned}\Gamma_1' = {} & (\varepsilon/2)\Gamma_1(x,y(x),\Theta_k) - (n+2)\Gamma_1(x,y(x),\Theta_k)\Big|_{y(x)\to 0}\Gamma_2(x,y(x),\Theta_k)\Big|_{y(x)\to 0} \\ & + n\Gamma_1(x,y(x),\Theta_k)\Big|_{y(x)\to 0} U(x,y(x),\Theta_k)\Big|_{y(x)\to 0} \\ & + 2\Gamma_1(x,y(x),\Theta_k)\Big|_{y(x)\to 0} \langle U(x,y(x),\Theta_k)\rangle_{\rho(\Theta_k)}, \quad \varepsilon \to +0,\end{aligned} \tag{2.4.8}$$

$$\begin{aligned}\Gamma_2' = {} & \varepsilon\Gamma_2(x,y(x),\Theta_k) - (n+8)\Gamma_2^2(x,y(x),\Theta_k)\Big|_{y(x)\to 0} \\ & + 12\Gamma_2(x,y(x),\Theta_k)\Big|_{y(x)\to 0} \langle U(x,y(x),\Theta_k)\rangle_{\rho(\Theta_k)} \\ & - 4\langle U^2(x,y(x),\Theta_k)\rangle_{\rho(\Theta_k)}, \quad \varepsilon \to +0,\end{aligned} \tag{2.4.9}$$

$$\begin{aligned}U' = {} & \varepsilon U(x,y(x),\Theta_k) + nU^2(x,y(x),\Theta_k)\Big|_{y(x)\to 0} \\ & - 2(n+2)\Gamma_2(x,y(x),\Theta_k)\Big|_{y(x)\to 0} U(x,y(x),\Theta_k)\Big|_{y(x)\to 0} \\ & + 4U(x,y(x),\Theta_k)\Big|_{y(x)\to 0} \langle U(x,y(x),\Theta_k)\rangle_{\rho(\Theta_k)}, \quad \varepsilon \to +0\end{aligned} \tag{2.4.10}$$

[$\rho(\Theta_k)$ is the distribution function over angles between the direction of the dislocation wave vector and the crystallographic axes, and $y(x) = q$ is a function of the variable x.] The initial value of the amplitude U is

$$U_0(\mathbf{q}) = (1/2V)\langle\Gamma_{30}(\mathbf{q})\Gamma_{30}(-\mathbf{q})\rangle_{(D^\nu)} . \tag{2.4.11}$$

The symbol $\langle \dots \rangle_{(D^\nu)}$ indicates averaging over the dislocation distribution in the crystal. The diagrams for Γ_1, Γ_2, and U are presented in Fig. 2.14, where the undulating line corresponds to the

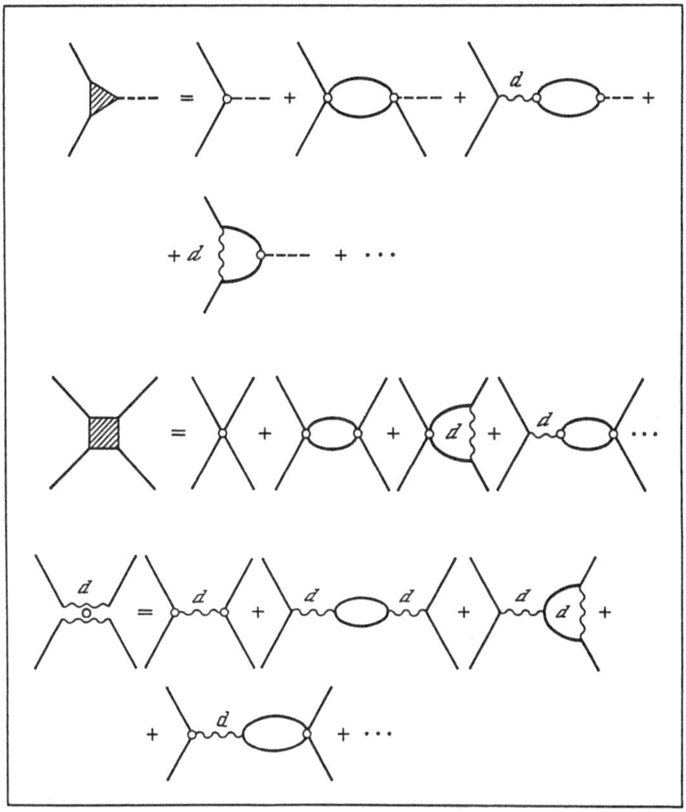

Fig. 2.14. Schematic representation of the solutions to equations for the amplitudes Γ_1, Γ_2, and U in an antiferromagnet with dislocations in a strong magnetic field

correlator (2.4.11). Let us consider ring dislocations of radius R_ν. In this case,

$$U_0(\mathbf{q}) = \frac{\pi^2}{q^2 v_0 V} \sum_{i,k,\ell,\ell',n,n',\nu} \varphi_{ik\ell n}(\mathbf{q}_0)\varphi_{ik\ell'n'}(\mathbf{q}_0)\delta_{\ell\ell'} b_\ell^\nu b_{\ell'}^\nu$$

$$\cdot\, R_\nu^2 (\delta_{nn'} - q_0^n q_0^{n'}) \int_0^1 dz\, J_2(2qzR_\nu) \ , \qquad (2.4.12)$$

$J_2(2qzR_\nu)$ is a second-order Bessel function, and v_0 is the volume of a unit cell. Since $U_0(q)$ does not have a singularity in q, the main contribution to the initial interaction will come from $U_0(q_0)$, the zero-order term in the expansion of $U_0(q)$ in q. In this case, (2.4.7-10) can be reduced to

$$\lambda' = -(3/2)n\Gamma_1^2(x,\Theta_k) \ , \varepsilon \to +0 \ , \qquad (2.4.13)$$

$$\Gamma_1' = (\varepsilon/2)\Gamma_1(x,\Theta_k) - n\Gamma_1(x,\Theta_k)\tilde{\Gamma}_2(x,\Theta_k)$$

$$- 2\Gamma_1(x,\Theta_k)\langle\tilde{\Gamma}_2(x,\Theta_k)\rangle_{\rho(\Theta_k)}, \quad \varepsilon \to +0, \tag{2.4.14}$$

$$\tilde{\Gamma}_2' = \varepsilon\tilde{\Gamma}_2(x,\Theta_k) - n\tilde{\Gamma}_2^2(x,\Theta_k) - 4\tilde{\Gamma}_2(x,\Theta_k)\langle\tilde{\Gamma}_2(x,\Theta_k)\rangle_{\rho(\Theta_k)}$$

$$- 4\langle\tilde{\Gamma}_2^2(x,\Theta_k)\rangle_{\rho(\Theta_k)}, \quad \varepsilon \to +0, \tag{2.4.15}$$

$$\tilde{\Gamma}_2 = \Gamma_2 - U. \tag{2.4.16}$$

It is interesting to note that (2.4.13-16) can also effectively take into account the anisotropy of the crystal lattice and are identical to (2.2.8-10) except for the exchange interaction anisotropy. The dislocation potential has the form

$$U_0(\mathbf{q}) = \frac{2\pi^2\eta^4}{3v_0 V}\sum_{k,\nu} R_\nu^4 b_k^{\nu 2}(1 - \mathbf{q}_0^{k\,2}). \tag{2.4.17}$$

Let us consider the isotropic dislocation potential $b_k = b^\nu$, $U_0(\Theta_k) = \langle U_0(\Theta_k)\rangle_{\rho(\Theta_k)}$. In this case, (2.4.13-16) are simplified and coincide with (2.2.13-15) except for the initial interactions. Hence the stability boundary x* is defined by (2.2.21), the only difference being that $\tilde{\Gamma}_{20} = \Gamma_{20} - (3\Gamma_{10}^2/\lambda_0 + 2k) - U_0$. Consequently, x* decreases with increasing U_0, and the discontinuity in the order parameter ℓ_0 increases. However, the critical fluctuations decrease at the transition point and hence the probability of phase transition at this point decreases. Thus ring dislocations intensify the magneto-elastic coupling near the phase transition point, but at the same time increase the exchange rigidity of the system. The phase transition in this case will be of hysteresis type (Fig. 2.15). The magnitude γ of hysteresis will be proportional to $\Delta^2\tau_0$ as a function of the dislocation density n_d. The possibility of such transitions was experimentally demonstrated in [2.1].

For a weakly anisotropic dislocation potential

$$\delta U_0(\Theta_k)/\tilde{\Gamma}_{20} \ll 1, \quad \tilde{\Gamma}_{20} = \Gamma_{20} - (3\Gamma_{10}^2/\lambda_0 + 2\varkappa) - \langle U_0(\Theta_k)\rangle_{\rho(\Theta_k)}$$

the solutions of (2.4.13-16) for amplitudes Γ_1 and Γ_2 have the form

$$\Gamma_1(x,\Theta_k) = \Gamma_{10}\exp\left[\frac{4-n}{2(n+8)}\varepsilon x\right]\left[\varepsilon/\{\varepsilon e^{-\varepsilon x} + (n+8)\tilde{\Gamma}_{20}(1 - e^{-\varepsilon x})\}\right]^{\frac{n+2}{n+8}}$$

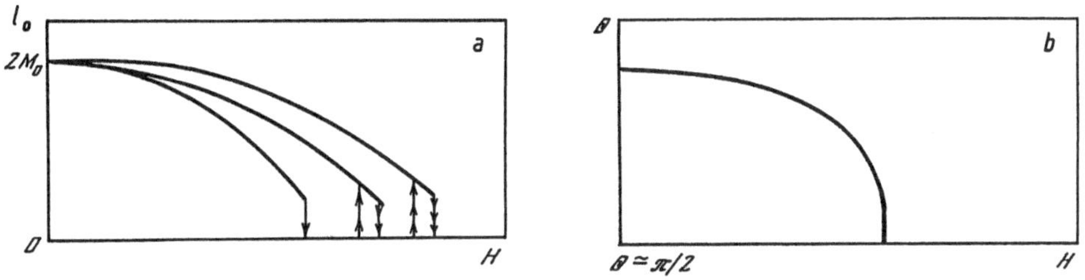

Fig. 2.15. Dependence of the magnitude of the equilibrium antiferromagnetism vector in an antiferromagnet with dislocations (a), and of the angle θ between the direction of the external magnetic field and the magnetic moment of the sublattice without line defects (b)

$$\cdot \exp\left[-\frac{n\delta U_0(\Theta_k)}{(4-n)\tilde{\Gamma}_{20}}\left\{\exp[\tfrac{4-n}{n+8}\varepsilon x]\left[\varepsilon/\{\varepsilon e^{-\varepsilon x}+(n+8)\tilde{\Gamma}_{20}(1-e^{-\varepsilon x})\}\right]^{\frac{n+2}{n+8}} - 1\right\}\right] , \qquad (2.4.18)$$

$$\Gamma_{20}(x,\Theta_k) = \left[\varepsilon\tilde{\Gamma}_{20}/\{\varepsilon e^{-\varepsilon x}+(n+8)\tilde{\Gamma}_{20}(1-e^{-\varepsilon x})\}\right] + \delta U_0 \exp[\tfrac{4-n}{n+8}\varepsilon x]\left[\varepsilon/\{\varepsilon e^{-\varepsilon x}+(n+8)\tilde{\Gamma}_{20}(1-e^{-\varepsilon x})\}\right]^{\frac{2(n+2)}{n+8}} ,$$
$$\varepsilon \to +0 . \qquad (2.4.19)$$

Since $\delta U_0(\Theta_k) < 0$, it follows from (2.4.18,19) that the dislocation potential anisotropy leads to a further intensification of the magnetoelastic coupling near the phase transition point and to an even larger increase in the order parameter discontinuity. For $2\lambda\delta U_0(\Theta_k)/3\Gamma_{10}^2 << 1$, we obtain

$$x^* = \frac{1}{\varepsilon}\ln\left[1+\frac{\varepsilon}{(n+8)\tilde{\Gamma}_{20}}\left\{\left(2\lambda\tilde{\Gamma}_{20}/3\Gamma_{10}^2\right)^{\frac{n+8}{4-n}} \cdot\left[1+\frac{(4+n)(n+8)}{(4-n)^2}\,\frac{2\lambda\delta U_0(\Theta_k)}{3\Gamma_{10}^2}\right]-1\right\}\right] , \quad \varepsilon\to +0 . \qquad (2.4.20)$$

Thus, an analysis of an inhomogeneous antiferromagnet with linear defects shows that the dislocations are capable of strengthening the magnetoelastic coupling near the phase transition point, which results in an increase in the order parameter discontinuity. An increase in the dislocation density n_d leads to an enhanced exchange rigidity of the system, which decreases the probability of a phase transition at this point. In this section, we have not considered the anisotropy of the crystal lattice. However, it is quite obvious that a consideration of this anisotropy causes an even more significant increase in the amplitude Γ, i.e., leads to an even stronger manifestation of the exchange enhancement effect [2.22,2.16]. consequently, the phase transition in an antiferromagnet in a strong magnetic field will still be determined by the strong magnetoelastic coupling. If the dislocation density $n_d \lesssim 10^7 - 10^8$ cm^{-2}, $R_\nu \simeq 10^{-4}$ cm, then $U_0 >> 1/\lambda_0$, and the phase transition will be determined by the presence of dislocations in the crystal.

The introduction of dislocations in a magnetically ordered crystal may lead to other interesting effects. For example, bound states of two magnons may be formed on dislocations in antiferromagnets [2.23].

3. Phase Lamination and the Exchange Enhancement Effect

3.1 Phase Transitions in Complex Magnetic Structures and the Exchange Enhancement Effect in Spin-Orbit Interaction

It was shown in Chap. 2 that the exchange enhancement effect plays a significant role in the phase transitions in ferromagnets and antiferromagnets. The exchange enhancement effect changes the nature of a phase transition: instead of a second-order phase transition, the system undergoes a first-order transition close to a second-order one. However, we considered only the interaction of the fluctuations of the order parameter of ferromagnetism or antiferromagnetism vector ℓ with quadrupole moment fluctuations or acoustic phonons, longwave bulk deformations or optical phonons. In all these cases, the effective Hamiltonian contained a term corresponding to the striction-type interaction. Integration over the fluctuating quadrupole moment field, over the phonon field and longwave bulk strain field in the presence of the exchange enhancement effect led to an instability in the magnetic subsystem and to a first-order phase transition close to the second. Anomalies like jumps in the order parameter, bulk modulus and soft ferroelectric mode ω are found to be exponentially small in a number of cases in spite of the exchange enhancement due to the fact that the quadrupole or magnetoelectric interaction constants are small, and also due to the fact that the effective dimensionality d of the space turns out to be equal to four.

In the last two decades [3.1], a large body of experimental data has been acquired on the investigation of phase transitions in magnets. According to these data, there exists a large class of magnetically ordered crystals with a complex magnetic structure, where instead of second-order magnetic phase transitions, only first-order transitions or first-order transitions close to the second are observed. In some of these materials (e.g., in MnO, MnS_2, Eu), the jumps in the order parameter reached about 50% of the initial value. Table 3.1 contains a list of the most characteristic types of magnetically ordered crystals with a complex magnetic structure. In most of these

Table 3.1

Material	Dimensionality of order parameter	Space group	Magnetic structure for $T<T_{C(N)}$
$TbAu_2$, DyC_2	2m = 4	I4/mmm	Sinusoidal, $\mathbf{K}\parallel[110]$ ($TbAu_2$), $\mathbf{K}\parallel[100]$ (DyC_2), $\mathbf{m}\parallel[001]$
NbO_2	2m = 4	$P4_2/mnm$	$\mathbf{K}(1/4, 1/4, 2)$
TbD_2	2m = 6	Fm3m	Sinusoidal, $\mathbf{m}\parallel\mathbf{K}\parallel[100]$
Nd	2m = 6	$P6_3/mmc$	Sinusoidal, $\mathbf{m}\parallel\mathbf{K}\parallel[100]$
K_2IrCl_6	2m = 6	Fm3m	Antiferromagnetic, type III, $\mathbf{K} = (1/2, 0, 1)$, $\mathbf{m}\parallel[1,0,0]$
MnO, MnSe NiO, ErSb	8	Fm3m	Antiferromagnetic, type III, $\mathbf{m}\perp\mathbf{K}$
TbAs, TbP, TbSb	4	Fm3m	Antiferromagnetic, type II, $\mathbf{m}\parallel\mathbf{K}$
UO_2	6	Fm3m	Antiferromagnetic, type I, $\mathbf{m}\perp\mathbf{K}$
Cr, Eu	12	Im3m	Helicoidal, $\mathbf{S}\perp\mathbf{K}$(Eu); Sinusoidal, $\mathbf{S}\parallel\mathbf{K}$(Cr)
Er	2m = 6	$P6_3/mmc$	Sinusoidal, $\mathbf{S}\parallel\mathbf{K}$
Dy,Tb,Ho	2m = 6	$P6_3/mmc$	Helicoidal, $\mathbf{S}\perp\mathbf{K}$
Er-Dy,	2m = 6	$P6_3/mmc$	Tilted spiral, $T< T_{N_2}$ ($\mathbf{S}\hat{\psi}\mathbf{K}$)
Er-Tb, Er-Ho, Er,Ho	2×2m = 12		

materials, first-order phase transitions from paramagnetic to magnetically ordered states were experimentally observed. The Table also contains the spatial symmetry groups of these materials. At temperatures of the order of transition temperature to the paramagentic state, the striction interaction in these crystals is weak, and the quadrupole exchange interaction has not been observed experimentally [3.1]. Hence a consideration of these interactions for explaining phase transitions in this class of materials is not justified. The explanation for such phase transitions should be sought in the peculiarities of the magnetic structure of a magnetically ordered crystal on the basis of magnetic symmetry.

In this section, we shall construct a general pattern for describing phase transitions in magnetically ordered crystals with a complex magnetic structure in the framework of the modern fluctuation theory of phase transitions. The order parameter in crystals with a complex magnetic structure is generally an N-dimensional vector in the Euclidean space.

The effective Hamiltonian as a function of the N-dimensional order parameter can be written in the form

$$\mathbb{H}^{eff} = \int d\mathbf{x}\ \mathbb{H}\left(S_i, \frac{\partial S_i}{\partial \chi_k}, \ldots\right), \qquad (3.1.1)$$

where $\mathbb{H}$ is invariant to the corresponding symmetry group transformations. The statistical sum of the system can be presented as a functional integral with respect to the fluctuating fields S_i:

$$Z = \prod_{i=1,\mathbf{x}}^{N} \int DS_i(\mathbf{x}) \exp(-\mathbb{H}^{eff}) . \qquad (3.1.2)$$

The density of the effective Hamiltonian can be presented as a sum of two components

$$\mathbb{H}\left(S_i, \frac{\partial S_i}{\partial \chi_k}\right) = \mathbb{H}^{(1)}(S_i) + \mathbb{H}^{(2)}\left(S_i, \frac{\partial S_i}{\partial \chi_k}\right), \qquad (3.1.3)$$

$$\mathbb{H}^{(1)}(S_i) = \frac{1}{2}\sum_{i=1}^{N} \tau_i S_i^2 + \sum_{p;i,j,k,\ell=1}^{N} \delta_p \beta^p_{ijk\ell} S_i S_j S_k S_\ell + \ldots, \qquad (3.1.4)$$

$$\mathbb{H}^{(2)} = \sum_{i,j=1}^{N} \sum_{k=1}^{d} A_{ijk} S_i \frac{\partial S_j}{\partial \chi_k} + \sum_{i,j=1}^{N} \sum_{\ell=1}^{d} B_{ijk\ell} \frac{\partial S_i}{\partial \chi_k} \frac{\partial S_j}{\partial \chi_\ell} + \ldots, \qquad (3.1.5)$$

where τ_i is the effective temperature equal to $\tau_i = \tau + a_i$; $\tau = T/T_{N(C)} - 1$; a_i is the anisotropy constant, the sum over p indicates summation of all fourth-order invariants over the magnetic symmetry group G of the crystal, and d is the dimensionality of the effective space. Hence, in order to construct the effective Hamiltonian of the system under consideration, we must write down all the necessary invariants of the crystal symmetry group G.

Phase transitions from the paramagnetic to the ordered state in most of the crystals listed in Table 3.1 were studied in detail by **Bak, Mukamel**, and **Krinsky** [3.2-4]. Using the Wilson-Fisher fluctuation theory of phase transitions [3.5-7], these authors were able to determine from (3.1.3-5) possible phase transitions and their character in a given type of material. They obtained and analyzed systems of nonlinear RG equations for effective amplitudes $\Gamma_{ijk\ell} = \sum_p \delta_p \beta^p_{ijk\ell}$ for the corresponding fourth-order invariants. By analyzing the system of differential equations for the quantity $\Gamma_{ijk\ell}$, Bak, Krinsky and Mukamel showed that the more complex the magnetic structure of a crystal and the less significant the role of relativistic interactions in its formation, the more probable the first-order phase transition from the paramagnetic phase to the ordered state. Conversely, the stronger the relativistic interaction and the smaller the number of fluctuating fields, the less probable is a first-order phase transition. Hence a second-order phase transition should be anticipated in the system in this case. For example, in Er and Ho (space group $P6_3/mmc$), whose magnetic structure at high temperatures is a spin-density wave (sinusoidal for Er and helicoidal for Ho) with a wave vector q oriented along the preferred axis z of the crystal, a second-order phase transition must be observed due to a strong magnetic anisotropy [3.2-4].

In [3.2-4], the authors did not take into account the renormalization of relativistic interactions upon a phase transition in the systems they considered. Hence they predicted phase transitions to states which were not observed experimentally [3.1].

For the group of rare-earth metals and their alloys (space group $P6_3/mmc$), the magnetic anisotropy of the systems was experimentally found to depend strongly on temperature [3.8], and changed sharply during phase transitions. The renormalization of the magnetic anisotropy in rare-earth metals and their alloys is associated with a change in the topology of the Fermi surface of conduction electrons which are responsible for the effective exchange interaction in this class of materials (indirect s-f exchange) [3.9]. This means that during phase transitions in rare-earth metals and their alloys, the renormalization of the spin-orbit interaction parameter must lead to a change in the magnetic structure of the system. Consequently, the phase transitions in this class of materials from the paramagnetic phase to the ordered state must have a much more complicated nature, and the spin-orbit interaction will play a decisive role in the formation of a magnetic structure. This means that if one of the components of the order parameter has large fluctuations in the paramagnetic phase, it still does not mean that this component will play a decisive role in forming the magnetic structure of the ordered state. On account of fluctuations, the anisotropy of the system is renormalized by the exchange parameter, and this may lead to the emergence of a new type of magnetic structure upon a phase transition. This is a manifestation of the exchange enhancement effect during phase transitions in complex magnetic structures. In this section, we shall consider high-temperature phase transitions in rare-earth alloys of the type Er-Dy, Er-Tb, and Er-Ho ($T_N \approx$ 100 K) taking into account the exchange enhancement effect.

On the basis of neutron scattering experiments in rare-earth compounds Er-Dy and Er-Tb, **Woods** [3.10], **Stringfellow** [3.11], and **Milhause** and **Koehler** [3.12] established that phase transitions from paramagnetic phase to ordered state in these materials take place as follows (Fig. 3.1). If the concentration of Dy and Tb in an alloy exceeds the concentration of Er ($C_x > C_y$), the system goes over from the paramagnetic phase to a normal spiral (NS). As the temperature is lowered, the system moves to the state of a tilted spiral (TS), i.e., its magnetic structure is a helical wave with a wave vector **q** oriented along the preferred axis z (hexagonal axis) of the crystal, whose polarization plane is inclined at a certain angle ψ to the direction of its wave vector **q**. Upon a further decrease in temperature, the system undergoes a phase transition to a conic spiral (cone), which is found to be stable down to zero temperature ($T \to 0$). However, another intermediate state, called a tilted cone, may be observed between tilted and conic spiral states [3.8]. This state is found to be metastable. If the concentration of Dy and Tb in the alloy is lower than the concentration of Er ($C_x < C_y$), the system goes over from the paramagnetic phase to a c-sin state with a longitudinal sinusoidal wave, after which it behaves in the same way as in the case considered above.

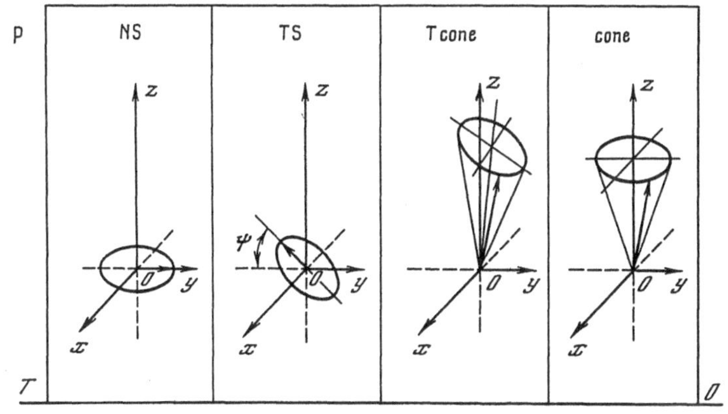

Fig. 3.1. Phase transitions in magnetically ordered alloys of rare-earth metals such as Er-Dy and Er-Tb

Let us analyze the phase transitions from the paramagnetic phase to the ordered state up to a tilted spiral (TS) by taking into account the exchange enhancement effect for the spin-orbit interaction. We shall not consider phase transitions from TS to a cone, since a strong static magnetostriction is observed experimentally in the temperature region corresponding to this transition [3.13], and the short-range fluctuations are considerably suppressed. Hence the exchange enhancement effect in this case will not be manifested together with the interaction of fluctuations of various components of a multicomponent order parameter. It is more appropriate to explain these transitions in the framework of the self-consistent field theory by a magneto-striction mechanism or mechanisms associated with a strong static temperature dependence of the second- and fourth-order anisotropy constants (K_2, K_4) [3.11].

Thus, the magnetic structure of the alloys Er-Dy, Er-Tb, and Er-Ho (space group $P6_3/mmc\text{-}D^4_{6h}$) in the lowest ordered state is a tilted spiral spin density wave with its periodicity along the preferred axis z of the crystal. A tilted spiral wave may be imagined as a superposition of two spin-density (longtitudinal sinusoidal and a helicoidal) waves which are periodic along the preferred axis z of the crystal and which have the same wave vector q. Since the complex conjugate of the wave vector q of both longitudinal and transverse waves has the form q = 0, 0, q_3, 0, ..., we can introduce two complex spin-density vectors [3.14] (Fig. 3.2):

$$\mathbf{S}_{\parallel} = \mathbf{s}^+_{\parallel} + i\mathbf{s}^-_{\parallel} \ , \quad \mathbf{s}^+_{\parallel} = \mathbf{s}^+_{\parallel\, o}\mathrm{Re}\,\{\exp[i(\mathbf{q}\mathbf{z} + \varphi^+)]\}$$

$$\mathbf{s}^-_{\parallel} = \mathbf{s}^-_{\parallel\, o}\, \mathrm{Im}\{\exp[i(\mathbf{q}\mathbf{z} + \varphi^-)]\} \ , \quad \mathbf{s}^+_{\parallel\, o} \parallel \mathbf{s}^-_{\parallel\, o} \ ; \tag{3.1.6}$$

$$\mathbf{S}_{\perp} = \mathbf{s}^+_{\perp} + i\mathbf{s}^-_{\perp} \ , \quad \mathbf{s}^+_{\perp} = \mathbf{s}^+_{\perp o}\ \mathrm{Re}\{\exp(i\mathbf{q}\mathbf{z})\} \ ,$$

$$\mathbf{s}^-_{\perp} = \mathbf{s}^-_{\perp o}\ \mathrm{Im}\{\exp(i\mathbf{q}\mathbf{z})\} \ , \quad \mathbf{s}^+_{\perp oi} \perp \mathbf{s}^-_{\perp ok} \ . \tag{3.1.7}$$

Using vectors $\mathbf{S}_{\parallel}$ and $\mathbf{S}_{\perp}$, we can compose the following second- and fourth-order invariants of exchange and relativistic type:

$$(\mathbf{S}_{\parallel}\, \mathbf{S}^*_{\parallel}) \ , \quad (\mathbf{S}_{\perp}\mathbf{S}^*_{\perp}) \ , \quad S^z_{\parallel}\, S^{z*}_{\parallel} \ , \quad S^z_{\perp}S^{z*}_{\perp} \ ,$$

$$(\mathbf{S}_{\parallel}\, \mathbf{S}^*_{\parallel})^2 \ , \quad (\mathbf{S}_{\perp}\mathbf{S}^*_{\perp})^2 \ , \quad S^2_{\parallel}\, S^{2*}_{\parallel} \ , \quad S^2_{\perp}S^{2*}_{\perp} \ ; \tag{3.1.8}$$

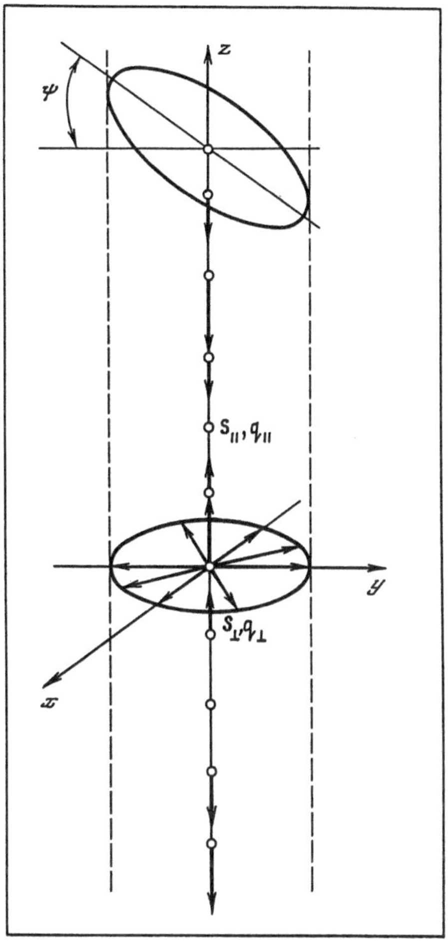

Fig. 3.2. Tilted spiral (TS) magnetic structure

$$(\mathbf{S}_{\parallel}\ \mathbf{S}_{\parallel}^{*})(\mathbf{S}_{\perp}\mathbf{S}_{\perp}^{*})\ ,\quad (\mathbf{S}_{\parallel}\ \mathbf{S}_{\parallel})(\mathbf{S}_{\parallel}^{*}\ \mathbf{S}_{\perp}^{*}) + (\mathbf{S}_{\parallel}\ \mathbf{S}_{\perp}^{*})(\mathbf{S}_{\parallel}^{*}\ \mathbf{S}_{\perp})\ ,$$

$$S_{\parallel}^{z^2}\ S_{\parallel}^{z^{2*}}\ ,\quad S_{\perp}^{z^2}\ S_{\perp}^{z^{2*}}\ ,\quad S_{\parallel}^{z}\ S_{\parallel}^{z*}\ S_{\perp}^{z}S_{\perp}^{z*}\ . \tag{3.1.9}$$

Henceforth, we shall neglect the fourth-order relativistic invariants since the fourth-order anisotropy constant describing the angle of inclination ψ of the tilted spiral is small in comparison with the second-order anisotropy constant: $K_4 \ll K_2$ [3.8]. Using the reduced invariants, we can write the following expression for the free energy of the system in the paramagnetic region:

$$F = (1/2)\tau_{\parallel}(s_{\parallel}^{+2} + s_{\parallel}^{-2}) + (1/2)\tau_{\perp}(s_{\perp}^{+2} + s_{\perp}^{2}) + (1/2)a_{\parallel}(s_{\parallel z}^{+2} + s_{\parallel z}^{-2})$$
$$+(1/2)a_{\perp}(s_{\perp z}^{+2} + s_{\perp z}^{-2}) + (1/8)\Gamma_{1\parallel}(s_{\parallel}^{+4} + s_{\parallel}^{-4}) + (1/8)\Gamma_{1\perp}(s_{\perp}^{+4} + s_{\perp}^{-4})$$
$$+(1/4)\Gamma_{2\parallel} s_{\parallel}^{+2} s_{\parallel}^{-2} + (1/4)\Gamma_{2\perp} s_{\perp}^{+2} s_{\perp}^{-2} + (1/2)\Gamma_{3\parallel}(\mathbf{s}_{\parallel}^{+}\mathbf{s}_{\parallel}^{-})^2$$
$$+(1/2)\Gamma_{3\perp}(\mathbf{s}_{\perp}^{+}\mathbf{s}_{\perp}^{-})^2 + (1/4)\Gamma_4(s_{\parallel}^{+2} s_{\perp}^{+2} + s_{\parallel}^{+2} s_{\perp}^{-2} + s_{\parallel}^{-2} s_{\perp}^{+2} + s_{\parallel}^{-2} s_{\perp}^{-2})$$
$$+(1/2)\Gamma_5[(\mathbf{s}_{\parallel}^{+}\mathbf{s}_{\perp}^{+})^2 + (\mathbf{s}_{\parallel}^{+}\mathbf{s}_{\perp}^{-})^2 + (\mathbf{s}_{\parallel}^{-}\mathbf{s}_{\perp}^{+})^2 + (\mathbf{s}_{\parallel}^{-}\mathbf{s}_{\perp}^{-})^2]. \quad (3.1.10)$$

In this formula, $a_{\parallel} < 0$ and $a_{\perp} > 0$ are the anisotropy constants while the initial values τ_{jo} of temperature and Γ_{ijo} of amplitude are given by

$$\tau_{\parallel o} = \tau_{\perp o} = \tau, \quad \Gamma_{1\parallel o} = \Gamma_{1\perp o} = \Gamma_{1o},$$
$$\Gamma_{2\parallel o} = \Gamma_{2\perp o} = \Gamma_{1o} - 2\Gamma_{3o}, \quad \Gamma_{3\parallel o} = \Gamma_{3\perp o} = \Gamma_{3o}. \quad (3.1.11)$$

In accordance with the experimental data [3.8,12-19], we shall consider the case of a strong anisotropy, both longitudinal and transverse ($|a_{\parallel}| \lesssim 1$, $|a_{\perp}| \lesssim 1$). In this case, the invariant corresponding to the amplitude Γ_5 vanishes ($s_{\parallel} \perp s_{\perp}$). The effective Hamiltonian of the system will then have the form

$$\mathbb{H}^{eff} = (1/V)\sum_{\mathbf{q}} \{(1/2)(\tau_{\parallel}+q^2)[(\mathbf{s}_{\parallel}^{+}(\mathbf{q})\mathbf{s}_{\parallel}^{+}(-\mathbf{q}))$$
$$+(\mathbf{s}_{\parallel}^{-}(\mathbf{q})\mathbf{s}_{\parallel}^{-}(-\mathbf{q}))] + (1/2)a_{\parallel}[s_{\parallel z}^{+}(\mathbf{q})s_{\parallel z}^{+}(-\mathbf{q}) + s_{\parallel z}^{-}(\mathbf{q})s_{\parallel z}^{-}(-\mathbf{q})]$$
$$+(1/2)(\tau_{\perp}+q^2)[(\mathbf{s}_{\perp}^{+}(\mathbf{q})\mathbf{s}_{\perp}^{+}(-\mathbf{q})) + (\mathbf{s}_{\perp}^{-}(\mathbf{q})\mathbf{s}_{\perp}^{-}(-\mathbf{q}))]$$
$$+ (1/2)a_{\perp}[s_{\perp z}^{+}(\mathbf{q})s_{\perp z}^{+}(-\mathbf{q}) + s_{\perp z}^{-}(\mathbf{q})s_{\perp z}^{-}(-\mathbf{q})]\}$$
$$+(1/V^3)\sum_{\mathbf{q}_1+\mathbf{q}_2+\mathbf{q}_3+\mathbf{q}_4=0} \{(1/8)\Gamma_{1\parallel}[(\mathbf{s}_{\parallel}^{+}(\mathbf{q}_1)\mathbf{s}_{\parallel}^{+}(\mathbf{q}_2))(\mathbf{s}_{\parallel}^{+}(\mathbf{q}_3)\mathbf{s}_{\parallel}^{+}(\mathbf{q}_4))$$
$$+(\mathbf{s}_{\parallel}^{-}(\mathbf{q}_1)\mathbf{s}_{\parallel}^{-}(\mathbf{q}_2))(\mathbf{s}_{\parallel}^{-}(\mathbf{q}_3)\mathbf{s}_{\parallel}^{-}(\mathbf{q}_4))] + (1/4)\Gamma_{2\parallel}(\mathbf{s}_{\parallel}^{+}(\mathbf{q}_1)\mathbf{s}_{\parallel}^{+}(\mathbf{q}_2))$$
$$\times(\mathbf{s}_{\parallel}^{-}(\mathbf{q}_3))(\mathbf{s}_{\parallel}^{-}(\mathbf{q}_4)) + (1/4)\Gamma_{2\perp}(\mathbf{s}_{\perp}^{+}(\mathbf{q}_1)\mathbf{s}_{\perp}^{+}(\mathbf{q}_2))(\mathbf{s}_{\perp}^{-}(\mathbf{q}_3)\mathbf{s}_{\perp}^{-}(\mathbf{q}_4))$$
$$+(1/2)\Gamma_{3\parallel}(\mathbf{s}_{\parallel}^{+}(\mathbf{q}_1)\mathbf{s}_{\parallel}^{-}(\mathbf{q}_2))(\mathbf{s}_{\parallel}^{+}(\mathbf{q}_3)\mathbf{s}_{\parallel}^{-}(\mathbf{q}_4))$$

$$+(1/2)\Gamma_{3\perp}(\mathbf{s}_\perp^+(\mathbf{q}_1)\mathbf{s}_\perp^-(\mathbf{q}_2))(\mathbf{s}_\perp^+(\mathbf{q}_3)\mathbf{s}_\perp^-(\mathbf{q}_4))$$

$$+(1/4)\Gamma_4[(\mathbf{s}_\|^+(\mathbf{q}_1)\mathbf{s}_\|^+(\mathbf{q}_2))(\mathbf{s}_\perp^+(\mathbf{q}_3)\mathbf{s}_\perp^+(\mathbf{q}_4))$$

$$+(\mathbf{s}_\|^+(\mathbf{q}_1)\mathbf{s}_\|^+(\mathbf{q}_2))(\mathbf{s}_\perp^-(\mathbf{q}_3)\mathbf{s}_\perp^-(\mathbf{q}_4))$$

$$+(\mathbf{s}_\|^-(\mathbf{q}_1)\mathbf{s}_\|^-(\mathbf{q}_2))(\mathbf{s}_\perp^+(\mathbf{q}_3)\mathbf{s}_\perp^+(\mathbf{q}_4))+(\mathbf{s}_\|^-(\mathbf{q}_1)\mathbf{s}_\|^-(\mathbf{q}_2))(\mathbf{s}_\perp^-(\mathbf{q}_3)\mathbf{s}_\perp^-(\mathbf{q}_4))]\}. \quad (3.1.12)$$

As in Sect. 2.1, we introduce Green functions

$$G^{\pm\pm}_{j\alpha\beta}(\mathbf{q}) = \langle s^{\pm}_{j\alpha}(\mathbf{q})s^{\pm}_{j\beta}(-\mathbf{q})\rangle_o = \frac{\delta_{\alpha\beta}}{\tau_j + q^2},$$

$$\tilde{G}^{\pm\pm}_{j\alpha\beta}(\mathbf{q}) = \langle s^{\pm}_{j\alpha}(\mathbf{q})s^{\pm}_{j\beta}(-\mathbf{q}) = \frac{\delta_{\alpha\beta}}{\tau_j+q^2+\Sigma_j(\tau_\|,\tau_\perp,\mathbf{q})},$$

$$\alpha,\ \beta = x,\ y,\ j = \|,\ \perp,$$

$$G^{\pm\pm}_{jzz}(\mathbf{q}) = \langle s^{\pm}_{jz}(\mathbf{q})s^{\pm}_{jz}(-\mathbf{q})\ \rangle_o = \frac{1}{\tau_j + a_j + q^2},$$

$$\tilde{G}^{\pm\pm}_{jzz}(\mathbf{q}) = \langle s^{\pm}_{jz}(\mathbf{q})s^{\pm}_{jz}(-\mathbf{q})\rangle = \frac{1}{\tau_j + a_j + q^2 + \Sigma_j(\tau_\|,\tau_\perp,\mathbf{q})}. \quad (3.1.13)$$

For the case when all the fields fluctuate strongly, the RG equations can be written in the form

$$-\Gamma'_{1j} = (n_j+8)\Gamma^2_{1j} + n_j\Gamma^2_{2j} + 4\Gamma_{2j}\Gamma_{3j} + 4\Gamma_{3j} + 2n_k\ \Gamma^2_4,$$

$$-\Gamma'_{2j} = 2(n_j+2)\Gamma_{1j}\Gamma_{2j} + 4\Gamma_{1j}\Gamma_{3j} + 4\Gamma^2_{2j} + 4\Gamma^2_{3j} + 2n_k\Gamma^2_4,$$

$$-\Gamma'_{3j} = 4\Gamma_{1j}\Gamma_{3j} + 8\Gamma_{2j}\Gamma_{3j} + 2(n_j+2)\Gamma^2_{3j},$$

$$-\Gamma'_4 = \sum_i [(n_i+2)\Gamma_{1i} + n_i\Gamma_{2i} + 2\Gamma_{3i}]\Gamma_4 + 4\Gamma^2_4,$$

$$-\tau'_j = \tau_j[(n_j+2)\Gamma_{1j} + n_i\Gamma_{ij} + 2\Gamma_{3j}] + 2n_k\tau_k\Gamma_4,$$

$$i,j,k = \|,\ \perp,\quad j \neq k, \quad (3.1.14)$$

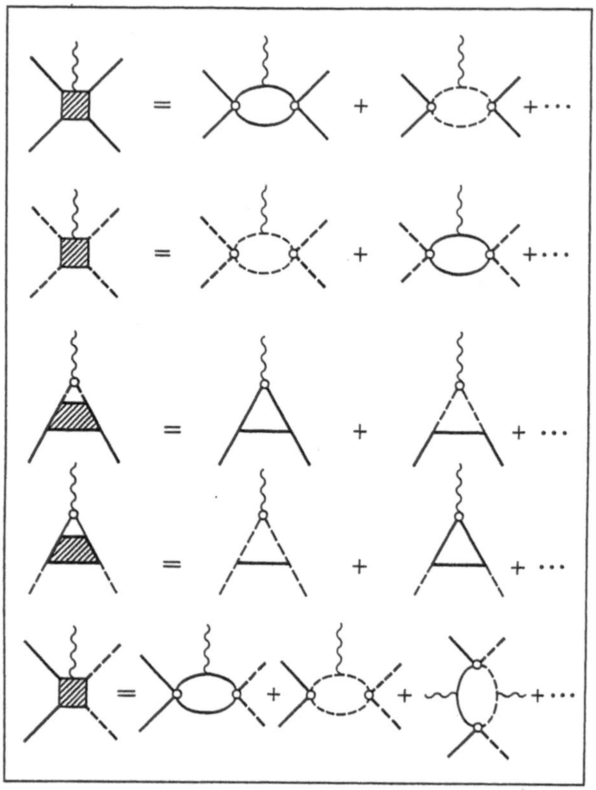

Fig. 3.3. Renormalized group equations for the quantities τ_j and Γ_{ij}

while the ε-expansion method variable has the form

$$x = \frac{2}{\varepsilon(\Lambda^2)^{\varepsilon/2}} \left\{ \left(\frac{\Lambda^2}{\max[\lambda^2(\tau_{\parallel},\tau_{\perp}), s_{\perp o}^2, s_{\parallel o}^2]} \right)^{\varepsilon/2} - 1 \right\}, \quad \varepsilon \to 0 \; ; \tag{3.1.15}$$

$\lambda^2(\tau_{\parallel},\tau_{\perp})$ is the scaling factor as a function of $\tau_{\parallel},\tau_{\perp}$; $s_{\parallel o}^2 = s_{\parallel o}^{+2} + s_{\parallel o}^{-2}$, $s_{\perp o}^2 = s_{\perp o}^{+2} + s_{\perp o}^{-2}$; $n_{\parallel}$, $n_{\perp}$ and $n_{\parallel}$ and $n_{\perp}$ are the dimensionalities of the fluctuating fields $s_{\parallel}^{\pm}$ and $s_{\perp}^{\pm}$.

The diagrams corresponding to the quantities Γ_{ij} and τ_j are shown in Fig. 3.3. The points in the diagrams correspond to the initial interactions Γ_{ijo} and τ_{jo}, while the solid and dashed lines correspond to the functions $G_{\parallel\alpha\beta}^{\pm\pm}(q_k)$ and $G_{\perp\alpha\beta}^{\pm\pm}(q_k)$; α, $\beta \to (x,y,z)$. Summation is carried out over all internal momenta q_k. The undulating line indicates differentiation. After deriving the general RG equations, we can go over to an analysis of specific magnetic structures.

It follows from (3.1.10) that in the case of a strong anisotropy $a_\perp$, two helical polarizations $(s^+_{\perp x}, s^-_{\perp y})$ and $(s^+_{\perp y}, s^-_{\perp x})$ coexist in the basal plane (x,y). This means that together with the phase transition to the tilted spiral, a phase transition to a state with a tilted sinusoidal wave may also take place in the system. However, such a state was not observed experimentally in the alloys investigated, and hence we shall consider the case of only one helicoidal polarization in the (x,y) plane. For a large anisotropy, the expression for the free energy of the system has the form

$$F = (1/2)\tau_{\parallel}(s^{+2}_{\parallel} + s^{-2}_{\parallel}) + (1/2)\tau_{\perp}(s^{+2}_{\perp} + s^{-2}_{\perp})$$

$$+ (1/8)\Gamma_{1\parallel}(s^{+2}_{\parallel} + s^{-2}_{\parallel})^2 + (1/8)\Gamma_{1\perp}(s^{+4}_{\perp} + s^{-4}_{\perp})$$

$$+ (1/4)\Gamma_{2\perp}s^{+2}_{\perp}\, s^{-2}_{\perp} + (1/4)\Gamma_4(s^{+2}_{\parallel} + s^{-2}_{\parallel})(s^{+2}_{\perp} + s^{-2}_{\perp}), \quad (3.1.16)$$

where

$$\tau_{\parallel 0} = \tau + a_{\parallel}, \quad \tau_{\perp 0} = \tau + a_{\perp}, \quad a_{\parallel} < 0, \quad a_{\perp} < 0,$$

$$\Gamma_{1\parallel 0} = \Gamma_{1\perp 0} = \Gamma_{10}, \quad \Gamma_{2\perp 0} = \Gamma_{10} - 2\Gamma_{30}; \; \Gamma_{10} > 0, \; \Gamma_{40} > 0, \; \Gamma_{30} < \Gamma_{40},$$

$$\mathbf{s}^+_{\parallel} = (0, 0, s^+_{\parallel z}), \quad \mathbf{s}^+_{\perp} = (\mathbf{s}^+_{\perp x}, 0, 0),$$

$$\mathbf{s}^-_{\parallel} = (0, 0, s^-_{\parallel z}), \quad \mathbf{s}^-_{\perp} = (0, s^-_{\perp y}, 0).$$

Let us study the behavior of the system depending on the relation between $|a_{\parallel}|$ and $|a_\perp|$.

1. $|a_\perp| < |a_{\parallel}|$. In this case, the field $\mathbf{S}_{\parallel}$ experiences the strongest fluctuations and the intrinsic fluctuations of the field $\mathbf{S}_\perp$ are weaker. The RG equations for $\Gamma_{ij}(x)$, $\tau_j(x)$ can be derived easily from (3.1.14):

$$-\Gamma'_{1\parallel} = 10\Gamma^2_{1\parallel}, \quad -\Gamma'_{2\perp} = 2\Gamma^2_4, \quad -\tau'_{\parallel} = 4\tau_{\parallel}\Gamma_{1\parallel},$$

$$-\Gamma'_{1\perp} = 2\Gamma^2_4, \quad -\Gamma'_4 = 4\Gamma_{1\parallel}\Gamma_4, \quad -\tau'_\perp = 2\tau_{\parallel}\Gamma_4. \quad (3.1.17)$$

In this case, the variable of the RG method is found to be equal to

$$x = \frac{2}{\varepsilon(\Lambda^2)^{\varepsilon/2}}\left\{\left(\frac{\Lambda^2}{\max[\lambda^2(\tau_{\parallel}), s^2_{\perp 0}]}\right)^{\varepsilon/2} - 1\right\}, \quad \varepsilon \to 0. \quad (3.1.18)$$

Equations (3.1.17) can be solved easily. Their solutions have the form

$$\Gamma_{1\parallel}(x) = \frac{\Gamma_{10}}{1 + 10\Gamma_{10}x}, \quad \Gamma_{1\perp}(x) = \Gamma_{10} - \frac{\Gamma_{40}^2}{\Gamma_{10}}[(1 + 10\Gamma_{10}x)^{1/5} - 1],$$

$$\Gamma_{2\perp}(x) = \Gamma_{1\perp}(x) - 2\Gamma_{30}, \quad \Gamma_4(x) = \Gamma_{40}/(1 + 10\Gamma_{10}x)^{2/5},$$

$$\tau_{\parallel}(x) = \tau_{\parallel 0}/(1 + 10\Gamma_{10}x)^{2/5}, \quad \tau_{\perp}(x) = \tau_{\perp 0} - \frac{\tau_{\parallel 0}\Gamma_{40}}{\Gamma_{10}}[(1+10\Gamma_{10}x)^{1/5}-1]. \tag{3.1.19}$$

It follows from (3.1.19) that the transverse subsystem becomes unstable since its fluctuations are enhanced by the fluctuations of the longitudinal magnetic subsystem. This is just a manifestation of the exchange enhancement effect. Consequently, a first-order phase transition will take place to a state with a helicoidal magnetic structure NS. The stability boundary of the paramagnetic phase is determined from the condition $(1/2)(\Gamma_{1\perp} + \Gamma_{2\perp}) = 0$ or $\Gamma_{1\perp}(x) - \Gamma_{30} = 0$, and is found to be equal to

$$x_h^* = \frac{1}{10\Gamma_{10}}\left\{\left[1 + \frac{\Gamma_{10}(\Gamma_{10} - \Gamma_{30})}{\Gamma_{40}^2}\right]^5 - 1\right\}. \tag{3.1.20}$$

In this case, the jump in the order parameter $s_{\perp 0}^2$ and the temperature of transition to the helicoidal state are respectively given by

$$\tau_{\perp h} = 2b(x_h^*)\left(\frac{1 - \varepsilon/2}{1 + \varepsilon x^*/2}\right)^{2/\varepsilon}, \quad s_{\perp 0}^2 = \tau_{\perp h}/2b(x_h^*), \tag{3.1.21}$$

where

$$b(x_h^*) = (1/4)\Gamma_4^2(x_h^*).$$

Following a phase transition to the normal spiral state, the temperature $\tau_{\parallel}$ of the longitudinal subsystem receives an increment $(1/2)\Gamma_4(x_h^0)s_{\perp 0}^2$, where

$$x_h^0 = \frac{1}{10\Gamma_{10}}\left\{\left[\frac{(\tau_{\perp 0} - \tau_{\perp h})\Gamma_{10}}{\tau_{\parallel 0}\Gamma_{40}} + 1\right]^5 - 1\right\}. \tag{3.1.22}$$

This means that the magnetic anisotropy constant is renormalized, it receives an exchange type increment and is reduced as a result. The fluctuations of the longitudinal subsystem freeze out and the subsystem remains in the paramagnetic state. The free energy of the longitudinal subsystem can be expressed through the relation

$$F = (1/2)\,\tau_{\parallel}(s_{\parallel}^{+2} + s_{\parallel}^{-2}) + (1/8)\,\Gamma_{1\parallel}(s_{\parallel}^{+2} + s_{\parallel}^{-2})^2 . \qquad (3.1.23)$$

Upon a subsequent decrease in temperature, the effective anisotropy constant gradually decreases. When the order parameter $s_{\perp 0}$ becomes saturated, a phase transition to the tilted spiral state becomes possible. After transition to the TS state, the problem should be solved with initial conditions $x = x_{0h}$, which indicates that the system undergoes a second-order phase transition [3.14]. Solving the equations for $\Gamma_{\parallel}$ and $\tau_{\parallel}$, we can determine the critical susceptibility index for a second-order phase transition. This index is found to be $\gamma = 1 + \varepsilon/5 + \dots$.

2. $|a_{\parallel}| < |a_{\perp}|$. In this case, the transverse magnetic subsystem experiences the strongest fluctuations, and the fluctuations of the longitudinal magnetic subsystem increase. The RG equations have the form

$$-\Gamma'_{1\parallel} = 2\Gamma_4^2 , \quad -\Gamma'_{2\perp} = 6\Gamma_{1\perp}\Gamma_{2\perp} + 4\Gamma_{2\perp}^2 ,$$

$$-\Gamma'_{1\perp} = 9\Gamma_{1\perp}^2 + \Gamma_{2\perp}^2 , \quad -\Gamma'_4 = (3\Gamma_{1\perp} + \Gamma_{2\perp})\Gamma_4 ,$$

$$-\tau'_{\parallel} = 2\tau_{\perp}\Gamma_4 , \quad -\tau'_{\perp} = \tau_{\perp}(3\Gamma_{1\perp} + \Gamma_{2\perp}) \qquad (3.1.24)$$

and hence

$$x = \frac{2}{\varepsilon(\Lambda^2)^{\varepsilon/2}}\left\{\left(\frac{\Lambda^2}{\max[\lambda^2(\tau),\, s_{\parallel 0}^2]}\right)^{\varepsilon/2} - 1\right\}, \varepsilon \to 0 . \qquad (3.1.25)$$

Introducing the new variables

$$y_{1\parallel} = \Gamma_{1\parallel}/\Gamma_{1\perp} , \quad y_{2\perp} = \Gamma_{2\perp}/\Gamma_{1\perp} , \quad y_4 = \Gamma_4/\Gamma_{1\perp} ,$$

$$z = -\ln \Gamma_{1\perp} , \qquad (3.1.26)$$

we can reduce the system of equations for the amplitudes to the form

$$\frac{dy_{1\parallel}}{dz} = \frac{y_{1\parallel}(9 + y_{2\perp}^2) - 2y_4^2}{9 + y_{2\perp}^2} ,$$

$$\frac{dy_{2\perp}}{dz} = y_{2\perp}\,\frac{(1 - y_{2\perp})(3 - y_{2\perp})}{9 + y_{2\perp}^2} ,$$

$$\frac{dy_4}{dz} = y_4 \frac{6 - y_{2\perp}(1 - y_{2\perp})}{9 + y_{2\perp}^2} . \qquad (3.1.27)$$

Let us consider in detail the second equation in this system. This equation has three stationary points $y_{2\perp}^{(1)} = 0$, $y_{2\perp}^{(2)} = 1$, and $y_{2\perp}^{(3)} = 3$. The point $y_{2\perp}^{(2)} = 1$ is found to be stable while the remaining two points are unstable (Fig. 3.4). If the initial values of $\Gamma_{1\perp 0}$ and $\Gamma_{2\perp 0}$ are such that $0 < y_{2\perp 0} < 3$, i.e., if $-\Gamma_{10} < \Gamma_{30} < (\frac{1}{2})\Gamma_{10}$, then $y_{2\perp}$ tends to a stable stationary point $y_{2\perp}^{(2)} = 1$, $\Gamma_{1\perp} \to \Gamma_{2\perp} \to 1/10x$. This means that the system can undergo a first-order phase transition to a state with a longitudinal sinusoidal wave. Indeed, reducing (3.1.24) and (3.1.27) to quadratures, we obtain

$$\frac{\Gamma_{1\parallel}(x)}{\Gamma_{10}} = 1 - \frac{2y_{40}^2}{y_{2\perp 0}(3 - y_{2\perp 0})} \; \frac{y_{2\perp} - y_{2\perp 0}}{1 - y_{2\perp}} ,$$

$$\frac{\Gamma_{1\perp}(x)}{10} = \left(\frac{3 - y_{2\perp 0}}{3 - y_{2\perp}}\right)^3 \left(\frac{1 - y_{2\perp}}{1 - y_{2\perp 0}}\right)^5 \left(\frac{y_{2\perp 0}}{y_{2\perp}}\right)^3 ,$$

$$\frac{\Gamma_4(x)}{\Gamma_{40}} = \left(\frac{3 - y_{2\perp 0}}{3 - y_{2\perp}}\right)\left(\frac{1 - y_{2\perp}}{1 - y_{2\perp 0}}\right)^2 \left(\frac{y_{2\perp 0}}{y_{2\perp}}\right) , \qquad (3.1.28)$$

$$\frac{\tau_{\parallel}(x)}{\tau_{\parallel 0}} = 1 - \frac{2y_{40}\tau_{\perp 0}}{\tau_{\parallel 0}y_{2\perp 0}(3 - y_{2\perp 0})} \left(\frac{y_{2\perp} - y_{2\perp 0}}{1 - y_{2\perp}}\right) ,$$

$$\frac{\tau_{\perp}(x)}{\tau_{\perp 0}} = \left(\frac{3 - y_{2\perp 0}}{3 - y_{2\perp}}\right)\left(\frac{1 - y_{2\perp}}{1 - y_{2\perp 0}}\right)^2 \left(\frac{y_{2\perp 0}}{y_{2\perp}}\right). \qquad (3.1.29)$$

In this case, the stability boundary is determined by the condition

$$\Gamma_{1\parallel}(x)/\Gamma_{10} = 0 , \qquad (3.1.30)$$

whence

$$\Gamma_{10}x_s^* = \frac{(1 - y_{2\perp 0})^5}{y_{2\perp 0}^3(3 - y_{2\perp 0})^3} w(y_{2\perp s}^* , y_{2\perp 0}) , \qquad (3.1.31)$$

$$w(y_{2\perp s}^* , y_{2\perp 0}) = \int_{y_{2\perp 0}}^{y_{2\perp s}^*} dy_{2\perp} \frac{y_{2\perp}^2(3 - y_{2\perp})^2}{(1 - y_{2\perp})^6} , \qquad (3.1.32)$$

$$y^*_{2\perp s} = y_{2\perp 0}\,\frac{2y_{40}^2 + 3 - y_{2\perp 0}}{2y_{40}^2 + y_{2\perp 0}(3 - y_{2\perp 0})} . \tag{3.1.33}$$

With the help of (3.1.31-33), we can easily obtain the value of the jump in the order parameter upon a first-order phase transition, as well as the critical temperature of transition to a state with a longitudinal sinusoidal wave. Indeed,

$$\tau_{\parallel s} = 2b(x*)\left(\frac{1 - \varepsilon/2}{1 + x*\varepsilon/2}\right)^{2/\varepsilon} , \quad \varepsilon \to 0 ,$$

$$s_{\parallel 0}^2 = \tau_{\parallel s}/2b(x_s^*) , \tag{3.1.34}$$

where

$$b(x_s^*) = (1/4)\Gamma_4^2(x_s^*) . \tag{3.1.35}$$

After a transition to the c-sin state, the anisotropy of the transverse subsystem is also renormalized and decreases by an amount $(1/4)\Gamma_4(x_s) \times s_{\parallel 0}^2$. The anisotropy of the longitudinal system in this case is a slowly varying function of temperature. The expression for the free energy of the helicoidal subsystem has the form

$$F_\perp = (1/2)\tau_\perp(s_\perp^{+2} + s_\perp^{-2}) + (1/8)\,\Gamma_{1\perp}(s_\perp^{+4} + s_\perp^{-4})$$
$$+ (1/4)\Gamma_{2\perp}s^{+2}s^{-2} . \tag{3.1.36}$$

The equations for the amplitudes $\Gamma_{1\perp}$ and $\Gamma_{2\perp}$ remain unchanged, and the system tends to reach the stable point $y_2^{(2)} = 1$ as before in view of the fact that $0 < y_{2\perp 0} < 3$. Only the initial condition is now different, and is determined from the equation

$$\frac{\tau_{\parallel s}}{\tau_{\parallel 0}} = 1 - \frac{2y_{40}\tau_{\perp 0}}{\tau_{\parallel 0}y_{2\perp 0}(3 - y_{2\perp 0})}\,\frac{y_{2\perp s}^0 - y_{2\perp 0}}{1 - y_{2\perp s}^0} . \tag{3.1.37}$$

Consequently,

$$\Gamma_{10}x_{0s} = \frac{(1 - y_{2\perp 0})^5}{y_{2\perp 0}^3(3 - y_{2\perp 0})^3}\,w(y_{2\perp s}^0 , y_{2\perp 0}) , \tag{3.1.38}$$

where

$$y^0_{2\perp s} = y_{2\perp 0}\,\frac{2y_{40}\tau_{\perp 0} + \tau_{\parallel 0}(3 - y_{2\perp 0})(1 - \tau_{\parallel s}/\tau_{\parallel 0})}{2y_{40}\tau_{\perp 0} + \tau_{\parallel 0}y_{2\perp 0}(3 - y_{2\perp 0})(1 - \tau_{\parallel s}/\tau_{\parallel 0})} . \quad (3.1.39)$$

Thus, the system undergoes a second-order phase transition to a tilted spiral (TS). The critical index γ remains the same as for the case $|a_\perp| < |a_\parallel|$ (see above), viz., $\gamma = 1 + \varepsilon/5 + \ldots$.

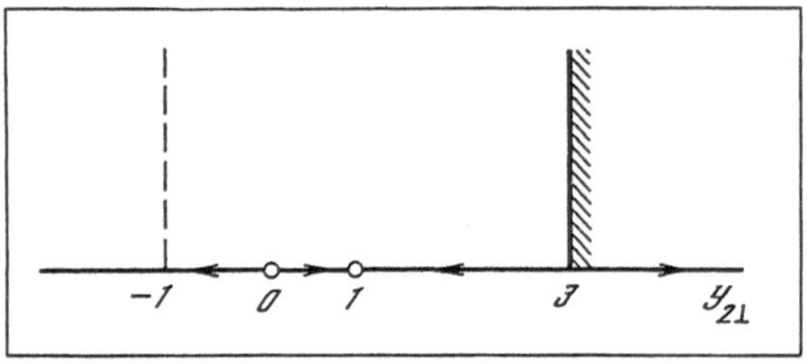

Fig. 3.4. Phase trajectories of the equation

$$\frac{dy_{2\perp}}{dz} = y_{2\perp}\frac{(1 - y_{2\perp})(3 - y_{2\perp})}{9 + y^2_{2\perp}}$$

Let us consider the case when the initial values of the amplitudes Γ_{ij0} are such that $-1 < y_{2\perp 0} < 0$. This means that the function $y_{2\perp}(z)$ falls in the region of unstable solutions of the second equation in the system (3.1.27) (Fig. 3.4). The helicoidal system becomes unstable and undergoes a first-order phase transition. Its stability boundary is determined by the condition $y_{2\perp} = -1$. The value of $y_{2\perp}$ at the transition point, viz., $y^0_{2\perp h}$, is then equal to

$$y^0_{2\perp h} = \frac{2 + 3a + (9a^2 + 8a)^{\frac{1}{2}}}{2(1 + a)} , \quad (3.1.40)$$

$$a = \frac{\tau^0_{\perp h}}{\tau_{\perp 0}}\,\frac{(1 - y_{2\perp 0})^2}{y_{2\perp 0}(3 - y_{2\perp 0})} , \quad (3.1.41)$$

($\tau^0_{\perp h}$ is the temperature of transition of the helicoidal subsystem to the ordered state). It follows from (3.1.40) that $y^0_{2\perp h}$ and $\tau^0_{\perp h}/\tau_{\perp 0}$ vary in the following intervals of values:

$$-3 < y^0_{2\perp h} < -1 , \quad (3.1.42)$$

$$-\frac{8y_{2\perp 0}(3 - y_{2\perp 0})}{9(1 - y_{2\perp 0})} < \frac{\tau^0_{\perp h}}{\tau_{\perp 0}} < -\frac{y_{2\perp 0}(3 - y_{2\perp 0})}{(1 - y_{2\perp 0})^2} . \quad (3.1.43)$$

The jump in the order parameter is determined from the expression

$$s^2_{\perp 0} = \tau^0_{\perp h}/2b(x^{*0}_{\perp h}) ,$$

where

$$b(x^{*0}_{\perp h}) = \frac{1}{16} \Gamma^2_{10} \frac{(3 - y_{2\perp 0})^6 y^6_{2\perp 0}(1 - y_{2\perp})^{11}}{y^5_{2\perp}(1 - y_{2\perp 0})^{10}(3 - y_{2\perp})^5}, \tag{3.1.44}$$

and $x^{*0}_{\perp h}$ is the renormalized value of $x^*_{\perp h}$ given by

$$\Gamma_{10} x^*_{\perp h} = \frac{(1 - y_{2\perp 0})^5}{y^3_{2\perp 0}(3 - y_{2\perp 0})^3} w(-1 , y_{2\perp 0}) . \tag{3.1.45}$$

Accordingly, $x^{*0}_{\perp h} > x^*_{\perp h}$. The quantity $x^{*0}_{\perp h}$ is determined from the condition

$$\tau^0_{\perp h} = 2b(x^{*0}_{\perp h}) \left\{\frac{1 - (\varepsilon/2)[1 + f(x^{*0}_{\perp h})]}{1 + x^{*0}_{\perp h} \varepsilon/2}\right\}^{2/\varepsilon} ,\varepsilon \to 0 , \tag{3.1.46}$$

where

$$f(x^{*0}_{\perp h}) = \frac{1 + y_{2\perp}}{\Gamma_{10} y_{2\perp 0}(3 - y_{2\perp 0})(1 - y_{2\perp})} \left(\frac{3 - y_{2\perp}}{3 - y_{2\perp 0}}\right)^2$$

$$\times \left(\frac{1 - y_{2\perp 0}}{1 - y_{2\perp}}\right)^5 \left(\frac{y_{2\perp}}{y_{2\perp 0}}\right)^2 . \tag{3.1.47}$$

Let us now consider the behavior of the longitudinal sinusoidal subsystem. The inequalities $y^*_{2\perp s} < 0$ and $y^0_{2\perp s} < 0$ will hold if the following conditions are satisfied:

$$3/2 - (9/4 + 2y^2_{40})^{\frac{1}{2}} < y_{2\perp 0} < 0 , \tag{3.1.48}$$

$$3/2 - \left[\frac{9}{4} + \frac{2y_{40}\tau_{\perp 0}}{\tau_{\parallel 0}(1 - \tau_{\parallel s}/\tau_{\parallel 0})}\right]^{\frac{1}{2}} < y_{2\perp 0} < 0 . \tag{3.1.49}$$

It follows from these conditions that the quantity $1 - \tau_{\parallel s}/\tau_{\parallel 0}$ must change within the interval

$$\frac{\tau_{\perp 0}}{y_{40}\tau_{\parallel 0}} < 1 - \frac{\tau_{\parallel s}}{\tau_{\parallel 0}} < 1 . \tag{3.1.50}$$

As $y_{2\perp 0} \to -\varepsilon$, the quantities $\tau_{\parallel s}$ and $s^2_{\parallel 0}$ are defined by the expressions

$$\tau_{\parallel s} = 2b(x_s^*)\left(\frac{1-\varepsilon/2}{1+x_s^*\,\varepsilon/2}\right)^{2/\varepsilon}, \quad \varepsilon \to 0 , \qquad (3.1.51)$$

$$s_{\parallel 0}^2 = \tau_{\parallel s}/2b(x_s^*) ,$$

$$b(x_s^*) = \frac{1}{4}\,\Gamma_4^2(x_s^*) , \qquad b(x_s^*) = \Gamma_{40}^2\,\frac{y_{40}^4}{(2y_{40}^2+3)^2} . \qquad (3.1.52)$$

The parameters x_s^* , $y_{2\perp s}^*$ and $y_{2\perp s}^0$ are defined by the expressions

$$\Gamma_{10}x_s^* = \frac{1}{9}\left[\left(\frac{2y_{40}^2+3}{2y_{40}^2}\right)^3 - 1\right] , \qquad (3.1.53)$$

$$y_{2\perp s}^* = y_{2\perp 0}\,\frac{2y_{40}^2+3}{2y_{40}^2}, \qquad y_{2\perp s}^0 = y_{2\perp 0}\,\frac{3\tau_{\parallel 0}}{2\tau_{\perp 0}} . \qquad (3.1.54)$$

The analysis carried out for the case $-1 < y_{2\perp 0} < 0$ shows that the following phase transitions can take place in the system:

1. $-1 < y_{2\perp s}^0$. In this case, the system first undergoes a first-order phase transition to the c-sin state, followed by a first-order phase transition to a tilted spiral.

2. $-3 < y_{2\perp s}^0 < -1$. Under these conditions, a situation may arise when $y_{2\perp s}^0 \to y_{2\perp h}^0$. This means that a first-order phase transition from the paramagnetic phase directly to a tilted spiral is possible.

3. $y_{2\perp s}^0 < -3$. The system undergoes a first-order phase transition from the paramagnetic phase to a normal spiral, after which a first-order transition to a tilted spiral is possible.

4. $y_{2\perp 0} < (3/2) - (9/4 + 4\,y_{40}^2)^{\frac{1}{2}}$. The system undergoes a first-order phase transition to a normal spiral, followed by a second-order phase transition to a tilted spiral.

3.2 Phase Diagrams of Complex Magnetic Structures: Phase Lamination

The analysis carried out in Sect. 3.1 for the phase transitions from the paramagnetic phase to the ordered state in rare-earth metal alloys Er-Dy, Er-Tb, and Er-Ho with a complex magnetic structure led to the construction of a new type of phase diagrams in the temperature-aniso-

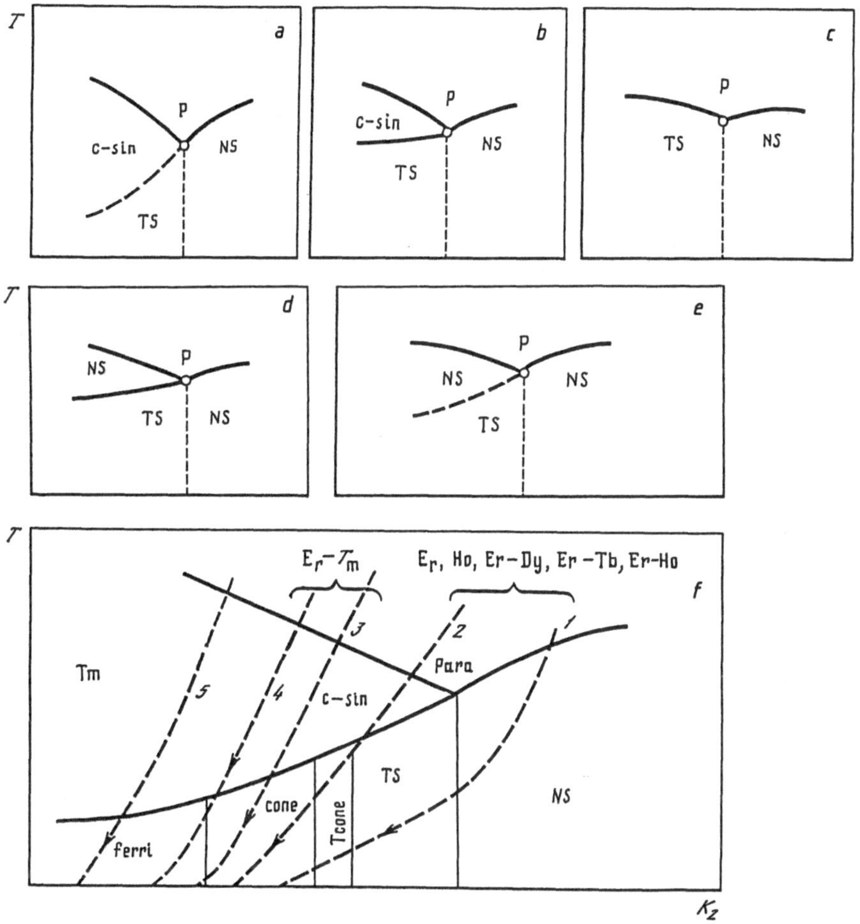

Fig. 3.5. Phase diagrams of rare-earth metals and their alloys having a complex magnetic structure in the lowest ordered state (K_2 is the renormalized magnetic anisotropy constant)

tropy variables (T, $|a_{||}|$). All possible modifications of these diagrams are presented in Figs. 3.5a-e. The diagram in Fig. 3.5a is a phase diagram of rare-earth metals and their alloys, obtained experimentally [3.8] (see Fig. 2.5f). The remaining diagrams in Fig. 3.5b-e correspond to cases (1)-(4) when $-1 < y_{2\perp 0} < 0$, and are modifications of the main phase diagram Fig. 3.5a. It follows from these phase diagrams that all of them have a singular point to which four or (Fig. 3.5c) three lines corresponding to first- and second-order phase transitions converge. Before analyzing the behavior of the system in the vicinity of the singular point, let us consider in detail the magnetic structure of the NS, c-sin, and TS phases. All these phases have the same space group $P6_3/mmc\text{-}D^4_{6h}$ and a hexagonal unit cell (Fig. 3.6).

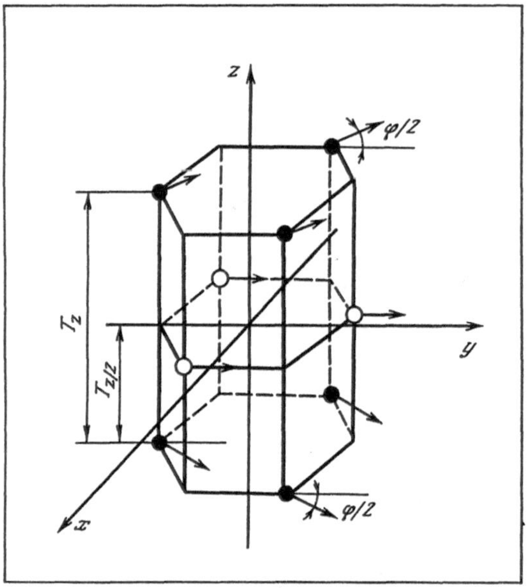

Fig. 3.6. Simplest unit cell for a magnetic structure of normal spiral (NS) type

1. Normal spiral (NS). In the normal spiral, the spins form ferromagnetic layers perpendicular to the hexagonal axis c(z), while the direction of spins along this axis changes from plane to plane; the spins in two adjacent planes being inclined toward each other at an angle $\varphi/2$. The spatial multiplicative group for this configuration is obtained from the crystallographic group by carrying out the following operations on its elements: (1) translation T_z, multiplied by $\Omega_{z\varphi}$ (Fig. 3.6), where $\Omega_{z\varphi}$ indicates rotation in the spin space by an angle φ about the c(z) axis; (2) translations T_x, T_y, leaving all vertical axes C_6^v and C_2^v of the sixth and second order free, as well as all diagonal slip planes relative to c(z); (3) S_d, S_c multiplied by $\Omega_{z\varphi/2}$, leaving as free all triad axes C_3, vertical mirror planes $\sigma_{\parallel}$ and vertical planes S_a and S_b with slip relative to a and b; (4) all horizontal spin-oriented symmetry planes $\sigma_{\perp}$ and second-order spin-oriented horizontal axes C_2^h, multiplied by $\Omega_{u,\pi}$, where u is also a spin-oriented vector indicating the direction of the rotational axis $(\hat{uu^*}) = \varphi/2$; and (5) the remaining horizontal axes and all centers of inversion I multiplied by $\Omega_{z,\varphi/2}\,\Omega_{u\pi}$. The set of elements obtained in this way is the required spatial multiplicative group for the normal spiral [3.15].

The order p rameter can then be presented in the form

$$\tilde{\boldsymbol{\gamma}} = \boldsymbol{\gamma}_0 \cos q(\mathbf{nx}) + [\mathbf{n}\,\boldsymbol{\gamma}_0] \sin q(\mathbf{nx}) \,. \qquad (3.2.1)$$

The topological degeneracy space (manifold) corresponding to the order parameter $\tilde{\gamma}$ (degeneracy space) is the rotational group [3.16]

$$R_{NS} = SO(3) = SU(2)/Z_2 . \qquad (3.2.2)$$

The first and second homotopic groups on the manifold are respectively equal to [3.17]

$$\pi_1(R_{NS}) = Z_2 , \quad \pi_2(R_{NS}) = 0 . \qquad (3.2.3)$$

2. Tilted Spiral (TS). The symmetry group of this structure can be easily obtained from the symmetry group of NS by replacing the orientational rotation axis z by d, $(d\hat{z}) = \psi$. In this case, we obtain

$$R_{TS} = SO(3) \times S^1 , \quad \pi_1(R_{TS}) = Z_2 + Z , \quad \pi_2(R_{TS}) = 0 . \qquad (3.2.4)$$

3. Longitudinal Sinusoidal Modulation (c-sin). In order to obtain the symmetry group of this phase, we must supplement the symmetry group of TS by an element $\Omega_{z,\delta}$ (δ is an arbitrary angle):

$$R_{c\text{-}sin} = S^2 \times S^1 , \quad \pi_1(R_{c\text{-}sin}) = Z , \quad \pi_2(R_{c\text{-}sin}) = Z . \qquad (3.2.5)$$

It follows from (3.2.2-5) that the following relations are satisfied between the topological degeneracy spaces corresponding to the order parameter in the NS, c-sin, and TS phases and their first homotopy groups:[1]

$$R_{c\text{-}sin}, \; R_{NS} \subset R_{TS} , \quad \pi_1(R_{c\text{-}sin}) , \quad \pi_1(R_{NS}) \subset \pi_1(R_{TS}) .$$

Phase transitions from one ordered phase to another in the space $(T, |a_{\|}|)$ are accompanied by a variation of the topological manifold of the order parameter. For example, the transition TS $\to$ NS involves the mapping $R_{TS} \xrightarrow{W_1} R_{NS}$, and the transition TS $\to$ c-sin is accompanied by the mapping $R_{TS} \xrightarrow{W_2} R_{c\text{-}sin}$.

[1] While writing the degeneracy spaces corresponding to the magnetic structures, we did not take into consideration the fact that the strong anisotropy fixes the direction of the wave vector q of the structure. Hence it cannot describe the sphere S^2. The degeneracy space in q is nothing but two points on the sphere S^2: ±1. Consequently, the form of $R_{c\text{-}sin}$, R_{NS} and R_{TS} is considerably simplified. However, the relation (3.2.5) remains unchanged in this case, and hence we shall consider also degeneracy spaces of the most general type.

Let us consider these mappings in greater detail. In order to carry out the mapping $R_{TS} \xrightarrow{W_1} R_{NS}$, $R_{TS} = SO(3) \times S^1$, $R_{NS} = SO(3)$, we must "remove" a layer $SO(2) \times S^1$ from the surface corresponding to the manifold $SO(3) \times S^1$. This gives a normal two-dimensional sphere in a three-dimensional space. In order to obtain SO(3), we must contract S^1 to a point and cover the sphere S^2 by the manifold SO(2). This can be represented schematically as follows:

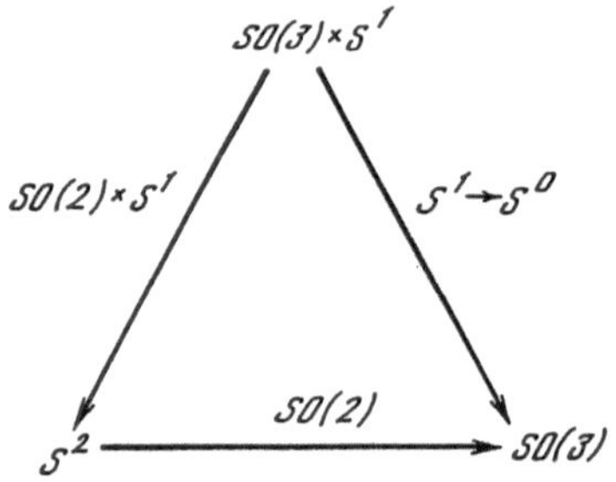

(3.2.6)

For the mapping $R_{TS} \xrightarrow{W_2} R_{c\text{-}sin}$, we obtain the following sequence:

$$SO(3) \times S^1 \xrightarrow{S^1 \to S^0} SO(3) \xrightarrow{SO(2)} S^2 \xrightarrow{S^0 \to S^1} S^2 \times S^1 . \tag{3.2.7}$$

In algebraic topology, the mappings (3.2.6) and (3.2.7) are called fibrations (laminations) [3.17]. Hence it can be concluded that the sequence of phase transitions in the (T, $|a_{\parallel}|$)-space from the lowest-ordered state to the paramagnetic phase can be described by a sequence of fibres. Such a sequence of phase transitions will be called a phase fibration (lamination) in the (T, $|a_{\parallel}|$)-space. It can be easily concluded from the above analysis that the mapping $SO(3) \to S^2 \times S^1$ is also possible for a fixed phase of the order parameter in the c-sin phase, i.e., for a fixed value of the wave vector of the structure. This means that the singular point in the phase diagram may be stable, but allows a second-order phase transition from the paramagnetic phase directly to the tilted spiral phase. In the vicinity of the singular point in the phase diagram, the fields $S_{\parallel}$ and $S_{\perp}$ are found to fluctuate strongly. This corresponds to the case $|a_{\parallel}| \to |a_{\perp}|$. The RG equations have the form

$$-\Gamma'_{1\parallel} = 10\Gamma^2_{1\parallel} + 2\Gamma^2_4 , \quad -\Gamma_{2\perp} = 6\Gamma_{1\perp}\Gamma_{2\perp} + 4\Gamma^2_{2\perp} + 2\Gamma^2_4 ,$$

$$-\Gamma'_{1\perp} = 9\Gamma^2_{1\perp} + \Gamma_{2\perp} + 9\,\Gamma^2_4 ,$$

$$-\Gamma'_4 = (4\Gamma_{1\parallel} + 3\Gamma_{1\perp} + \Gamma_{2\perp})\Gamma_4 + 4\Gamma^2_4 ,$$

$$-\tau'_{\parallel} = 4\tau_{\parallel}\Gamma_{1\parallel} + 2\tau_{\perp}\Gamma_4 , \quad -\tau'_{\perp} = \tau_{\perp}(3\Gamma_{1\perp} + \Gamma_{2\perp}) + 2\tau_{\parallel}\Gamma_4 ; \tag{3.2.8}$$

and the variable x is defined by (3.1.15).

In the variables $y_{1\parallel}$, $y_{2\perp}$, y_4 , and z , the equations for the amplitudes are written in the form

$$\frac{dy_{1\parallel}}{dz} = \frac{y_{1\parallel}(9 + y^2_{2\perp} - 10y_{1\parallel}) - 2y^2_4(1 - y_{1\parallel})}{9 + y^2_{2\perp} + 2y^2_4} ,$$

$$\frac{dy_{2\perp}}{dz} = (1 - y_{2\perp})\frac{y_{2\perp}(3 - y_{2\perp}) - 2y^2_4}{9 + y^2_{2\perp} + 2y^2_4} ,$$

$$\frac{dy_4}{dz} = y_4\frac{6 - 4y_{1\parallel} - (1 - y_{2\perp})y_{2\perp} - 4y_4 + 2y^2_4}{9 + y^2_{2\perp} + 2y^2_4} . \tag{3.2.9}$$

These equations have seven stationary points:

$$A(1,1,1) ,$$

$$B_1(0,0,0) , \quad B_2(0,3,0) , \quad B_3(9/5, 3,0)$$

$$C_1(9/10, 0,0), \quad C_2(0,1,0) , \quad C_3(1,1,0) .$$

The phase trajectories of the system are presented in Fig. 3.7. Points B_1-B_3 and C_1-C_3 are unstable, while point A is weakly stable. This means that if the initial values of $y_{1\parallel 0}$, $y_{2\perp 0}$, and y_{40} are such that $y_{1\parallel 0} > 0$, $0 < y_{2\perp 0} < 3$, $y_{40} > 0$ (Fig. 3.7a), the system may take an infinitely long time to tend to the given point x.

Figure 3.7a shows the phase trajectories in the plane $y_4 = 0$ and Fig. 3.7b in the $y_4 = 1$ plane. Let us consider the behavior of the system when $y_4 = 1$. When $y_{1\parallel} > 0$, $0 < y_{2\perp 0} < 3$, $y_{40} > 0$, the system will tend to fall in the given plane. While in the plane $y_4 = 1$, the system will arrive at point A if it appears in its δ_ε-neighborhood. In this case, a second-order phase transition to a TS takes place with a critical index $\gamma = 1 + \varepsilon/4 + \ldots$. Outside the δ_ε-neighborhood of the point A in the plane $y_4 = 1$, the system will wander for a long time and may reach the stability boundary of the paramagnetic phase

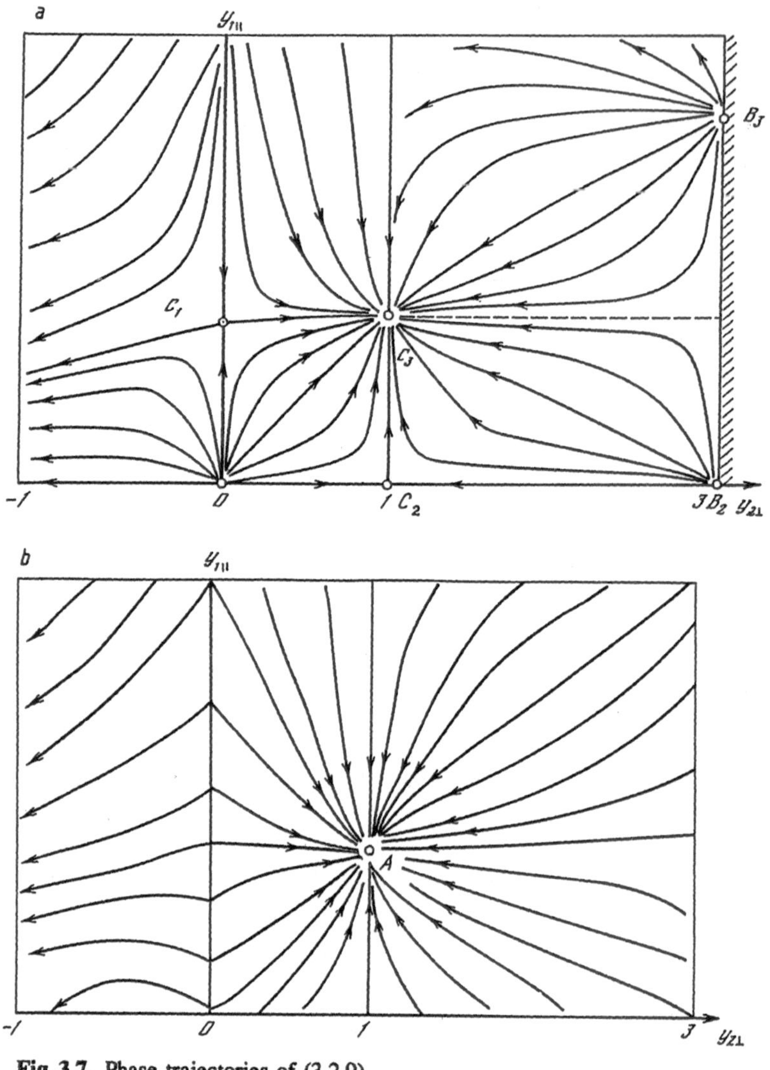

Fig. 3.7. Phase trajectories of (3.2.9)

(straight lines (1) $y_{1\parallel} = 0$, $-1 < y_{2\perp} < 3$; (2) $y_{1\parallel} > 0$, $y_{2\perp} = -1$). In this case, a first-order phase transition will occur to the c-sin state or to the NS phase, respectively. If $y_{1\parallel 0} > 0$, $-1 < y_{2\perp 0} < 0$, $y_{40} > 0$, the system cannot appear in the δ_ε-neighborhood of the point A and will move to the stability boundary of the paramagnetic phase (planes (1) $y_{1\parallel} = 0$, $-1 < y_{2\perp} < 0$, $y_4 > 0$; (2) $y_{1\parallel} > 0$, $y_{2\perp} = -1$, $y_4 > 0$). In this case, a first-order phase transition to the states shown in the phase diagrams of Fig 3.5b-d will take place, after which the system will behave in the same way as described in Sect. 3.1.

This analysis shows that the interaction mechanism for fluctuations of different components of the multi-component order parameter during phase transitions from the paramagnetic phase to the ordered state, based on the exchange enhancement of spin-orbit interaction, and allows us to isolate all high-temperature states of a complex magnetic structure. This makes it possible to construct a new type of phase diagram in T and $|a_{||}|$ variables, in which phase lamination takes place. Such a diagram can be used to predict the nature of phase transitions at high temperatures for a whole range of materials. Indeed, the anisotropy constants $a_{||}$ and $a_{\perp}$ depend on the concentration of the ions of a certain rare-earth element in the alloy, and hence the phase diagram in the temperature-anisotropy variables can be juxtaposed to the temperature-concentration phase diagram (T, x). In addition to the phase transitions in alloys of rare-earth metals, the phase diagram obtained by us describes phase transitions in pure Er, Ho [3.8]. The behavior of Ho corresponds to the case $|a_{\perp}|<|a_{||}|$, and that of Er to the case $|a_{||}|<|a_{\perp}|$ In other words, the nature of phase transitions in Er and Ho is quite different from that in [3.4].

A correct treatment of fluctuations within the framework of the ε-expansion method allows us to correctly describe all possible transitions in the magnetic structures (tilted spiral) considered. It is interesting to note that all transitions from the paramagnetic region turn out to be first-order fluctuation phase transitions. This is due to the fact that in spite of a strong anisotropy, the number of fluctuating fields is found to be sufficient for producing instabilities leading to a first-order phase transition. After the first-order phase transition, when one of the subsystems (helicoidal or sinusoidal) is condensed, the number of fluctuating fields decreases. Hence transitions to the TS phase are, as a rule, second-order phase transitions. The critical indices of these transitions have similar values. The phase diagrams shown in Fig. 3.8 have a fourfold critical point. In

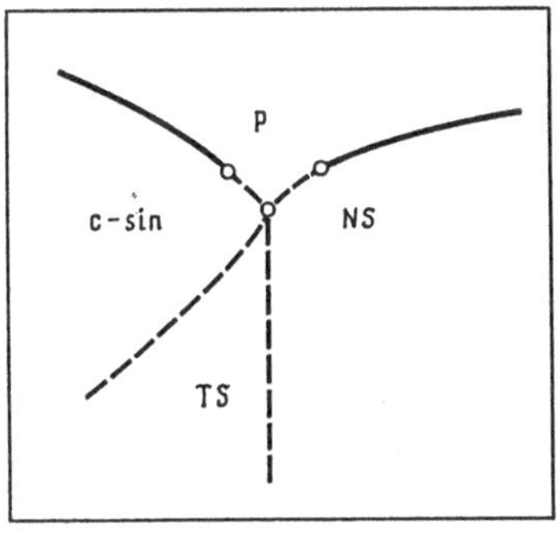

Fig. 3.8. The structure of a singularity point of a phase diagram of a complex magnetic structure of tilted spiral (TS) type. Solid lines correspond to first order phase transitions; dashed lines correspond to second order phase transitions

order to correctly describe the behavior of the system in its neighborhood for the case $y_{1\parallel} > 0$, $0 < y_{2\perp 0} < 3$, $y_{40} > 0$, we must write the RG equations including terms up to ε^2. As a result, the first-order phase transition lines near the singular point are transformed into second-order phase transition lines, i.e., a stable fourfold critical point appears in the system. This is in agreement with the above topological analysis of the phase diagram of the class of materials under consideration.

3.3 Phase Transitions in Ferroelectro-antiferromagnets

Finally we consider another example of a magnetically ordered crystal, including a ferroelectro-antiferromagnet, in which the exchange enhancement mechanism is equally effective. This allows us to obtain the phase diagram of a system with phase lamination.

According to the experimental data [3.18], a decrease in temperature ($T \to 0$ K) in a ferroelectro-antiferromagnet leads to a phase transition to the ferroelectric and antiferromagnetic states, the transition may be first ferroelectric and then magnetic (as, for example, in $PbFe_{\frac{1}{2}}Nb_{\frac{1}{2}}O_3$), or first magnetic and then ferroelectric (as, for example, in $PbFe_{2/3}W_{1/3}O_3$). As in Sects. 3.1 and 3.2, we can use the fluctuation theory of phase transitions to show that the sequence of phase transitions in ferroelectro-antiferromagnets may be different for different relations between the coefficients in the expression for the free energy. The effective Hamiltonian of a ferroelectro-antiferromagnet can be written in the form [3.19]

$$\mathbb{H}^{\text{eff}} = \int dx \left\{ \frac{1}{2}\tau_1 \ell^2 + \frac{1}{2}\tau_2 p^2 + \frac{1}{8}\Gamma_1 \ell^4 \right.$$
$$+ \frac{1}{8}\Gamma_2 p^4 + \frac{1}{4}\Gamma_3 p^2 \ell^2 + \frac{1}{2}\lambda_{iklm} u_{ik} u_{\ell m}$$
$$+ \frac{1}{2}\alpha_{ik}\left(\frac{\partial \boldsymbol{l}}{\partial x_i}\frac{\partial \boldsymbol{l}}{\partial x_k}\right) + \frac{1}{2}\beta_{ik}\frac{\partial p}{\partial x_i}\frac{\partial p}{\partial x_k}$$
$$\left. + \frac{1}{2\sqrt{2}}\Gamma_{4ik} u_{ik} \ell^2 + \frac{1}{2\sqrt{2}}\Gamma_{5ik} u_{ik} p^2 \right\}, \qquad (3.3.1)$$

where $\tau_1 = 1 - (T_{c1}/T)$; $\tau_2 = 1 - (T_{c2}/T)$; T_{c1} , T_{c2} are the initial transition temperatures to the ferroelectric and antiferromagnetic states, p is the spontaneous polarization, $\boldsymbol{l}$ the antiferromagnetism vector, u_{ik} the deformation tensor, λ_{iklm} the tensor of elastic constants, and Γ_j and Γ_{jik} the interaction amplitudes. We shall consider the isotropic model. Using Fourier transforms and integrating with respect to the variables u_{ik}, we obtain the effective Hamiltonian in the form

$$\mathbb{H}^{eff} = \frac{1}{V}\sum_{\mathbf{q}}\left[\frac{1}{2}(\tau_1 + \alpha q^2)\ell_i(\mathbf{q})\ell_i(-\mathbf{q}) + \frac{1}{2}(\tau_2 + \beta q^2)p_i(\mathbf{q})p_i(-\mathbf{q})\right]$$

$$+ \frac{1}{V^3}\sum_{\mathbf{q}_1+\mathbf{q}_2+\mathbf{q}_3+\mathbf{q}_4=0}\left[\frac{1}{8}\tilde{\Gamma}_1\ell_i(\mathbf{q}_1)\ell_i(\mathbf{q}_2)\ell_j(\mathbf{q}_3)\ell_j(\mathbf{q}_4) + \frac{1}{8}\Gamma_2\, p_i(\mathbf{q}_1)p_i(\mathbf{q}_2)p_j(\mathbf{q}_3)p_j(\mathbf{q}_4)\right]$$

$$+ \frac{1}{V^3}\sum_{\mathbf{q}_1+\mathbf{q}_2+\mathbf{q}_3+\mathbf{q}_4=0}\frac{1}{4}\Gamma_3 p_i(\mathbf{q}_1)p_i(\mathbf{q}_2)\ell_j(\mathbf{q}_3)\ell_j(\mathbf{q}_4)$$

$$+ \frac{1}{V^3}\frac{1}{8}\tilde{\Gamma}_4\Big[\sum_{\mathbf{q}}\ell_i(\mathbf{q})\ell_i(-\mathbf{q})\Big]^2 + \frac{1}{V^3}\frac{1}{8}\tilde{\Gamma}_5\Big[\sum_{\mathbf{q}}p_i(\mathbf{q})p_i(-\mathbf{q})\Big]^2;$$

$$\tilde{\Gamma}_{10} = \Gamma_{10} - 3\Gamma^2_{40}/(\lambda_0 + 2\varkappa), \quad \tilde{\Gamma}_{20} = \Gamma_{20} - 3\Gamma^2_{50}/(\lambda_0 + 2\varkappa),$$

$$\Gamma_{40} = -3\Gamma^2_{40}/2(\lambda_0 + 2\varkappa), \quad \tilde{\Gamma}_{50} = -3\Gamma^2_{50}/2(\lambda_0 + 2\varkappa), \varkappa \to 0. \qquad (3.3.2)$$

In this case, the RG equations for a single-component model $n = 1$ can be written

$$-\Gamma_1' = 9\tilde{\Gamma}_1^2 + \Gamma_3^2, \quad -\Gamma_2' = 9\tilde{\Gamma}_2^2 + \Gamma_3^2,$$

$$-\Gamma_3' = 3\tilde{\Gamma}_1\Gamma_3 + 3\tilde{\Gamma}_2\Gamma_3 + 4\Gamma_3^2,$$

$$-\tilde{\Gamma}_4' = \tilde{\Gamma}_4^2 + 6\tilde{\Gamma}_1\tilde{\Gamma}_4, \quad -\tilde{\Gamma}_5' = \tilde{\Gamma}_5^2 + 6\tilde{\Gamma}_2\tilde{\Gamma}_5, \qquad (3.3.3)$$

while the RG method variable is defined by the standard expression

$$x = \frac{2}{\varepsilon(\Lambda^2)^{\varepsilon/2}}\left\{\left(\frac{\Lambda^2}{\max[\lambda^2(\tau_1,\tau_2), \ell_0^2, p_0^2]}\right)^{\varepsilon/2} - 1\right\}, \varepsilon \to 0. \qquad (3.3.4)$$

In order to determine the thermodynamic potential of the system, we must evaluate the functional integral

$$F = -\ln \prod_{i,\mathbf{q}} D\ell_i^{(1)}(\mathbf{q}) Dp_i^{(1)}(\mathbf{q}) \exp[-\mathbb{H}^{\mathrm{eff}}(\mathbf{q}) - \;] \tag{3.3.5}$$

($\ell_i^{(1)}$ and $p_i^{(1)}$ are nonequilibrium values of the antiferromagnetism vector and polarization, respectively). Thus

$$F = (1/2)\tau_1 \ell_0^2 + (1/2)\tau_2 p_0^2 + (1/8)\tilde{\Gamma}_1 \ell_0^4 + (1/8)\tilde{\Gamma}_2 p_0^4 + (1/4)\Gamma_3 \ell_0^2 p_0^2 , \tag{3.3.6}$$

where $\tilde{\Gamma}_1 = \tilde{\Gamma}_1 + \tilde{\Gamma}_4$; $\tilde{\Gamma}_2 = \tilde{\Gamma}_2 + \tilde{\Gamma}_5$; ℓ_0 and p_0 are the equilibrium values of the antiferromagnetism vector and polarization, respectively.

Let us now consider two different cases.

1. $\tau_{10} < \tau_{20}$. The RG equations for this case have the form

$$-\tilde{\Gamma}_1' = 9\tilde{\Gamma}_1^2 , \quad -\Gamma_2' = \Gamma_3^2 , \quad -\Gamma_3' = 3\tilde{\Gamma}_1\Gamma_3 ,$$

$$-\tilde{\Gamma}_4' = \tilde{\Gamma}_4^2 + 6\tilde{\Gamma}_1\tilde{\Gamma}_4 , \quad -\tilde{\Gamma}_5' = 0 ; \tag{3.3.7}$$

$$x = \frac{1}{(\varepsilon/2)(\Lambda^2)^{\varepsilon/2}} \left\{ \left(\frac{\Lambda^2}{\max[\lambda^2(\tau_1), p_0^2]} \right)^{\varepsilon/2} - 1 \right\}, \varepsilon \to 0. \tag{3.3.8}$$

The solutions of (3.3.7) are

$$\tilde{\Gamma}_1 = \frac{\tilde{\Gamma}_{10}}{1 + 9\tilde{\Gamma}_{10}x} , \quad \Gamma_3 = \frac{\Gamma_{30}}{(1+9\tilde{\Gamma}_{10}x)^{1/3}} ,$$

$$\tilde{\Gamma}_2 = \tilde{\Gamma}_{20} - \frac{\Gamma_{30}^2}{3\tilde{\Gamma}_{10}}[(1 + 9\tilde{\Gamma}_{10}x)^{1/3} - 1] ,$$

$$\Gamma_4 = \frac{3\tilde{\Gamma}_{10}\Gamma_{40}}{\tilde{\Gamma}_{40}(1 + 9\tilde{\Gamma}_{10}x) + (3\tilde{\Gamma}_{10} - \tilde{\Gamma}_{40})(1 + 9\tilde{\Gamma}_{10}x)^{2/3}} .$$

If $\tilde{\Gamma}_5 = \tilde{\Gamma}_{50}$, a first-order phase transition to the ferroelectric state is caused by the interaction of magnetons, magnons, and acoustic phonons.

The stability boundary of the free energy (3.3.6) has the form

$$x_p^* = \frac{1}{9\tilde{\Gamma}_{10}} \left\{ \left[1 + \frac{3\tilde{\Gamma}_{10}(\Gamma_2 + \tilde{\Gamma}_{50})}{\Gamma_{30}^2} \right] - 1 \right\} . \tag{3.3.10}$$

The order parameter jump and the transition temperature are determined from

$$p_0^2 = [(2-\varepsilon)/(2+x_p^*\varepsilon)]^{2/\varepsilon} \quad ; \quad \tau_{2p} = 2b(x_p^*)p_0^2 \ , \qquad (3.3.11)$$

$$b(x_p^*) = (1/8)\Gamma_3^2 \ (x_p^*) \ . \qquad (3.3.12)$$

After transition to the ferroelectric state, the expression for the free energy is

$$F = (1/2)\tau_1\ell^2 + (1/8)\tilde{\Gamma}_1\ell^4 \ , \quad \tilde{\Gamma} = \tilde{\Gamma}_1 + \tilde{\Gamma}_4 \ . \qquad (3.3.13)$$

Here also, the magnetic subsystem temperature τ_1 has been renormalized. For $\tilde{\Gamma}_1 + \tilde{\Gamma}_4 = 0$, we obtain a first-order phase transition to the antiferromagnetic state, caused by the interaction of magnons with acoustic phonons:

$$x_\ell^* = (1/9\tilde{\Gamma}_{10})\left\{ [(\tilde{\Gamma}_{40} - 3\Gamma_{10})/4\tilde{\Gamma}_{40}]^3 - 1\right\} . \qquad (3.3.14)$$

If

$$-\frac{\tilde{\Gamma}_{10} + \tilde{\Gamma}_{40}}{4\tilde{\Gamma}_{40}} < \frac{\tilde{\Gamma}_{10}(\tilde{\Gamma}_{20} + \tilde{\Gamma}_{50})}{\Gamma_{30}^2}, \qquad (3.3.15)$$

a decrease in temperature T would lead to a first-order phase transition initially to the antiferromagnetic state, and then to the ferroelectric state.

2. $\tau_{20} < \tau_{10}$. In this case, a first-order phase transition to the antiferromagnetic state, caused by the interaction of magnons, magnetons, and acoustic phonons, is followed by a first-order phase transition to the ferroelectric state, caused by the interaction of magnetons and acoustic phonons. If

$$-\frac{\tilde{\Gamma}_{20} + \tilde{\Gamma}_{50}}{4\tilde{\Gamma}_{50}} < \frac{(\tilde{\Gamma}_{10} + \tilde{\Gamma}_{40})\tilde{\Gamma}_{20}}{\Gamma_{30}^2} \ , \qquad (3.3.16)$$

a decrease in temperature T ($T \to 0$) would first lead to a first-order phase transition to the ferroelectric state, followed by a first-order phase transition to the antiferromagnetic state.

Equations (3.3.3) have seven stationary points, and all of them are unstable:

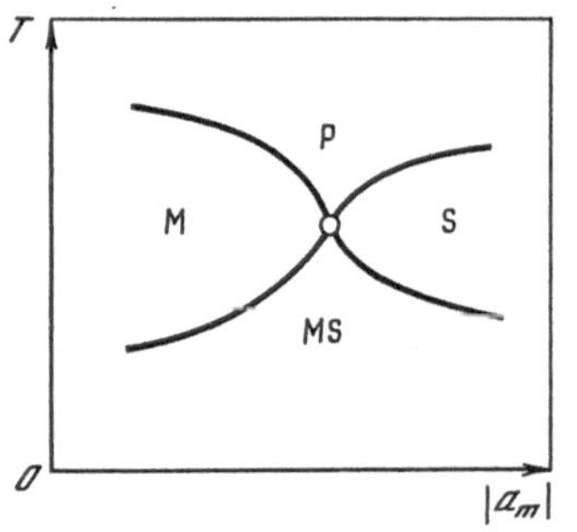

Fig. 3.9. Phase diagram of a segnetoantiferromagnet, constructed in the variables temperature and magnetic subsystem anisotropy (T, $|a_m|$). The symbols P, M, S and MS denote the paramagnetic, magnetic, ferroelectric, and segnetoantiferromagnetic states

$M_1(\varepsilon/9,\ \varepsilon/9,\ 0,\ \varepsilon/3)$, $M_2(0,\ \varepsilon/9,\ 0,\ 0,\ \varepsilon/3)$, $M_3(\varepsilon/9,\ 0,\ 0, \varepsilon/3, 0)$,

$M_4(0,0,0,0,0)$, $M_5(\varepsilon/18,\ \varepsilon/18,\ \varepsilon/6,\ 0,\ 2\varepsilon/3)$,

$M_6(\varepsilon/18,\ \varepsilon/18,\ \varepsilon/6,\ 2\varepsilon/3,\ 0)$, $M_7(\varepsilon/18,\ \varepsilon/18,\ \varepsilon/6,\ 0,\ 0)$.

Exactly in the same way as in Sect.3.2, the results of computations can be used to construct the phase diagram of ferroelectro-antiferromagnet in T, $|a_m|$ variables, where $|a_m|$ is the absolute value of the anisotropy of the magnetic subsystem (Fig. 3.9). This figure also contains a fourfold critical point. All transitions in the phase diagram are first-order transitions. This is due to the interaction of magnetons and magnons with the crystal lattice. A topological analysis of the phase diagram shows that a phase lamination takes place in the system, and the fourfold critical point may be stable. However, (3.3.3) do not have stable stationary points, which means that the critical point is unstable and a second-order phase transition directly to the ferroelectric state is impossible. This is not in contradiction with the topological analysis, since the RG equations take into account the interaction of ferroelectric and magnetic subsystems with acoustic phonons and longwave bulk deformations.

4. Superconducting Phases in Rare-Earth Metal Compounds: Phenomenological Theory

4.1 Superconducting Compounds of Rare-Earth Metals: Formulation of the Problem

In this and the next chapters, we shall construct the fluctuation theory of superconducting compounds of rare-earth metals by using the concept of phase lamination occurring in phase transitions in complex systems and based on the exchange enhancement effect.

Experimental investigations in recent years have led to the conclusion that there exists a wide range of rare-earth metal compounds in which phases can coexist during a phase transition to the ordered state. One of these phases is found to be superconducting, while the remaining phases are purely magnetic [4.1-20]. These compounds can be classified into the following two groups: $RERh_4B_4$ (Rh_6B_6) and $REMo_6S_8(Se_8)$, RE = Nd, Sm, Gd, Tb, Dy, Ho, Er, Tm and Lu. Recently, a new class of superconducting compounds of rare-earth metals has been discovered: $RERh_2Si_2$, RE_2RhSi_2, RE = La, Ce, Nd, Sm, Gd, Tb, Dy, Ho, Er [4.16,19]. All these materials behave identically upon phase transitions to the ordered state and a superconducting phase is formed in all of them. These compounds differ in the position of this phase on the temperature scale relative to the paramagnetic and magnetically ordered phases, as well as in the superconducting transition temperatures. Among the most thoroughly investigated materials of this class are the compounds $Er_{1-x}Ho_xRh_4B_4$ and the particular case $ErRh_4B_4$. We shall describe the entire body of the available experimental data and then explore the ways of constructing the theory of phase transitions in the superconducting compounds of rare-earth metals.

Let us consider the compound $ErRh_4B_4$. The following parameters have been experimentally investigated for this compound: heat capacity C as a function of temperature T, magnetic susceptibility χ_{ac} and electrical resistivity ρ as functions of temperature and external magnetic field H, the coefficient of thermal expansion, and the magnetostriction $\epsilon(T,H)$. In addition, neutron diffraction studies

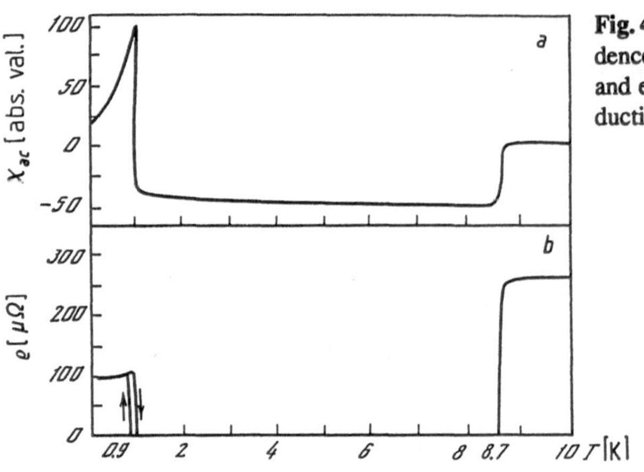

Fig. 4.1. Experimental temperature dependences of the magnetic susceptibility (**a**) and electric resistance (**b**) of the superconducting compound $ErRh_4B_4$

of the magnetic structure of the magnetically ordered phase of this compound were also performed.

If the dependence of magnetic susceptibility χ_{ac} of $ErRh_4B_4$ is studied in the absence of an external magnetic field, we obtain a dependence characteristic of the paramagnetic state at temperatures $T > 8.7$ K (Fig. 4.1a). At a temperature $T_{c1} = 8.7$ K, the susceptibility sharply drops to values below zero, which corresponds to a phase transition to the superconducting state. Upon a further decrease in temperature, the superconducting state is preserved down to a temperature $T_{c2} \simeq 0.9$ K. At $T = T_{c2}$, the susceptibility experiences a sharp increase which is followed by a decrease in the value of χ_{ac}, corresponding to a phase transition to the magnetically ordered state. The superconductivity is destroyed in this case. The neutron diffraction studies of the magnetic structure of the normal state [4.1] indicate that the magnetic structure is close to ferromagnetic.

Indeed, one neutron reflection can be clearly seen in the crystallographic directions (101) and (102). However, two reflections of different amplitudes are seen in the directions (110) and (002). Hence the results of neutron diffraction studies fail to provide unambiguous information on the magnetic structure of the normal state. The dependence $\rho(T,H)$ of the electrical resistivity of $ErRh_4B_4$ is presented in Fig. 4.1b [4.4]. The dependence $\rho(T)$ for $H = 0$ is somewhat similar to the dependence $\chi_{ac}(T)$. The only difference between the two is that the $\rho(T)$ dependence clearly shows a hysteresis of magnitude $T \simeq 0.1$ K during the phase transition at $T = T_{c2}$. This means that superconductivity and magnetism can coexist in this tem-

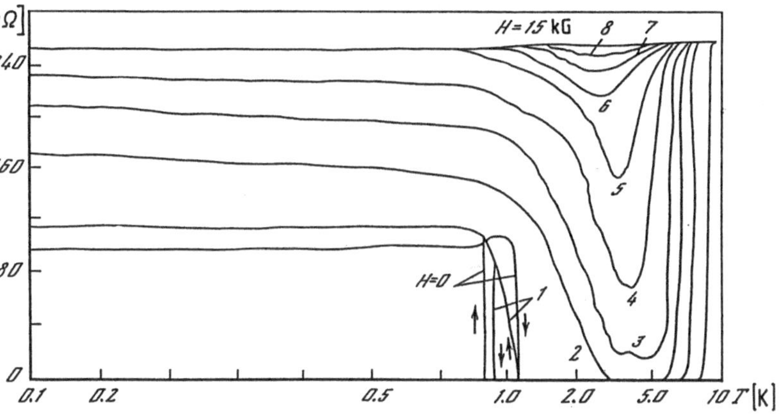

Fig. 4.2. Experimental dependence of the electrical resistance of $ErRh_4B_4$ on temperature and magnetic field strength

perature region. Transitions at T_{c1} and T_{c2} are clearly visible (Fig. 4.2) in magnetic fields of 0.1 and 2 kG, while the hysteresis still persists in the temperature region $T \sim T_{c2}$. However, superconductivity disappears at H = 3 kG, and a steep fall in resistivity is observed in the temperature region where superconductivity had existed before. The fact that hysteresis disappears together with the superconductivity confirms the possibility of coexistence of superconductivity and magnetism at $T \sim T_{c2}$. Moreover, the sign of the magnetoresistance indicates that the magnetic ordering below T_{c2} is different from ferromagnetic. This is possible if the superconducting fluctuations penetrate the temperature region below T_{c2}.

Hence, a long-range magnetic order exists in the normal phase. The positive value of the Curie-Weiss temperature Θ, obtained from the measurement of static magnetic susceptibility, confirms the existence of ferromagnetic ordering. However, the magnetoresistance in the normal phase below T_{c2} is opposite in sign to that expected for a ferromagnet. Magnetization of the system is practically independent of temperature in a constant magnetic field. This becomes possible if we have an antiferromagnetic order or an "easy plane" type of ferromagnet with a strong anisotropy in the basal plane. A similar situation arises for the conic spiral structure which is observed in pure Er at low temperatures. It is interesting to note that the measurement of χ_{ac} or ρ alone is not sufficient for determining the superconducting transition temperatures T_{c1} and T_{c2}. We must measure both χ_{ac} and ρ since the values of T_{c1} and T_{c2} in these

Table 4.1 Upper and lower superconducting transition temperatures for two samples of Er. The values of T_c and ΔT_c obtained from the resistivity measurements are shown in the denominators. The remaining values are obtained from susceptibility measurements [4.6,7].

Sample	T_{c1}, K	ΔT_{c1}, K	T_{c2}, K		ΔT_{c2}, K	
			Heating	Cooling	Heating	Cooling
A	$\frac{8.70}{8.73}$	$\frac{0.03}{0.03}$	$\frac{0.93}{0.91}$	$\frac{0.85}{0.83}$	$\frac{0.03}{0.02}$	$\frac{0.04}{0.02}$
B	8.66	0.03	0.94	0.86	0.02	0.04

experiments are not identical. Table 4.1 shows the extent to which these quantities differ [4.6,7].

Measurements of the magnetostriction $\epsilon(H)$ of $ErRh_4B_4$ for the cases $\epsilon(H) \parallel H$ and $\epsilon(H) \perp H$ [4.6] show that the strongest magnetostriction is observed in the temperature range $T \lesssim 1$ K. Such a behavior of magnetostriction at low temperatures is characteristic of magnetically ordered crystals with a complex magnetic structure and, above all, for conic spirals [4.21].

Let us now consider the compounds $Er_{1-x}Ho_xRh_4B_4$ [4.5,9]. The nature of phase transitions from paramagnetic phase to the ordered state in these compounds is much more complicated than in $ErRh_4B_4$. Indeed, in the concentration range of Ho ions $0 \leq x \lesssim 0.9$, a decrease in temperature first leads to a transition to the superconducting state. Upon a further lowering of temperature, a phase transition to the magnetically ordered phase is observed. However, with increasing concentration x, the temperature interval for the existence of the superconducting phase becomes narrower, and for $x \to x_{cr} \simeq 0.9$, the superconducting phase disappears altogether. A purely magnetic transition is observed at concentrations $0.9 \lesssim x \leq 1$. It is interesting to note that the hysteresis, which is observed at $T \sim T_{c2}$, gradually disappears with increasing concentration and is not observed experimentally for $x = 0.813$. All that has been stated above can be clearly illustrated by considering the temperature dependence of χ_{ac} for different values of the concentration x (Fig. 4.3). From the results of measurement of susceptibility χ_{ac} and heat capacity C, the authors of [4.5,9] constructed a phase diagram of the compound $Er_{1-x}Ho_xRh_4B_4$ in temperature-concentration variables (Fig. 4.4).

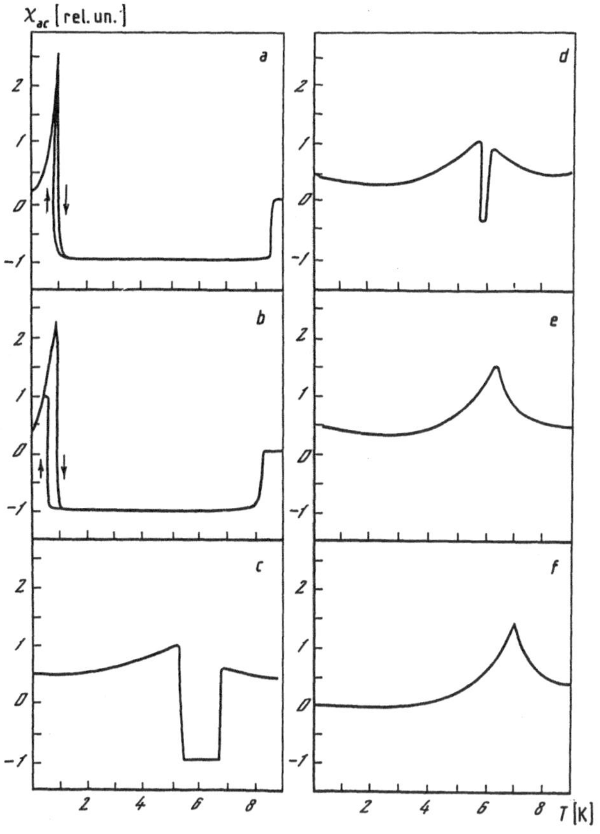

Fig. 4.3. Experimental temperature dependence of the magnetic susceptibility of the compounds $Er_{1-x}Ho_xRh_4B_4$ for different concentrations of Ho ions, x = 0 (a), 0.270 (b), 0.813 (c), 0.890 (d), 0.915 (e), and 1.00 (f)

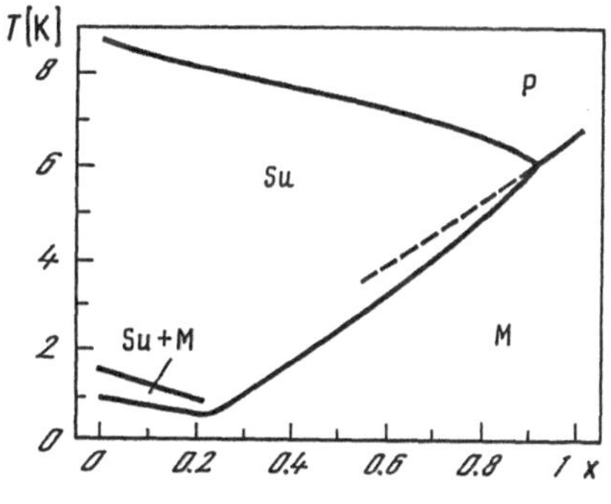

Fig. 4.4. Phase diagram of the superconducting compounds of rare-earth metals $Er_{1-x}Ho_xRh_4B_4$, constructed from experimental data

All the states arising during phase transitions from the paramagnetic phase to the ordered state are clearly shown in this diagram.

Let us consider the temperature dependence of the heat capacity of these compounds for different values of x (Fig. 4.5). For x = 1, the C(T) dependence clearly shows two peaks (see Fig. 4.5), which indicates the possibility of two magnetic transitions, the second of which must be accompanied by a change in the magnetic structure.

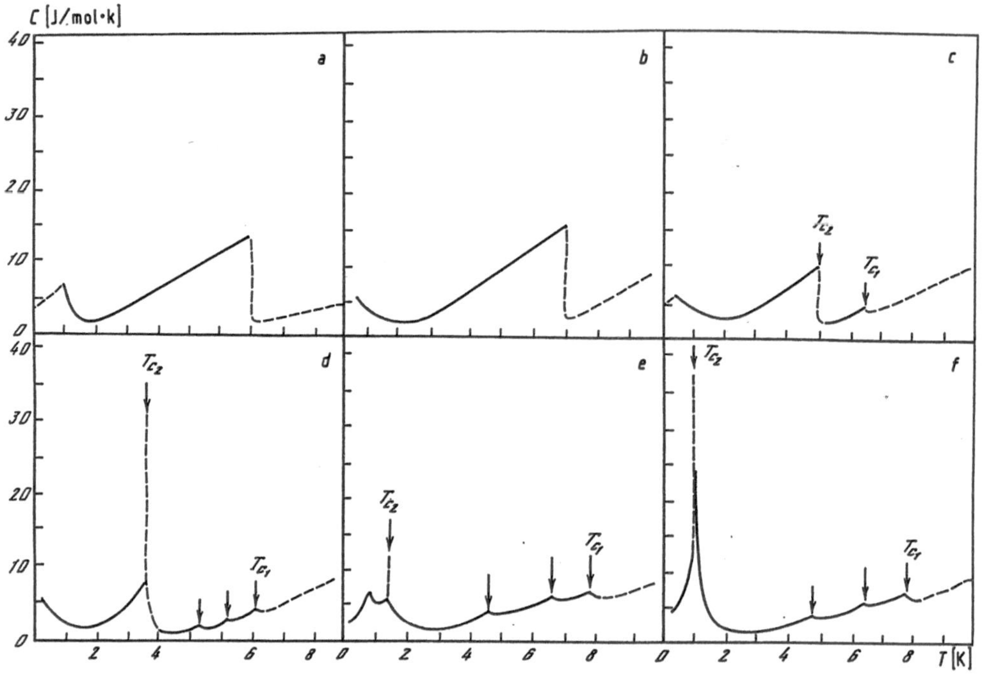

Fig. 4.5. Experimental temperature dependence of the specific heat of the compounds $Er_{1-x}Ho_xRh_4B_4$ for different concentrations of Ho ions, x = 1.00 (a), 0.912 (b), 0.813 (c), 0.60 (d), 0.30 (e), and 0 (f)

As the concentration of Ho ions decreases to x = 0.3, the magnetically ordered phase continues to exhibit two peaks. For x = 0, however, the second peak is not observed experimentally. The existence of a second magnetic phase transition points towards the complex magnetic structure of the normal phase. This fact is also noted by **Maple** [4.9]. Recent neutron diffraction patterns obtained for the magnetic structures $ErRh_4B_4$ and $HoRh_4B_4$, which are limiting cases of the compounds $Er_{1-x}Ho_xRh_4B_4$, show [4.22] that the former corresponds to an easy plane type of ferromagnet, while the latter corresponds to an easy axis type of ferromagnet. This leads to the conclusion that the magnetic structure of the compounds of this class is determined by two types of anisotropy, viz., axial anisotropy and planar anisotropy ($a_{\parallel}$, $a_{\perp}$). However, an exactly similar situation is also observed in the magnetic structure of a tilted spiral (TS) which is the most general type of magnetic symmetry (Sects. 3.1,2). hence we shall use this structure as the basis for constructing the phase diagrams of superconducting compounds of rare-earth metals of the type $Er_{1-x}Ho_xRh_4B_4$. Moreover, this magnetic structure has the best agreement with experimental observations.

4.2 Fluctuation Free Energy: Renormalization Group Equations

Having identified the most general type of magnetic symmetry in the magnetically ordered phase of the superconducting compounds $Er_{1-x}Ho_xRh_4B_4$ of rare-earth metals, we can now construct the phenomenological theory for the emergence of a superconducting phase in these compounds. We shall construct this theory on the basis of the Wilson-Fisher fluctuation theory of phase transitions [4.23,24]. As in Chap. 3, we write down an expression for the free energy of a system in the paramagnetic region. We introduce the following three order parameters [4.25,26]:

$$\mathbf{S}_{\parallel} = \mathbf{s}^{+}_{\parallel} + i\mathbf{s}^{-}_{\parallel} \ , \quad \mathbf{S}_{\perp} = \mathbf{s}^{+}_{\perp} + i\mathbf{s}^{-}_{\perp} \ , \Delta \ . \tag{4.2.1}$$

Vectors $\mathbf{S}_{\parallel}$ and $\mathbf{S}_{\perp}$ correspond to two spin-density waves of a magnetic subsystem of the TS type (Sects. 3.1,2), while Δ is the order parameter of the superconducting subsystem (band gap in the conduction electron spectrum). The second- and fourth-order invariants of exchange and relativistic type for the case of a strong magnetic anisotropy, which characterizes the experimental situation [4.5,9], can be written in the form

$$(\mathbf{S}_{\parallel} \mathbf{S}^*_{\parallel}), \quad (\mathbf{S}_{\perp} \mathbf{S}^*_{\perp}), \quad S^z_{\parallel} S^{z*}_{\parallel} \ , \quad S^z_{\perp} S^{z*}_{\perp} \ , \quad \Delta\Delta^*$$

$$(\mathbf{S}_{\parallel} \mathbf{S}^*_{\parallel})^2, \quad (\mathbf{S}_{\perp}\mathbf{S}^*_{\perp})^2, \quad \mathbf{S}^2_{\parallel} \mathbf{S}^{2*}_{\parallel} \ , \quad S^2_{\perp} S^{2*}_{\perp} \ ,$$

$$(\mathbf{S}_{\parallel} \mathbf{S}^*_{\parallel})(\mathbf{S}_{\perp}\mathbf{S}^*_{\perp}), \quad (\Delta\Delta^*)^2, \quad (\Delta\Delta^*)(\mathbf{S}_{\parallel} \mathbf{S}^*_{\parallel}), \quad (\Delta\Delta^*)(\mathbf{S}_{\perp}\mathbf{S}^*_{\perp}). \tag{4.2.2}$$

In this case, the expression for the free energy of the system will have the form [4.25,26]

$$\begin{aligned} F = \ & (1/2)\tau_{\parallel}(s^{+2}_{\parallel} + s^{-2}_{\parallel}) + (1/2)\tau_{\perp}(s^{+2}_{\perp} + s^{-2}_{\perp}) + (1/2)\tau_{\delta}|\Delta|^2 \\ & + (1/2)a_{\parallel}(s^{+2}_{\parallel z} + s^{-2}_{\parallel z}) + (1/2)a_{\perp}(s^{+2}_{\perp z} + s^{-2}_{\perp z}) + (1/8)\Gamma_{1\parallel}(s^{+4}_{\parallel} + s^{-4}_{\parallel}) \\ & + (1/8)\Gamma_{1\perp}(s^{+4}_{\perp} + s^{-4}_{\perp}) + (1/8)\Gamma_{1\delta}|\Delta|^4 + (1/4)\Gamma_{2\parallel}s^{+2}_{\parallel} s^{-2}_{\parallel} \\ & + (1/4)\Gamma_{2\perp}s^{+2}_{\perp} s^{-2}_{\perp} + (1/2)\Gamma_{3\parallel}(\mathbf{s}^{+}_{\parallel} \mathbf{s}^{-}_{\parallel})^2 + (1/2)\Gamma_{3\perp}(\mathbf{s}^{+}_{\perp} \mathbf{s}^{-}_{\perp})^2 \\ & + (1/4)\Gamma_4(s^{+2}_{\parallel} + s^{-2}_{\parallel})(s^{+2}_{\perp} + s^{-2}_{\perp}) + (1/4)\Gamma_5|\Delta|^2(s^{+2}_{\parallel} + s^{-2}_{\parallel}) \\ & + (1/4)\Gamma_6|\Delta|^2(s^{+2}_{\perp} + s^{-2}_{\perp}) \ . \end{aligned} \tag{4.2.3}$$

If the initial values of the temperature are such that $\tau_{\parallel i} \to \tau_{\perp i} \to \tau^{o}_{\delta i}$, all the fields are found to be strongly fluctuating, and the RG equations can be written in the form

$$-\Gamma_1' = (n_{||}+8)\Gamma_{1||}^2+n_{||}\Gamma_{2||}^2+4\Gamma_{2||}\Gamma_{3||}+4\Gamma_{3||}^2+2n_{\perp}\Gamma_4^2+n_{\delta}\Gamma_5^2 ,$$

$$-\Gamma_{2||}' = 2(n_{||}+2)\Gamma_{1||}\Gamma_{2||}+4\Gamma_{2||}^2+4\Gamma_{2||}\Gamma_{3||}+4\Gamma_{3||}^2+2n_{\perp}\Gamma_4^2+n_{\delta}\Gamma_5^2 ,$$

$$-\Gamma_{3||}' = 8\Gamma_{2||}\Gamma_{3||}+4\Gamma_{1||}\Gamma_{3||}+2(n_{||}+2)\Gamma_{3||}^2 ,$$

$$-\Gamma_{1\perp}' = (n_{\perp}+8)\Gamma_{1\perp}^2+n_{\perp}\Gamma_{2\perp}^2+4\Gamma_{2\perp}\Gamma_{3\perp}+4\Gamma_{3\perp}^2+2n_{||}\Gamma_4^2+n_{\delta}\Gamma_6^2 ,$$

$$-\Gamma_{2\perp}' = 2(n_{\perp}+2)\Gamma_{1\perp}\Gamma_{2\perp}+4\Gamma_{2\perp}^2+4\Gamma_{2\perp}\Gamma_{3\perp}$$
$$+4\Gamma_{3\perp}^2+2n_{||}\Gamma_4^2+n_{\delta}\Gamma_6^2 ,$$

$$-\Gamma_{3\perp}' = 8\Gamma_{2\perp}\Gamma_{3\perp}+\Gamma_{1\perp}\Gamma_{3\perp}+2(n_{\perp}+2)\Gamma_{3\perp}^2 ,$$

$$-\Gamma_4' = (n_{||}+2)\Gamma_{1||}\Gamma_4+(n_{\perp}+2)\Gamma_{1\perp}\Gamma_4+n_{||}\Gamma_{2||}\Gamma_4$$
$$+n_{\perp}\Gamma_{2\perp}\Gamma_4+2\Gamma_{3||}\Gamma_4+2\Gamma_{3\perp}\Gamma_4+4\Gamma_4^2+n_{\delta}\Gamma_5\Gamma_6 ,$$

$$-\Gamma_{1\delta}' = (n_{\delta}+8)\Gamma_{1\delta}^2+2n_{||}\Gamma_5^2+2n_{\perp}\Gamma_6^2 ,$$

$$-\Gamma_5' = (n_{||}+2)\Gamma_{1||}\Gamma_5+n_{||}\Gamma_{2||}\Gamma_5$$
$$+2\Gamma_{3||}\Gamma_5+(n_{\delta}+2)\Gamma_{1\delta}\Gamma_5+4\Gamma_5^2+2n_{\perp}\Gamma_4\Gamma_6 ,$$

$$-\Gamma_6' = (n_{\perp}+2)\Gamma_{1\perp}\Gamma_6+n_{\perp}\Gamma_{2\perp}\Gamma_6+2\Gamma_{3\perp}\Gamma_6+4\Gamma_6^2$$
$$+2n_{||}\Gamma_5\Gamma_4+(n_{\delta}+2)\Gamma_{1\delta}\Gamma_6 ,$$

$$-\tau_{||}' = \tau_{||}[(n_{||}+2)\Gamma_{1||}+n_{||}\Gamma_{2||}+2\Gamma_{3||}]+2n_{\perp}\tau_{\perp}\Gamma_4+n_{\delta}\tau_{\delta}\Gamma_5 ,$$
$$-\tau_{\perp}' = \tau_{\perp}[(n_{\perp}+2)\Gamma_{1\perp}+n_{\perp}\Gamma_{2\perp}+2\Gamma_{3\perp}]+2n_{||}\tau_{||}\Gamma_4+n_{\delta}\tau_{\delta}\Gamma_6 ,$$
$$-\tau_{\delta}' = (n_{\delta}+2)\tau_{\delta}\Gamma_{1\delta}+2n_{||}\tau_{||}\Gamma_5+2n_{\perp}\tau_{\perp}\Gamma_6 . \qquad (4.2.4)$$

Here, $n_{||}$, $n_{\perp}$, and n_{δ} are the dimensions of the fields $s_{||}^{\pm}$, $s_{\perp}^{\pm}$ and Δ, respectively. The variable used in the RG method is defined by the formula

$$x = \frac{2}{\varepsilon(\Lambda^2)^{\varepsilon/2}}\left\{\left(\frac{\Lambda^2}{\max[\lambda^2(\tau_{||},\tau_{\perp},\tau_{\delta}),s_{||o}^2,s_{\perp o}^2,|\Delta_o|^2]}\right)^{\varepsilon/2}-1\right\},\ \varepsilon\to 0. \qquad (4.2.5)$$

Analysis of the set of equations (4.2.4) is complicated because of their cumbersomeness. Hence we shall use the tilted spiral model in which only one helicoidal polarization exists (Sects. 3.1,2). In this case, the expression (4.2.3) for free energy becomes simplified and can be written in the form

$$
\begin{aligned}
F = {} & (1/2)\tau_{\parallel}(s_{\parallel}^{+2} + s_{\parallel}^{-2}) + (1/2)\tau_{\perp}(s_{\perp}^{+2} + s_{\perp}^{-2}) \\
& + (1/2)\tau_{\delta}|\Delta|^{2} + (1/8)\Gamma_{1\parallel}(s_{\parallel}^{+2} + s_{\parallel}^{-2})^{2} + (1/8)\Gamma_{1\perp}(s_{\perp}^{+4} + s_{\perp}^{-4}) \\
& + (1/4)\Gamma_{2\perp}s_{\perp}^{+2}s_{\perp}^{-2} + (1/8)\Gamma_{1\delta}|\Delta|^{4} \\
& + (1/4)\Gamma_{4}(s_{\parallel}^{+2} + s_{\parallel}^{-2})(s_{\perp}^{+2} + s_{\perp}^{-2}) \\
& + (1/4)\Gamma_{5}|\Delta|^{2}(s_{\parallel}^{+2} + s_{\parallel}^{-2}) + (1/4)\Gamma_{6}|\Delta|^{2}(s_{\perp}^{+2} + s_{\perp}^{-2}) .
\end{aligned}
\tag{4.2.6}
$$

The following relations are satisfied for the initial values of amplitudes:

$$
\begin{aligned}
& \Gamma_{1\perp 0} = \Gamma_{1\parallel 0} = \Gamma_{10} ,\ \Gamma_{2\perp 0} = \Gamma_{1\perp 0} - 2\Gamma_{30},\ \Gamma_{40} = \Gamma_{50} = \Gamma_{60} = \Gamma_{20} \\
& \Gamma_{10} - \Gamma_{30} < \Gamma_{1\delta 0} < \Gamma_{10} < \Gamma_{20} ,
\end{aligned}
\tag{4.2.7}
$$

while the initial values of temperatures $\tau_{\parallel 0}, \tau_{\perp 0}, \tau_{\delta 0}$ are functions of the concentration of Ho ions x:

$$
\begin{aligned}
& \tau_{\parallel 0} = \tau - |a_{\parallel}(x)| ,\ \tau_{\perp 0} = \tau - |a_{\perp}(x)| , \\
& \tau_{\delta 0} = \tau - f(|a_{\parallel}(x)| ,\ |a_{\parallel}(x)|) .
\end{aligned}
\tag{4.2.8}
$$

As $\tau_{\parallel 0} \to \tau_{\perp 0} \to \tau_{\delta 0}$, all fields are found to be strongly fluctuating and the RG equations for amplitudes and temperatures can be easily obtained from (4.2.4):

$$
\begin{aligned}
-\Gamma'_{1\parallel} &= 10\Gamma_{1\parallel}^{2} + 2\Gamma_{4}^{2} + \Gamma_{5}^{2} ,\quad -\Gamma'_{1\perp} = 9\Gamma_{1\perp}^{2} + \Gamma_{2\perp}^{2} + 2\Gamma_{4}^{2} + \Gamma_{6}^{2} , \\
-\Gamma'_{2\perp} &= 6\Gamma_{1\perp}\Gamma_{2\perp} + 4\Gamma_{2\perp}^{2} + 2\Gamma_{4}^{2} + \Gamma_{6}^{2} ,\quad -\Gamma'_{1\delta} = 9\Gamma_{1\delta}^{2} + 2\Gamma_{\delta}^{2} + 2\Gamma_{6}^{2} , \\
-\Gamma'_{4} &= 4\Gamma_{1\parallel}\Gamma_{4} + (3\Gamma_{1\perp} + \Gamma_{2\perp})\Gamma_{4} + 4\Gamma_{4}^{2} + \Gamma_{5}\Gamma_{6} , \\
-\Gamma'_{5} &= 4\Gamma_{1\parallel}\Gamma_{5} + 4\Gamma_{5}^{2} + 2\Gamma_{4}\Gamma_{6} + 3\Gamma_{1\delta}\Gamma_{5} , \\
-\Gamma'_{6} &= (3\Gamma_{1\perp} + \Gamma_{2\perp})\Gamma_{6} + 4\Gamma_{6}^{2} + 2\Gamma_{4}\Gamma_{5} + 3\Gamma_{1\delta}\Gamma_{6} ,
\end{aligned}
$$

$$-\tau_{\perp}' = \tau_{\perp}(3\Gamma_{1\perp} + \Gamma_{2\perp}) + 2\tau_{\parallel}\Gamma_4 + \tau_{\delta}\Gamma_6 \,,$$

$$-\tau_{\parallel}' = 4\tau_{\parallel}\Gamma_{1\parallel} + 2\tau_{\perp}\Gamma_4 + \tau_{\delta}\Gamma_5, \quad -\tau_{\delta}' = 3\tau_{\delta}\Gamma_{1\delta} + 2\tau_{\parallel}\Gamma_5 + 2\tau_{\perp}\Gamma_6 \,. \tag{4.2.9}$$

The variable x in the RG method is defined by (4.2.5). Using (4.2.6) for the free energy of the system in the paramagnetic phase, as well as (4.2.9), we can construct the phase diagram of the compound $Er_{1-x}Ho_xRh_4B_4$. We shall construct this diagram in temperature-anisotropy variables (T, $|a_{\parallel}(x)|$). It was shown in Sects. 3.1,2 that such a phase diagram is equivalent to the phase diagram in temperature-concentration variables, since the magnetic anisotropy of rare-earth metal alloys is determined by concentration of some element in the alloy. Consequently, the theoretical phase diagram will be equivalent to the phase diagram determined experimentally in [4.5,9] in (T,x) variables. While studying the phase diagram of these compounds, we shall use the mechanism of phase lamination based on the exchange enhancement effect (Sects. 3.1-3).

4.3 Phase Diagram of the Superconducting Compounds of Rare-Earth Metals

We shall construct the phase diagram of the system by proceeding from the nature of the phase transitions, which depends on the relation between the initial values of the temperatures $\tau_{\parallel 0}$, $\tau_{\perp 0}$, and $\tau_{\delta 0}$, as in Sects. 3.1-3.

1. $\tau_{\perp 0} < \min(\tau_{\parallel 0}, \tau_{\delta 0})$. In this case, the field $\mathbf{S}_{\perp}$ fluctuates most strongly upon a transition from the paramagnetic phase, while the fluctuations of the fields Δ and $\mathbf{S}_{\parallel}$ are weaker.

The RG equations have the form

$$-\Gamma_{1\parallel}' = 2\Gamma_4^2 \,, \quad -\Gamma_{1\perp}' = 9\Gamma_{1\perp}^2 + \Gamma_{2\perp}^2 \,, \quad -\Gamma_2' = 6\Gamma_{1\perp}\Gamma_{2\perp} + 4\Gamma_{2\perp}^2 \,,$$

$$-\Gamma_4' = (3\Gamma_{1\perp} + \Gamma_{2\perp})\Gamma_4 \,, \quad -\Gamma_5' = 2\Gamma_4\Gamma_6 \,, \quad -\Gamma_{\delta}' = 2\Gamma_6^2 \,,$$

$$-\Gamma_6' = (3\Gamma_{1\perp} + \Gamma_{2\perp})\Gamma_6 \,, \quad -\tau_{\parallel}' = 2\tau_{\perp}\Gamma_4 \,,$$

$$-\tau_{\perp}' = \tau_{\perp}(3\Gamma_{1\perp} + \Gamma_{2\perp}), \quad -\tau_{\delta}' = 2\tau_{\perp}\Gamma_6 \,; \tag{4.3.1}$$

$$x = \frac{2}{\varepsilon(\Lambda^2)^{\varepsilon/2}}\left\{\left(\frac{\Lambda^2}{\max[\lambda^2(\tau_{\perp}),\, s_{\parallel 0}^2,\, |\Delta_0|^2]}\right)^{\varepsilon/2} - 1\right\}, \varepsilon \to 0 \,. \tag{4.3.2}$$

Introducing the new variables

$$y_{1\parallel} = \Gamma_{1\parallel}/\Gamma_{1\perp}\,, \quad y_{2\perp} = \Gamma_{2\perp}/\Gamma_{1\perp}\,, \quad y_{1\delta} = \Gamma_{1\delta}/\Gamma_{1\perp}\,,$$

$$y_4 = \Gamma_4/\Gamma_{1\perp}\,, \quad y_5 = \Gamma_5/\Gamma_{1\perp}\,, \quad y_6 = \Gamma_6/\Gamma_{1\perp}\,,$$

$$z = -\ln\Gamma_{1\perp}\,,$$

we can reduce (4.3.1) to quadratures which can be easily calculated. Integrating, we obtain

$$\frac{\Gamma_{1\parallel}(x)}{\Gamma_{10}} = 1 - \frac{2y_{40}^2}{y_{2\perp0}(3 - y_{2\perp0})}\,\frac{y_{2\perp} - y_{2\perp0}}{1 - y_{2\perp}}\,,$$

$$\frac{\Gamma_{1\perp}(x)}{\Gamma_{10}} = \left(\frac{y_{2\perp0}}{y_{2\perp}}\right)^3\left(\frac{1 - y_{2\perp}}{1 - y_{2\perp0}}\right)^5\left(\frac{3 - y_{2\perp0}}{3 - y_{2\perp}}\right)^3\,,$$

$$\frac{\Gamma_{1\delta}(x)}{\Gamma_{10}} = y_{1\delta0} - \frac{2y_{40}^2}{y_{2\perp0}(3 - y_{2\perp0})}\,\frac{y_{2\perp} - y_{2\perp0}}{1 - y_{2\perp}}\,,$$

$$\frac{\Gamma_4(x)}{\Gamma_{10}} = y_{40}\left(\frac{y_{2\perp0}}{y_{2\perp}}\right)\left(\frac{1 - y_{2\perp}}{1 - y_{2\perp0}}\right)^2\left(\frac{3 - y_{2\perp0}}{3 - y_{2\perp}}\right)\,,$$

$$\frac{\Gamma_5(x)}{\Gamma_{10}} = y_{40} - \frac{2y_{40}^2}{y_{2\perp0}(3 - y_{2\perp0})}\,\frac{y_{2\perp} - y_{2\perp0}}{1 - y_{2\perp}}\,,$$

$$\frac{\tau_{\parallel}(x)}{\tau_{\parallel0}} = 1 - \frac{2\tau_{\perp0}\,y_{40}}{\tau_{\parallel0}y_{2\perp0}(3 - y_{2\perp0})}\,\frac{y_{2\perp} - y_{2\perp0}}{1 - y_{2\perp}}\,,$$

$$\frac{\tau_{\delta}(x)}{\tau_{\delta0}} = 1 - \frac{2\,\tau_{\perp0}y_{40}}{\tau_{\delta0}y_{2\perp0}(3 - y_{2\perp0})}\,\frac{y_{2\perp} - y_{2\perp0}}{1 - y_{2\perp}}\,,$$

$$\frac{\Gamma_6(x)}{\Gamma_{10}} = \frac{\Gamma_4(x)}{\Gamma_{10}}\,, \quad \frac{\tau_{\perp}(x)}{\tau_{\perp0}} = \frac{1}{y_{40}}\,\frac{\Gamma_4(x)}{\Gamma_{10}}\,. \qquad (4.3.3)$$

Let $0 < y_{2\perp0} < 3$. This means that the transverse magnetic subsystem tends to a stable stationary point $y_{2\perp} = 1$ (Sect. 3.1). The longitudinal magnetic and superconducting subsystems are unstable due to an interaction with the fluctuations of the transverse magnetic subsystem. However, in view of the initial conditions for the amplitudes (4.2.7), the vertex $\Gamma_{1\delta}$ is the first to lose positive definiteness. Consequently, the system will undergo a first-order phase transition to the superconducting state. The stability boundary for the first-order phase transition is defined by the relation

$$\Gamma_{10}x_\delta^* = \frac{(1 - y_{2\perp 0})^5}{y_{2\perp 0}^3(3 - y_{2\perp 0})} w(y_{2\perp\delta}^*, y_{2\perp 0}) , \tag{4.3.4}$$

where $w(y_{2\perp\delta}^*, y_{2\perp 0})$ is defined by (3.1.32), while

$$y_{2\perp\delta}^* = y_{2\perp 0} \frac{2y_{40}^2 + y_{1\delta 0}(3 - y_{2\perp 0})}{2y_{40}^2 + y_{1\delta 0}y_{2\perp 0}(3 - y_{2\perp 0})} . \tag{4.3.5}$$

The vertices of the cross-multiplication terms for the free energy and temperatures $\tau_{\parallel}$, $\tau_{\perp}$, τ_δ at the point $x = x_\delta^*$ are

$$\begin{aligned}
\Gamma_5(x_\delta^*)/\Gamma_{10} &= y_{40} - y_{1\delta 0} , \\
\Gamma_4(x_\delta^*)/\Gamma_{10} &= 2y_{40}^5/[2y_{40}^2 + y_{1\delta 0}(3 - y_{2\perp 0})][y_{40}^2 + y_{1\delta 0}y_{2\perp 0}], \\
\tau_{\parallel}(x_\delta^*)/\tau_{\parallel 0} &= 1 - (\tau_{\perp 0}y_{1\delta 0}/\tau_{\parallel 0}y_{40}) , \\
\tau_\delta(x_\delta^*)/\tau_{\delta 0} &= 1 - (\tau_{\perp 0}y_{1\delta 0}/\tau_{\delta 0}y_{40}) , \\
\tau_{\perp 0}(x_\delta^*)/\tau_{\perp 0} &= 2y_{40}^4/[2y_{40}^2 + y_{1\delta 0}(3 - y_{2\perp 0})][y_{40}^2 + y_{1\delta 0}y_{2\perp 0}].
\end{aligned} \tag{4.3.6}$$

The expression for the free energy of the superconducting subsystem may be obtained by summation of the loop graphs shown in Fig. 4.6, and has the standard form

$$\begin{aligned}
F_\Delta &= (1/2)\tau_\delta'|\Delta_0|^2 + (1/8)\tilde{\Gamma}_{1\delta}|\Delta_0|^4 = (1/2)\tau_\delta\,|\Delta_0|^2 \\
&\quad + b(x_\delta^*)\,|\Delta_0|^4 \ln(|\Delta_0|^2/\lambda_\delta^2(\tau_\perp)) , \\
b(x_\delta^*) &= (1/4)\Gamma_6^2(x_\delta^*) , \quad |\Delta_0|^2 = \tau_\delta^2/2b(x_\delta^*) , \\
\tau_\delta^0 &= 2b(x_\delta^*)[(2 - \varepsilon)/(2 + x_\delta^*\varepsilon)]^{2/\varepsilon} , \quad \varepsilon \to 0 .
\end{aligned} \tag{4.3.7}$$

Let x_δ^0 be the value of the variable x at the first-order superconducting transition point. Then the vertices Γ_5 and Γ_6 at the transition point are defined by the expressions

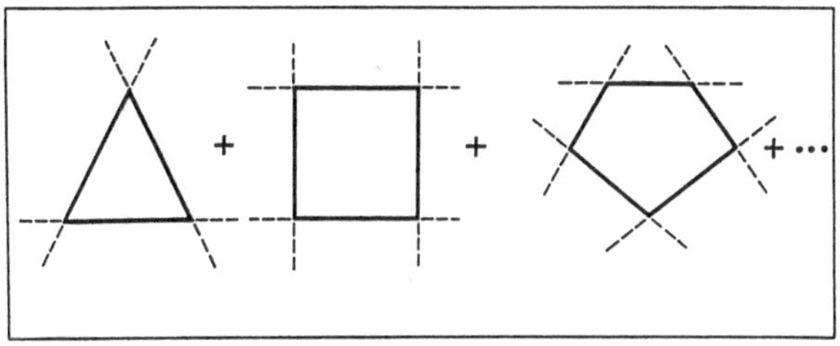

Fig. 4.6. Loop graphs whose summation allows us to determine the thermodynamic potential of the superconducting phase near the line of first-order phase transition to the paramagnetic state

$$\frac{\Gamma_5(x_\delta^0)}{\Gamma_{10}} = y_{40}\left(1 - \frac{\tau_{\delta 0}}{\tau_{\perp 0}}\Delta\tau_{\delta 0}\right), \qquad (4.3.8)$$

$$\frac{\Gamma_6(x_\delta^0)}{\Gamma_{10}} = \frac{2\tau_{\perp 0}^2 y_{40}^3}{[2\tau_{\perp 0}y_{40} + \tau_{\delta 0}\Delta\tau_{\delta 0}][\tau_{\perp 0}y_{40} + \tau_{\delta 0}y_{2\perp 0}\Delta\tau_{\delta 0}]},$$

$$\Gamma_{10}x_\delta^0 = \frac{(1 - y_{2\perp 0})^5}{y_{2\perp 0}^3(3 - y_{2\perp 0})^3}\, w(y_{2\perp\delta}^0, y_{2\perp 0}),$$

$$\Delta\tau_{\delta 0} = 1 - \tau_\delta^0/\tau_{\delta 0}, \qquad (4.3.9)$$

where the quantity $y^\delta_{2\perp 0}$ must satisfy the relations

$$0 < y_{2\perp 0} < 1, \quad y_{2\perp 0} < y_{2\perp\delta}^0 < y_{2\perp 0}\frac{2y_{40} + 3 - y_{2\perp 0}}{2y_{40} + y_{2\perp 0}(3 - y_{2\perp 0})}$$

$$1 < y_{2\perp 0} < 3, \quad y_{2\perp 0}\frac{2y_{40} + 3 - y_{2\perp 0}}{2y_{40} + y_{2\perp 0}(3 - y_{2\perp 0})} < y_{2\perp\delta}^0 < y_{2\perp 0}. \qquad (4.3.10)$$

Formulas (4.3.8,9) can be transformed as follows:

$$\frac{\Gamma_5(x_\delta^0)}{\Gamma_{10}} = y_{40}(1 - \zeta), \quad \frac{\Gamma_6(x_\delta^0)}{\Gamma_{10}} = \frac{y_{40}}{(1+\xi)(1+\eta)}, \qquad (4.3.11)$$

where

$$\zeta = Ay_{1\delta 0}/y_{40}, \quad \eta = Ay_{1\delta 0}y_{2\perp 0}/y_{40}^2,$$

$$\xi = Ay_{1\delta 0}(3 - y_{2\perp 0})/2y_{40}^2,$$

$$A = 2y_{40}^2(y_{2\perp\delta}^0 - y_{2\perp 0})/y_{1\delta 0}y_{2\perp 0}(3 - y_{2\perp 0})(1 - y_{2\perp\delta}^0). \qquad (4.3.12)$$

Following the transition to the superconducting state, the temperatures $\tau_{\parallel}$ and $\tau_{\perp}$ change. Their increments are given by

$$\Delta\tau_{\parallel} = (1/2)\Gamma_5(x_\delta^0)|\Delta_0|^2 , \quad \Delta\tau_{\perp} = (1/2)\Gamma_6(x_\delta^0)|\Delta_0|^2 . \tag{4.3.13}$$

The expression for the free energy of the magnetic subsystem is analogous to (3.1.16), except for the coefficients of $\tau_{\parallel}, \tau_{\perp}, \Gamma_{1\perp}, \Gamma_{2\perp}$ and Γ_4. The initial conditions for the amplitudes and temperatures are different, and are determined by the value of x_δ^0. Let $[\Gamma_6(x_\delta^0)/\Gamma_{10}] \lesssim [\Gamma_5(x_\delta^0)/\Gamma_{10}]$. This means that the inequality $\tau_{\perp}<\tau_{\parallel}$ between the temperatures is preserved. In this case, the transverse magnetic subsystem fluctuates strongly, while the longitudinal subsystem corresponding to the order parameter of $\mathbf{S}_{\parallel}$ becomes unstable due to interaction with the fluctuations of the transverse subsystem ($\mathbf{S}_{\perp}$). Since $0 < y_{2\perp\delta}^0 < 3$, it means that the transverse magnetic subsystem tends to the stable stationary point $y_{2\perp} = 1$ as before. Consequently, the system undergoes a first-order phase transition to the magnetically ordered state with a longitudinal sinusoidal spin-density wave. The expression for the effective free energy of the longitudinal magnetic subsystem can be written in the form

$$F_{\parallel} = \frac{1}{2}\tau_{\parallel} s_{\parallel 0}^2 + s_{\parallel 0}^4\, b(x_5^*) \ln \frac{s_{\parallel 0}^2}{\lambda_5^{2*}(\tau_{\perp})} . \tag{4.3.14}$$

The jump $s_{\parallel 0}^2$ in the order parameter and the transition temperature are respectively given by

$$s_{\parallel 0}^2 = \frac{\tau_{\parallel s}^0}{2b(x_s^*)} , \quad \tau_{\parallel s}^0 = 2b(x_s^*)\left(\frac{2-\varepsilon}{2+x_s^*\varepsilon}\right)^{2/\varepsilon}, \varepsilon \to 0 , \tag{4.3.15}$$

$$b(x_s^*) = (1/4)\Gamma_4^2(x_s^*) , \tau_{\parallel}(x_s^*) = \tau_{\parallel 0} - (\tau_{\perp 0} y_{1\parallel 0}/y_{40}) . \tag{4.3.16}$$

The temperature of transition to the c-sin state can be written in the same way:

$$\tau_{\parallel s}^0 = \tau_{\parallel 0} - \frac{2\tau_{\perp 0} y_{40}(y_{2\perp s}^0 - y_{2\perp 0})}{y_{2\perp 0}(3 - y_{2\perp 0})(1 - y_{2\perp s}^0)} , \tag{4.3.17}$$

where $y_{2\perp s}^0$ is defined by (3.1.39), or

$$\tau_{1s}^0 = \tau_{\parallel 0} - \tau_{\perp 0}(y_{1\parallel 0}/y_{40})D , \tag{4.3.18}$$

$$D = [2y_{40}^2/y_{1\,0}y_{2\perp 0}(3 - y_{2\perp 0})][(y_{2\perp s}^0 - y_{2\perp 0})/(1 - y_{2\perp s}^0)] . \tag{4.3.19}$$

Considering that

$$\tau_{\parallel 0} = \alpha_0 + \frac{1}{2}\Gamma_5(x_\delta^0)|\Delta_0|^2 , \quad \tau_{\perp 0} = \beta_0 + \frac{1}{2}\Gamma_6(x_\delta^0)|\Delta_0|^2 , \quad (4.3.20)$$

$$\alpha_0 < 0 , \quad \beta_0 < 0 , \quad |\alpha_0|<|\beta_0| ,$$

we obtain

$$\tau^0_{\parallel s} = [\alpha_0 - \beta_0(y_{1\parallel}/y_{40})D] + (1/2)[\Gamma_5(x_\delta^0) - \Gamma_6(x_\delta^0)(y_{1\parallel 0}/y_{40})D]|\Delta_0|^2 .$$

Thus, on the basis of the above analysis, we can write the general expression for energy when both the superconducting and magnetic order parameters $|\Delta_0|$ and $s_{\parallel 0}$ are nonzero [4.26]:

$$F_{\delta\parallel} = \frac{1}{2}a_\delta(\tau)|\Delta_0|^2 + A_\delta(\tau)\Phi_\delta\left[\frac{|\Delta_0|^2}{g_\delta(\tau)}\right]|\Delta_0|^4 + \frac{1}{2}a_\parallel(\tau)s^2_{\parallel 0}$$
$$+ A_\parallel(\tau)\Phi_\parallel\left[\frac{s^2_{\parallel 0}}{g_\parallel(\tau)}\right]s^4_{\parallel 0} + A_{\delta\parallel}(\tau)|\Delta_0|^2 s^2_{\parallel 0} . \quad (4.3.21)$$

The functions Φ_δ and $\Phi_\parallel$ stabilize the system during a first-order phase transition:

$$\lim_{\varepsilon\to 0}\Phi_\delta\left[\frac{|\Delta_0|^2}{g_\delta(\tau)}\right] \to \ln\frac{|\Delta_0|^2}{g_\delta(\tau)} ,$$

$$\lim_{\varepsilon\to 0}\Phi_\parallel\left(\frac{s^2_{\parallel 0}}{g_\parallel(\tau)}\right) \to \ln\frac{s^2_{\parallel 0}}{g_\parallel(\tau)} , \quad (4.3.22)$$

where $\tau = 1 - (T/T_{c1})$, T_{c1} is the temperature of the first superconducting transition, $0 \le \tau \le 1$; $a_\delta(0) = \tau^0_\delta$; $A_\delta(0) = (1/4)\Gamma^2_6(x^0_\delta)$; $g_\delta(0) = \lambda^{2*}_\delta(\tau_\perp)$; $g_\parallel(\tau_0) = \lambda^{2*}_\parallel(\tau_\perp)$; $a_\parallel(\tau_0) + A_{\delta\parallel}(\tau_0)|\Delta_0|^2 = \tau^0_{\parallel s}$; $A_\parallel(\tau_0) = (1/4)\Gamma^2_4(x^0_{\parallel s})$; $A_{\delta\parallel}(\tau_0) = (1/4)[\Gamma_5(x^0_\delta) - (Dy_{1\parallel}/y_{40})\times\Gamma_6(x^0_\delta)]$; $\tau_0 = 1 - T_\parallel/T_{c1}$, and $T_\parallel$ is the temperature of transition to the c-sin state.

The functions $g_\delta(\tau)$, $g_\parallel(\tau)$, $A_\delta(\tau)$, and $A_\parallel(\tau)$ are slowly varying functions of τ, while the nature of the dependences $a_\delta(\tau)$ and $a_\parallel(\tau)$ can be determined from the dependence of the free energy $F_\delta(F_\parallel)$ of the system on the order parameter $|\Delta_0|$ $(s_{\parallel 0})$ (Fig. 4.7). As a result, we find that the functions $a_\delta(\tau)$ and $a_\parallel(\tau)$ are monotonically decreasing functions. The appearance of magnetic order in the superconducting phase destroys the superconductivity, and if $a_\delta(\tau_0) + 2A_{\delta\parallel}(\tau_0)s^2_{\parallel 0}$

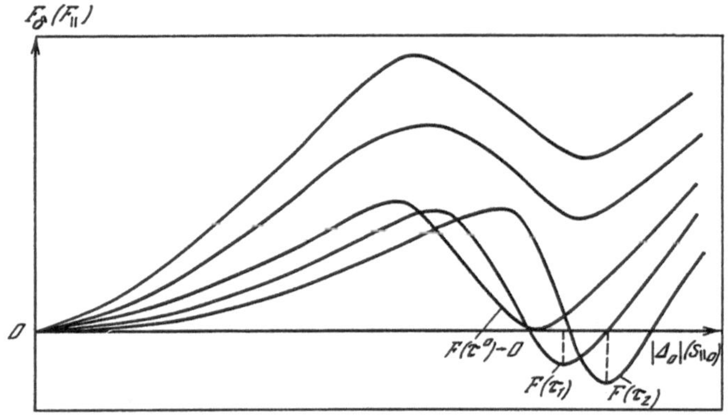

Fig. 4.7. Dependence of the free energy of a superconducting or magnetic (c-sin) phase on the magnitude of the order parameter. The existence of two minima indicates the possibility of a first-order phase transition to the corresponding state

$>\tau_\delta^0$, the free energy of the superconducting phase becomes unstable and the superconductivity disappears. For

$$a_\delta(\tau_0) + 2A_{\delta\,||}(\tau_0)s_{||0}^2 < \tau_\delta^0$$

a coexistence of superconductivity and magnetism in the form of a longitudinal sinusoidal spin density wave Su + c-sin becomes possible. As the temperature decreases, the order parameter $s_{||0}$ increases, and a first-order phase transition to the c-sin state takes place for

$$a_\delta(\tau) + 2A_{\delta\,||}(\tau)s_{||0}^2 = \tau_\delta^0 . \tag{4.3.23}$$

This equation allows us to determine the temperature of magnetic transition and the temperature interval $\Delta T_{\delta\,||}$ in which superconductivity and magnetism coexist [4.26]. A further decrease in temperature results in a second-order phase transition to the tilted spiral state TS (Sects. 3.1,2). If $[\Gamma_5(x_\delta^0)/\Gamma_{10}] < [\Gamma_6(x_\delta^0)/\Gamma_{10}]$, $\Delta\tau_{||} < \Delta\tau_\perp$, and $|\tau_{||0}| \ll \min[(1/2)\Gamma_5(x_\delta^0)|\Delta_0|^2, \ (1/2)\Gamma_6(x_\delta^0)|\Delta_0|^2]$, a situation may arise when $\tau_{||}(x_\delta^0) < \tau_\perp(x_\delta^0)$. This means that with decreasing temperature, the field $\mathbf{S}_{||}$ fluctuates strongly. The fluctuations of the order parameter of $\mathbf{S}_{||}$ stimulate fluctuations of the transverse magnetic subsystem $\mathbf{S}_\perp$, which becomes unstable and the system undergoes a first-order phase transition to a normal spiral phase in which superconductivity and magnetism coexist. Magnetic order destrits superconductivity, and a further decrease in temperature leads to a first-order phase transition to a normal spiral NS. A subsequent decrease in temperature results in a second-order phase transition to a tilted spiral TS. If $\tau_{||}(x_\delta^0) \to \tau_\perp(x_\delta^0)$, both $\mathbf{S}_{||}$ and

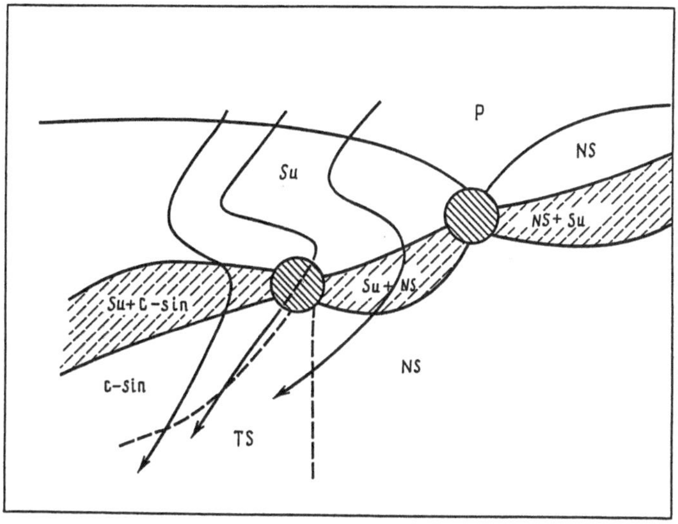

Fig. 4.8. Possible trajectories describing the behavior of a system during a phase transition from paramagnetic to ordered state in superconducting compounds of rare-earth metals $Er_{1-x}Ho_xRh_4B_4$. The *hatched circles* describe the singular points in the $(T, |a_{\|}(x)|$ space through which the trajectory (*dashed line*) cannot pass

$\mathbf{S}_{\perp}$ are strongly fluctuating fields. This means that the phase diagram acquires a singular point at which six lines of the first- and second-order phase transitions converge. Such a point will be unstable. These phase transitions are shown in Fig. 4.8.

2. $\tau_{\| 0} < \min(\tau_{\perp 0}, \tau_{\delta 0})$. In this case, the longitudinal magnetic subsystem oscillates most strongly. The RG equations have the form

$$
\begin{aligned}
&-\Gamma'_{1\|} = 10\Gamma^2_{1\|}\,, \quad -\Gamma'_{1\delta} = 2\Gamma^2_5\,, \quad -\Gamma'_{1\perp} = 2\Gamma^2_4\,,\\
&-\Gamma'_4 = 4\Gamma_{\|}\Gamma_4\,, \quad -\Gamma'_{2\perp} = 2\Gamma^2_4\,, \quad -\Gamma'_5 = 4\Gamma_{\|}\Gamma_5\,,\\
&-\Gamma'_6 = 2\Gamma_4\Gamma_5\,,\\
&-\tau'_{\perp} = 2\tau_{\|}\Gamma_4\,, \quad -\tau'_{\delta} = 2\tau_{\|}\Gamma_5\,,
\end{aligned}
\qquad (4.3.24)
$$

$$
x = \frac{2}{\varepsilon(\Lambda^2)^{\varepsilon/2}}\left\{\left(\frac{\Lambda^2}{\max[\lambda^2(\tau_{\|}),\ s^2_{\perp 0}\ ,\ |\Delta_0|^2]}\right)^{\varepsilon/2} - 1\right\}, \quad \varepsilon \to 0\,.
\qquad (4.3.25)
$$

Equations (4.3.24) can be solved easily and their solutions have the form

$$\Gamma_{1\parallel} = \frac{\Gamma_{1\parallel 0}}{1 + 10\Gamma_{1\parallel 0}x}, \quad \Gamma_6 = \Gamma_{60} - \frac{\Gamma_{40}\Gamma_{50}}{\Gamma_{1\parallel 0}}[(1 + 10\Gamma_{1\parallel 0}x)^{1/5} - 1],$$

$$\Gamma_{1\perp} = \Gamma_{1\perp 0} - (\Gamma_{40}^2/\Gamma_{1\parallel 0})[(1 + 10\Gamma_{1\parallel 0}x)^{1/5} - 1] ,$$

$$\Gamma_{2\perp} = \Gamma_{2\perp 0} - (\Gamma_{40}^2/\Gamma_{1\parallel 0})[(1 + 10\Gamma_{1\parallel 0}x)^{1/5} - 1] ,$$

$$\Gamma_4 = \Gamma_{40}/(1 + 10\Gamma_{1\parallel 0}x)^{2/5} , \quad \Gamma_5 = \Gamma_{50}/(1 + 10\Gamma_{1\parallel 0}x)^{2/5} ,$$

$$\tau_{\parallel} = \tau_{\parallel 0}/(1 + 10\Gamma_{1\parallel 0}x)^{2/5} , \quad \tau_{\perp} = \tau_{\perp 0} - (\tau_{\parallel 0}\Gamma_{40}/\Gamma_{1\parallel 0})$$

$$\times [(1 + 10\Gamma_{1\parallel 0}x)^{1/5} - 1] ,$$

$$\tau_{\delta} = \tau_{\delta 0} - (\tau_{\parallel 0}\Gamma_{50}/\Gamma_{1\parallel 0})[(1 + 10\Gamma_{1\parallel 0}x)^{1/5} - 1] . \tag{4.3.26}$$

It follows from the initial conditions (4.2.7) that the transverse magnetic subsystem is the first to be transformed into a "condensate", i.e., a first-order phase transition to the magnetically ordered state of a normal spiral takes place. The stability boundary for the first-order phase transition is given by

$$x_{\perp h}^* = \frac{1}{10\Gamma_{1\parallel 0}}\left\{\left[1 + \frac{\Gamma_{1\parallel 0}(\Gamma_{1\perp} + \Gamma_{2\perp 0})}{2\Gamma_{40}^2}\right]^5 - 1\right\}. \tag{4.3.27}$$

The amplitudes $\Gamma_4(x_{\perp h}^*)$ and $\Gamma_5(x_{\perp h}^*)$, $\Gamma_6(x_{\perp h}^*)$ are given by

$$\Gamma_4(x_{\perp h}^*) = \Gamma_5(x_{\perp h}^*) = \Gamma_{40}/[1 + \Gamma_{1\parallel 0}(\Gamma_{1\perp 0} + \Gamma_{2\perp 0})/2\Gamma_{40}]^2 ,$$

$$\Gamma_6(x_{\perp h}^*) = \Gamma_{40}[1 - (\Gamma_{1\perp 0} + \Gamma_{2\perp 0})/2\Gamma_{40}] . \tag{4.3.28}$$

The jump in the order parameter at the phase transition point is determined from the free energy minimum of the transverse subsystem:

$$F_{\perp} = \frac{1}{2}\tau_{\perp}s_{\perp 0}^2 + \frac{1}{16}(\Gamma_{1\perp} + \Gamma_{2\perp})s_{\perp 0}^4 = \frac{1}{2}\tau_{\perp}s_{\perp 0}^2 + b(x_{\perp h}^*)s_{\perp 0}^4 \ln\frac{s_{\perp 0}^2}{\lambda_{\perp}^{2*}(\tau_{\parallel})} \tag{4.3.29}$$

and is equal to

$$s_{\perp 0}^2 = \tau_\perp^0 / 2b(x_{\perp h}^*) \ ,$$

$$\tau_\perp^0 = 2b(x_{\perp h}^*) \left(\frac{2-\varepsilon}{2 + x_{\perp h}^* \varepsilon} \right)^{2/\varepsilon} , \ \varepsilon \to 0 \ ,$$

$$b(x_{\perp h}^*) = (1/4)\Gamma_4^2(x_{\perp h}^*) \ . \qquad (4.3.30)$$

If $\Gamma_4(x_{\perp h}^0) < \Gamma_6(x_{\perp h}^0)$, $x_{\perp h}^0 \approx x_{\perp h}^*$, then $\Delta\tau_{\parallel} < \Delta\tau_\delta$ and $\tau_{\parallel}(x_{\perp h}^0) < \tau_\delta < (x_{\perp h}^0)$. This means that upon a subsequent decrease in temperature, a first-order phase transition to a phase of coexistence of superconductivity and magnetism may take place in the system in the form of a helicoidal spin-density wave Su + NS. The free energy for this state can be described by

$$F_{\perp\delta} = \frac{1}{2} a_\perp(\tau) s_{\perp 0}^2 + A_\perp(\tau) \ln \left[\frac{s_{\perp 0}^2}{h_\perp(\tau)} \right] + \frac{1}{2} a_\delta(\tau) \ |\Delta_0|^2$$

$$+ A_\delta(\tau) \ |\Delta_0|^4 \ \ln \left[\frac{|\Delta_0|^2}{h\delta(\tau)} \right] + A_{\delta\perp}(\tau) s_{\perp 0}^2 |\Delta_0|^2 \ ;$$

$$\tau = 1 - (T/T_\perp) \ , \quad A_{\delta\perp} > 0 \ ,$$

where $T_\perp$ is the temperature of transition to the NS state. If the function $a_\perp(\tau)$ decreases more rapidly than $a_\delta(\tau)$, the parameter $s_{\perp 0}$ continues to increase with decreasing temperature, and this leads to a decrease in the order parameter $|\Delta_0|$, i.e., to a suppression of superconductivity. At

$$a_\delta(\tau) + 2A_{\perp\delta}(\tau) s_{\perp 0}^2 = \tau_\delta^0 \ , \qquad (4.3.31)$$

the system undergoes a first-order phase transition to a magnetically ordered state of a normal spiral. Equations (4.3.31) defines the lower boundary T_{c2} of the superconducting transition as well as the temperature interval $\Delta T_{\perp\delta}$ in which superconductivity and magnetism coexist. A further decrease in temperature leads to a second-order phase transition to the tilted spiral state.

The above analysis can be used to construct the phase diagram of superconducting compounds of rare-earth metals of the type $Er_{1-x}Ho_x Rh_4B_4$ in (T, $|a_{\parallel}(x)|$) variables. Figure 4.9 shows the modifications of this phase diagram. It is interesting to note that all singular points in the phase diagrams are unstable. This is due to the fact

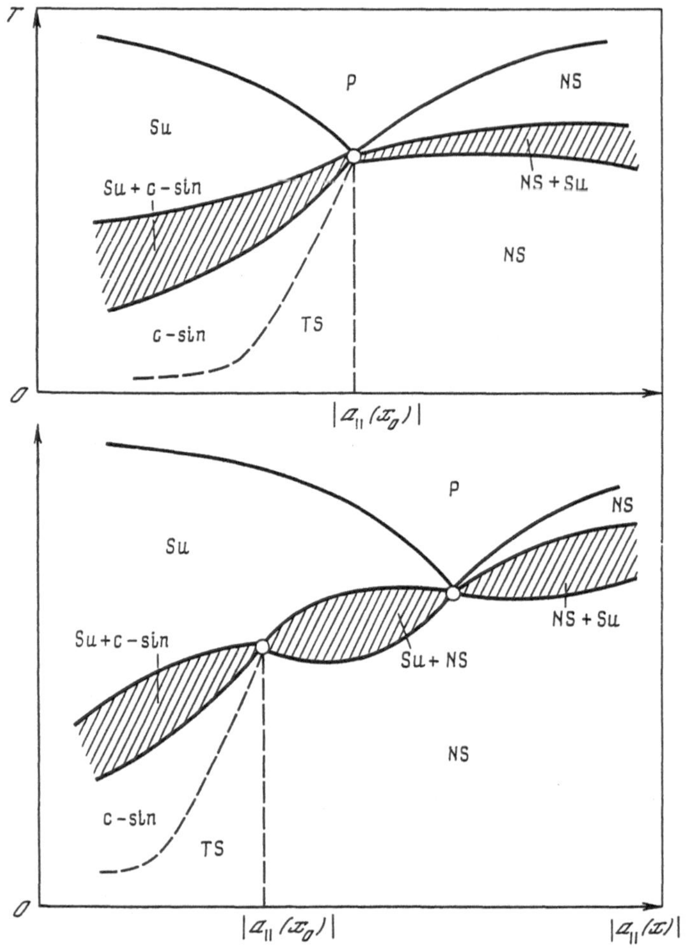

Fig. 4.9. Theoretical phase diagrams of magnetic superconductors $Er_{1-x}Ho_xRh_4B_4$

that superconductivity suppresses magnetic fluctuations in the vicinity of singular points and renders impossible a second-order phase transition from paramagnetic or superconducting phase directly to the lowest magnetically ordered state (Fig. 4.9).

4.4 Topological Analysis of the Phase Diagram of Superconducting Compounds of Rare-Earth Metals

While constructing the phenomenological theory of the emergence of superconducting phase in $Er_{1-x}Ho_xRh_4B_4$ compounds, we assumed that the amplitude of the order parameter corresponding to the superconducting subsystem $\Psi = |\Delta|e^{i\varphi}$ is a scalar function of temperature, which is uniform in space. This is in accordance with the fact that the phase transition to the superconducting state is due to longwave fluctuations of the superconducting component, which are stimulated by the longwave magnetic fluctuations (Sect. 4.3). It follows that longwave magnetic fluctuations must facilitate the attraction between conduction electrons and lead to a formation of bound electron states. The topological degeneracy space corresponding to Ψ is denoted by $R = S^1$. Hence such a superconductor may contain linear singularities of normal type (vortices) which can be classified by the elements of the fundamental group of the circle $\pi_1(S^1) = Z$. There are no point singularities in such a superconductor, since the second homotopic group on the manifold S^1 is trivial: $\pi_2(S^1) = 0$. Thus, within the framework of this model and away from the phase transition points when the magnetic subsystem fluctuations are suppressed to a considerable extent, the superconducting phase in this compound does not differ in any way from any ordinary type II superconductor.

In actual practice, magnetic fluctuations pierce the entire phase diagram of the compounds $Er_{1-x}Ho_xRh_4B_4$ (Chap. 1), i.e., they also exist in the superconducting phase. This must be reflected in the effective order parameter of the superconducting subsystem. Theoretically, two types of conduction electron-pairing are possible in a superconductor, viz., singlet pairing and triplet pairing [4.27,28]. In the case of triplet pairing, the order parameter is a tensor:

$$\Psi_{ik}(\mathbf{x}) = u(\mathbf{x})e^{i\varphi}R_{ik}(\mathbf{x}) .$$

The wave functions of electrons in the field of longwave spin-density fluctuations, defined in terms of the rotational matrix, were first introduced in [4.29]. Here, R_{ik} is the matrix of the rotational group SO(3) which, as in the B-phase of superfluid He^3, may be equal to

$$R_{ik} = \delta_{ik} + 2(n_i n_k - n^2\delta_{ik}) + 2e_{ik\ell}n_o n_\ell ,$$

$$\mathbf{n} = (\cos\xi \ , \mathbf{n}^0 \sin\xi \ ) \ ; \quad n^{02} = 1, \tag{4.4.1}$$

u(x) being a scalar function of coordinates.

A new class of superconducting compounds $CeCu_2Si_2$, UBe_{13}, UPt_3, and UPt_2C has been recently discovered [4.27,28]. All these compounds undergo a phase transition to the superconducting state at low temperatures $T_c = 1$ K. Measurements of the dependence of magnetic susceptibility, heat capacity, and electrical resistivity were carried out over a wide range of temperatures. It was found that in the superconducting state, the superconducting component of the electron heat capacity has a component proportional to $(T/T_c)^3$. The effective mass of carriers (5f-electrons) is found to be of the order of 200 μ_e, where μ_e is the electron mass. The authors of [4.27,28] explain these facts by assuming that a triplet conduction electron pairing takes place in UBe_{13}, and the order parameter is a tensor, although it differs from the order parameter of superfluid He^3 due to a strong spin-orbit interaction caused by the crystal lattice field.

It was shown in [4.30,31] that since magnetic fluctuations take place in the superconducting state, they modify the superconducting order parameter which is now a tensor and can be presented in the form

$$\hat{\Psi}(\mathbf{x}) = u(\mathbf{x})e^{iq\mathbf{x}}\hat{R}(\mathbf{x})\hat{a}, \quad \hat{a} = \left\| \begin{matrix} 0 \\ 1 \end{matrix} \right\|, \tag{4.4.2}$$

where $\hat{R}$ is a matrix of the SU(2) group. The symmetry group of the superconducting phase may be obtained by multiplying the space group G_p by the group formed by the elements of the SU(2) group: $G = G_p \times G_{Su}$, $G_{Su} \subset SU(2)$, $SO(3) \subset SU(2)$. Consequently, we can introduce superconducting classes for the magnetic superconductors under consideration. For this purpose, we must know the space R in which the order parameter of the superconducting state can be expressed. In this case, the superconducting class has the form $D_{4h} \times I_t \times R$ for tetgragonal symmetry, $D_{2h} \times I_t \times R$ for orthorhombic symmetry, and $O_h \times I_t \times R$ for cubic symmetry, where I_t indicates the time inversion. In the general case, the order parameter Ψ, determined on the basis of the fluctuation theory, may describe the singlet, as well as triplet pairing of conduction electrons. In particular, using the results of the fluctuation theory [4.30-32] we can determine the heat capacity in the superconducting phase. If gapless superconductivity is observed in the system, the low-temperature heat capacity contains a component $(T/T_c)^3$ in the same way as for compounds of the type UBe_{13}.

It follows from (4.4.2) that the order parameter is just an isospinor. The factor $\exp(iq\mathbf{x})$ indicates the possibility of observing superconducting structures and new types of currents and vortices in the superconducting phase. These are caused by the interaction of the superconducting subsystem with the spin-density wave fluctuations [4.30]. Since vortex structures can exist in a superconducting state and no point singularities are experimentally observed for superconductors, we shall choose the explicit form of matrix R by proceeding from the SU(2) group factorized by the center group Z_2: $SU(2)/Z_2 = SO(3)$, $\pi_1(SO(3)) = Z_2$, $\pi_2(SO(3)) = 0$. Consequently, all point singularities in the superconductors will be unstable.

We shall now describe the topological manifolds corresponding to the order parameter in the phase diagrams (Fig. 4.9), as well as their homotopic groups, in order to show the singularities that can appear in different states.

1. Superconducting Phase (Su).

$$R = S^1 , \quad \pi_1(R) = Z , \quad \pi_2(R) = 0 . \tag{4.4.3}$$

$$R_1 = SO(3) \times S^1 , \quad \pi_1(R_1) = Z_2 + Z , \quad \pi_2(R_1) = 0 . \tag{4.4.4}$$

There are no point singularities, although linear singularities exist.

2. The Phase of Coexistence of Superconductivity and Magnetism as a Longitudinal Sinusoidal Spin-Density Wave (Su + c-sin).

$$R = \begin{Vmatrix} S^2 \times S^1 \\ S^1 \end{Vmatrix} , \quad \pi_1(R) = \begin{Vmatrix} Z \\ Z \end{Vmatrix} , \quad \pi_2(R) = \begin{Vmatrix} Z \\ 0 \end{Vmatrix} , \tag{4.4.5}$$

$$\tilde{R} = \begin{Vmatrix} S^2 \times S^1/Z_2 \\ S^1 \end{Vmatrix} , \quad \pi_1(\tilde{R}) = \begin{Vmatrix} Z_4 \\ Z \end{Vmatrix} , \quad \pi_2(\tilde{R}) = \begin{Vmatrix} Z \\ 0 \end{Vmatrix} , \tag{4.4.6}$$

$$\tilde{R}_1 = \begin{Vmatrix} S^2 \times S^1 \\ SO(3) \times S^1 \end{Vmatrix} , \quad \pi_1(R_1) = \begin{Vmatrix} Z \\ Z_2 + Z \end{Vmatrix} , \quad \pi_2(R_1) = \begin{Vmatrix} Z \\ 0 \end{Vmatrix} , \tag{4.4.7}$$

$$\tilde{R}_1 = \begin{Vmatrix} S^2 \times S^1/Z_2 \\ SO(3) \times S^1 \end{Vmatrix} , \quad \pi_1(\tilde{R}_1) = \begin{Vmatrix} Z_4 \\ Z_2 + Z \end{Vmatrix} , \quad \pi_2(\tilde{R}_1) = \begin{Vmatrix} Z \\ 0 \end{Vmatrix} . \tag{4.4.8}$$

The matrix form of notation indicates that the superconducting and magnetic subsystems are independent even though they interact with each other. It can be seen from (4.4.7,8) that point singularities may exist in the Su + c-sin phase, but they must be purely magnetic. However, superconductivity exists in this phase and the singularity in the superconducting state must be associated with a nonzero superconducting phase has a gradient. At the point singularity, the superconducting current must be equal to zero in view of spherical symmetry. Hence the appearance of such a singularity is associated with the destruction of superconductivity.

3. c-sin Phase.

$$R = S^2 \times S^1 , \quad \pi_1(R) = Z , \quad \pi_2(R) = Z , \tag{4.4.9}$$

$$\tilde{R} = S^2 \times S^1/Z_2 , \quad \pi_1(\tilde{R}) = Z_4 , \quad \pi_2(\tilde{R}) = Z , \tag{4.4.10}$$

$$R_1 = S^2 , \quad \pi_1(R_1) = 0 , \quad \pi_2(R_1) = Z , \tag{4.4.11}$$

$$\tilde{R}_1 = S^2/Z_2 , \quad \pi_1(R_1) = Z_2 , \quad \pi_2(R_1) = Z . \tag{4.4.12}$$

This phase may contain point singularities as well as linear magnetic singularities (solitons). The emergence of point singularities is hampered by a strong magnetic anisotropy, and hence they can be observed, in all probability, for $|a_{\parallel}(x)| \to 0$, i.e., in the region of low concentrations (Fig. 4.9).

4. TS and NS Phases. The topological manifolds and homotopic groups for these phases are described in Sect. 3.2, from which it can be concluded that only magnetic linear singularities can exist in these phases.

5. The Phase of Coexistence of Superconductivity and Magnetism in the Form of a Normal Spiral (Su + NS).

$$R = \left\| \begin{matrix} SO(3) \\ S^1 \end{matrix} \right\| , \quad \pi_1(R) = \left\| \begin{matrix} Z_2 \\ Z \end{matrix} \right\| , \quad \pi_2(R) = \left\| \begin{matrix} 0 \\ 0 \end{matrix} \right\| , \tag{4.4.13}$$

$$\tilde{R} = \left\| \begin{matrix} S^3/Q \\ S^1 \end{matrix} \right\| , \quad \pi_1(\tilde{R}) = \left\| \begin{matrix} Q \\ Z \end{matrix} \right\| , \quad \pi_2(R) = \left\| \begin{matrix} 0 \\ 0 \end{matrix} \right\| , \tag{4.4.14}$$

$$R_1 = \left\| \begin{matrix} SO(3) \\ SO(3) \times S^1 \end{matrix} \right\| , \quad \pi_1(R_1) = \left\| \begin{matrix} Z_2 \\ Z_2 + Z \end{matrix} \right\| , \quad \pi_2(R) = \left\| \begin{matrix} 0 \\ 0 \end{matrix} \right\| , \tag{4.4.15}$$

$$\tilde{R}_1 = \left\| \begin{matrix} S^3/Q \\ SO(3) \times S^1 \end{matrix} \right\| , \quad \pi_1(\tilde{R}_1) = \left\| \begin{matrix} Q \\ Z_2 + Z \end{matrix} \right\| , \quad \pi_2(\tilde{R}_1) = \left\| \begin{matrix} 0 \\ 0 \end{matrix} \right\| . \tag{4.4.16}$$

Only linear singularities can exist in this case. While carrying out the above classification, we did not take into account the physical meaning of the spaces $\tilde{R}$ and $\tilde{R}_1$. We shall describe them in detail in the following chapters, when the specific form of linear and point singularities will be considered in the appropriate phase diagrams of the states of $Er_{1-x}Ho_xRH_4B_4$ (Fig. 4.9).

Before concluding this chapter, we observe that the following relations are satisfied for the spaces R, R_1, and the corresponding homotopic groups:

$$R_{Su}, \quad R_{Su+c-sin}, \quad R_{c-sin}, \quad R_{NS} \subset R_{TS}, \tag{4.4.17}$$

$$\pi_1(R_{Su}), \quad \pi_1(R_{Su+c-sin}), \quad \pi_1(R_{c-sin}), \quad \pi_1(R_{NS}) \subseteq \pi_1(R_{TS}), \tag{4.4.18}$$

$$R_{1Su}, \quad R_{NS}, \quad R_{c-sin}, \quad R_{TS} \subset R_{1NS+Su}, \quad R_{1c-sin+Su}, \tag{4.4.19}$$

$$\pi_1(R_{1Su}), \quad \pi_1(R_{NS}), \quad \pi_1(R_{c-sin}), \quad \pi_1(R_{TS}) \subset \pi_1(R_{1NS+Su}), \quad \pi_1(R_{1c-sin+Su}). \tag{4.4.20}$$

This means that phase transitions from the paramagnetic phase to the ordered state in this class of materials are accompanied by a phase lamination in the same way as in the alloys of rare-earth metals (Sects. 3.1,2).

It should be noted that if the dimensionality $n_{\|}(n_{\perp})$ of the order parameters $\mathbf{S}_{\|}$ $(\mathbf{S}_{\perp})$ is equal to 1, the tilted spiral state TS is transformed to a tilted ferromagnet (TF) state. Such a state was observed experimentally in the compounds $Er_{1-x}Ho_xRh_4B_4$ [4.22].

5. Theory of High-Temperature Superconducting Phases in Rare-Earth Metal Compounds

5.1 Lagrangian of the System near the Point of Phase Transition to Superconducting State

For the superconducting compounds of rare-earth metals considered in Chap. 4, the superconducting phase transition temperature did not exceed 10 K. In other words, all these compounds are low-temperature superconductors. We shall show that in magnetic superconductors there exists the possibility of the emergence of a high-temperature superconducting phase ($T_c >> 10$ K).

Let us consider the phase transition from the paramagnetic phase to the superconducting state. In magnetic superconductors, a phase transition from the paramagnetic phase to the superconducting state involves the mutual interaction of conduction electrons through virtual phonons, as well as the interaction of conduction electrons with the magnetic subsystem fluctuations (Chap. 1). We shall proceed by considering the symmetry of the system and the exchange enhancement effect, as well as the effect of the Fermi surface topology on the superconducting transition.

Most of the compounds $RERh_4B_4$ and $REMo_6S_8$ have an orthorhombic (tetragonal) symmetry. At present, 14 space symmetry groups corresponding to a tetragonal unit cell are known to exist:

P4, I4, P4/m, I4/m, P4mm, I4mm, P4/mmm,
P$\bar{4}$, I$\bar{4}$, P422, I422, P42m, I42m, I4/mmm.

Let us consider Rh (fcc symmetry) whose electron Fermi surface is smooth and lies near the geometrical center of the unit cell [5.1, p. 237]. On the other hand, the topology of the Fermi surface of rare-earth metals (e.g., Tb) is very complicated [5.1, p.245], and their magnetic properties depend significantly on the Fermi surface. Above all, this refers to the wave vector q of the magnetic structure [5.1, p. 247], which may be determined theoretically even from the parameters of Fermi surface singularities ("holes", "pockets", "arms").

On the basis of the neutron scattering experiments, **Moncton** [5.2] showed that in the phase of coexistence of superconductivity and magnetism, the magnetic structure considerably differs from the structure of the magnetically ordered phase. In particular, the period of the structure is found to be of the order of 100-200 Å ($HoMo_6S_8$) [5.3], which is much higher than the magnetic structure period in pure rare-earth metals. This leads to the conclusion that the superconducting transition must be inevitably linked with the variations of the topology of the Fermi surface. The topology of the Fermi surface of the compounds $REMo_6S_8$ is not known at present, but antiferromagnetic order has been observed experimentally in their magnetically ordered phase [5.4,5]. In the general case, the antiferromagnetic structure can be described with the help of three vectors:

$$\mathbf{S}_1 = \mathbf{S}_{1o}e^{i\pi x}\ , \quad \mathbf{S}_2 = \mathbf{S}_{2o}e^{i\pi y}\ , \quad \mathbf{S}_3 = \mathbf{S}_{3o}e^{i\pi z}\ . \tag{5.1.1}$$

While using the phenomenological approach in this case, we should take into consideration the following invariants:

$$1)\quad (\mathbf{S}_1\mathbf{S}_1^*),\quad (\mathbf{S}_2\mathbf{S}_2^*),\quad (\mathbf{S}_3\mathbf{S}_3^*),\quad S_1^zS_1^{z*}S_2^zS_2^{z*},\ S_3^z\ S_3^{z*};$$

$$\begin{aligned}2)\quad & (\mathbf{S}_1\mathbf{S}_1^*)^2,\ (\mathbf{S}_2\mathbf{S}_2^*)^2,\ (\mathbf{S}_3\mathbf{S}_3^*)^2,\ \ \mathbf{S}_1^2\mathbf{S}_1^{2*},\ \mathbf{S}_2^2\mathbf{S}_2^{2*},\ \mathbf{S}_3^2\mathbf{S}_3^{2*}\ ,\\ & (\mathbf{S}_1\mathbf{S}_1^*)(\mathbf{S}_2\mathbf{S}_2^*)^2,\ (\mathbf{S}_1\mathbf{S}_1^*)(\mathbf{S}_3\mathbf{S}_3^*),\ (\mathbf{S}_2\mathbf{S}_2^*)(\mathbf{S}_3\mathbf{S}_3^*)\ ,\\ & (\mathbf{S}_1\mathbf{S}_2)(\mathbf{S}_1^*\mathbf{S}_2^*) + (\mathbf{S}_1\mathbf{S}_2^*)(\mathbf{S}_1^*\mathbf{S}_2),\ (\mathbf{S}_1\mathbf{S}_3)(\mathbf{S}_1^*\mathbf{S}_3^*)\\ & + (\mathbf{S}_1\mathbf{S}_3^*)(\mathbf{S}_1^*\mathbf{S}_3),\ (\mathbf{S}_2\mathbf{S}_3)(\mathbf{S}_2^*\mathbf{S}_3^*) + (\mathbf{S}_2\mathbf{S}_3^*)(\mathbf{S}_2^*\mathbf{S}_3),\\ & \mathbf{S}_1^4 + \mathbf{S}_2^4 + \mathbf{S}_3^4\ .\end{aligned} \tag{5.1.2}$$

Since we are considering the case of an antiferromagnet, this means that the main contribution to the magnetic fluctuations is made by the fluctuations of the antiferromagnetism vector $\mathbf{I}_j = \mathbf{S}_j(\mathbf{x}) - \mathbf{S}_j(\mathbf{x}')$, while the fluctuations of the ferromagnetism vector $\mathbf{m}_j = \mathbf{S}_j(\mathbf{x}) + \mathbf{S}_j(\mathbf{x}')$ are negligibly small. Consequently, the term $\{\mathbf{S}_{jo}(\mathbf{x})\hat{\psi}^+(\mathbf{x})\boldsymbol{\sigma}\hat{\psi}(\mathbf{x})\}$ in the effective Lagrangian should be replaced by the term $\nabla_\alpha S_{jo}(\mathbf{x})\hat{\psi}_\sigma^+(\mathbf{x})\hat{\psi}_\sigma(\mathbf{x})$, where $\hat{\psi}_\sigma^+$ and $\hat{\psi}_\sigma$ are fermion operators. Then, the density of the Lagrangian system without taking into account the spin-phonon coupling and for the case of a strong anisotropy of the magnetic subsystem [5.6] can be presented in the form

$$\mathbb{L} = (1/2)(\nabla S_{\alpha o})^2 - (1/2)\tau_\alpha S_{\alpha o}^2 - (1/8)\Gamma_{1\alpha} S_{\alpha o}^4$$

$$- (1/8)\Gamma_{4\alpha\beta} S_{\alpha o}^2 S_{\beta o}^2 + (1/2)\hat{\psi}_\sigma^+ \hat{\pi} \hat{\psi}_\sigma$$

$$- (1/8)\mu_s^{d(f)} \hat{\psi}_\uparrow^+ \hat{\psi}_\downarrow^+ \hat{\psi}_\uparrow \psi_\downarrow - (1/2\sqrt{2})\nu \nabla_\alpha S_{\alpha o}\, \hat{\psi}_\sigma^+ \hat{\psi}_\sigma \quad , \tag{5.1.3}$$

where

$$\tau_{\alpha o} \to \tau_o \ , \ \Gamma_{\alpha o} \to \Gamma_{4\alpha\beta o} \to \Gamma_o \ , \quad \hat{\pi}(\mathbf{x}) = \frac{1}{\mu_\ell} \nabla_{\mathbf{x}}^2 - 2\hat{\tilde{g}}(\mathbf{x}) \ .$$

The statistical sum of the system may be determined in the form of the functional integral [5.7]

$$Z[\eta_\alpha^i(\mathbf{x},\tau) \ , \ \xi_\sigma^*(\mathbf{x},\tau), \quad \xi_\sigma(\mathbf{x},\tau)] = \int \prod_{\mathbf{x},\tau} DS_{\alpha o}^i(\mathbf{x},\tau) D\psi_\sigma(\mathbf{x},\tau) D\psi_\sigma^*(\mathbf{x},\tau)$$

$$\cdot \exp \{\int d\mathbf{x} \int_0^\beta d\tau [\mathbb{L}(\mathbf{x},\tau) + \eta_\alpha^i(\mathbf{x},\tau) S_{\alpha o}^i(\mathbf{x},\tau)$$

$$+ \xi_\sigma^*(\mathbf{x},\tau)\psi_\sigma(\mathbf{x},\tau) + \xi_\sigma(\mathbf{x},\tau)\psi_\sigma^*(\mathbf{x},\tau)]\} \ . \tag{5.1.4}$$

The Green functions of the system are determined by the variational derivatives of the generating functional (5.1.4):

$$G_{\alpha\beta}^{ik}(\mathbf{x}-\mathbf{x}', \tau-\tau') = \lim_{\eta_\alpha^i, \xi_\sigma, \xi_\sigma^* \to 0} \frac{1}{Z} \cdot \frac{\delta^2 Z}{\delta\eta_\alpha^i(\mathbf{x},\tau)\delta\eta_\beta^k(\mathbf{x}',\tau')}$$

$$= \langle \hat{T} S_\alpha^i(\mathbf{x},\tau) S_\beta^k(\mathbf{x}',\tau') \rangle , \tag{5.1.5}$$

$$G_{\sigma\sigma'}(\mathbf{x}-\mathbf{x}',\tau-\tau') = \lim_{\eta_\alpha^i, \xi_\sigma, \xi_\sigma^* \to 0} \frac{1}{Z} \cdot \frac{\delta^2 Z}{\delta\xi_\sigma^*(\mathbf{x},\tau)\delta\xi_{\sigma'}(\mathbf{x}',\tau')}$$

$$= \langle \hat{T}\psi_\sigma(\mathbf{x},\tau)\psi_{\sigma'}^*(\mathbf{x}',\tau')\rangle \ . \tag{5.1.6}$$

Near the phase transition point, the magnetic fluctuations are intensified and their correlation radius $\langle r_c \rangle \to \infty$. Consequently, the time dependence of the field S_α^i can be neglected, while the time dependence of electron operators ψ_σ and ψ_σ^* remains significant. A similar approach was used by **Moriya** [5.8] for studying the paramagnetic susceptibility and phase transition temperature in ferromagnetic and weakly ferro- and antiferromagnetic metals. In view of the symmetry of the initial values of the temperatures and amplitudes of the magnetic subsystem, the latter will tend to a stable

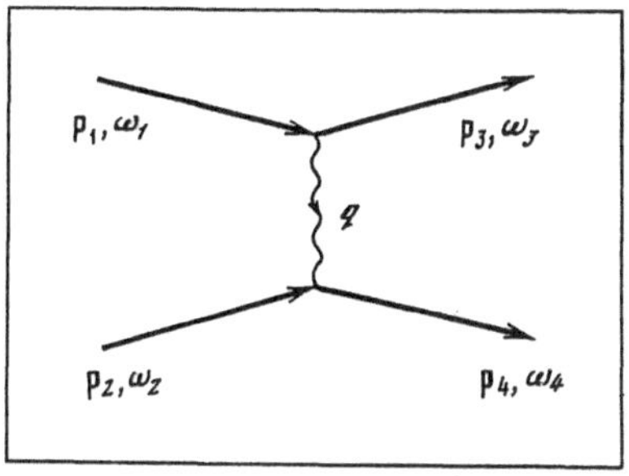

Fig. 5.1. Amplitude of interaction of through magnetic fluctuations

stationary point $\Gamma \to \varepsilon/11 + \dots$. In this case, the functional integral with respect to the fields $S^i_{\alpha o}(\mathbf{x})$ can be evaluated exactly and the Lagrangian (5.1.3) acquires a term (Fig. 5.1)

$$(1/8)\nu^2\tilde{G}(\mathbf{q})\psi_\uparrow(\mathbf{p}_1,\omega_1)\psi_\downarrow(\mathbf{p}_3,\omega_3)\psi^+_\uparrow(\mathbf{p}_2,\omega_2)\psi^+_\downarrow(\mathbf{p}_4,\omega_1)$$

$$\cdot\ \delta(\mathbf{p}_1+\mathbf{p}_3-\mathbf{q})\delta(\mathbf{q}-\mathbf{p}_2-\mathbf{p}_4)\delta(\omega_1+\omega_3-\omega_2-\omega_4)\ , \qquad (5.1.7)$$

$$\tilde{G}(\mathbf{q}) \ = \ \frac{q^2}{(q^{2-\eta}\ -\tau)}, \tau \to 0\ , \quad \eta = \frac{5}{242}\ \varepsilon^2 + \dots .$$

Since $\eta > 0$, $\tilde{G}(\mathbf{q}) \to 0$ as $\mathbf{q} \to 0$, i.e., interaction with the fluctuations cannot be responsible for the attraction between electrons and for the formation of bound electron states. However, the magnetic subsystem strongly influences the Fermi surface topology of conduction electrons. Magnetic fluctuations change the density of electron states on the Fermi surface [5.9,10] and suppress the Coulomb repulsion of electrons, thus ensuring an effective attraction of the conduction electrons through virtual phonons, formation of bound electron states, and the emergence of superconductivity.

Let us now write down the Lagrangian of the system by taking into account the interaction of electron and spin subsystems with the crystal lattice. In the general case, this Lagrangian has the form

$$\begin{aligned}\mathbb{L} = {} & (1/2)(\nabla S_{\alpha o})^2 - (1/2)\tau_\alpha S^2_{\alpha o}(\mathbf{x}) - (1/8)\Gamma_{1\alpha} S^4_{\alpha o}(\mathbf{x}) \\ & - (1/8)\Gamma_{4\alpha\beta} S^2_{\alpha o}(\mathbf{x}) S^2_{\beta o}(\mathbf{x}) + (1/2)\hat{\psi}^+_\sigma(x)\hat{\pi}(\mathbf{x})\hat{\psi}_\sigma(\mathbf{x}) \\ & - (1/8)\mu^{d(f)}_s(\mathbf{x}-\mathbf{x}')\hat{\psi}^+_\uparrow(\mathbf{x})\hat{\psi}^+_\downarrow(\mathbf{x}')\hat{\psi}_\uparrow(\mathbf{x})\hat{\psi}_\downarrow(\mathbf{x}') \\ & - (1/2\sqrt{2})\nu\nabla_\alpha S_{\alpha o}(\mathbf{x})\hat{\psi}^+_\sigma(\mathbf{x})\hat{\psi}_\sigma(\mathbf{x}) + \sum_\varkappa (1/2m_\varkappa)p^2_\varkappa(\mathbf{x})\end{aligned}$$

$$- (1/2) \sum_{\varkappa,\varkappa'} \lambda_{\alpha\beta\varkappa\varkappa'}(x-x')[u_{\alpha\varkappa}(x) - u_{\alpha\varkappa'}(x')][u_{\beta\varkappa}(x) - u_{\beta\varkappa'}(x')]$$

$$- (1/2\sqrt{2}) \sum_{\varkappa,\varkappa'} \Gamma_{2\alpha\gamma\varkappa\varkappa'}(x-x')[u_{\alpha\varkappa}(x) - u_{\alpha\varkappa'}(x')]S^2_{\gamma 0}(x),$$

$$(5.1.8)$$

where $\tilde{g}(x) = g_{ph}\hat{\varphi}(x)$; $\hat{\varphi}(x)$ is the phonon field operator.

Let us present the phonon subsystem in the form of a superposition of longwave oscillations coupled with the fluctuations of the
magnetic subsystem and virtual phonons participating in the exchange
between conduction electrons:

$$(1/2)\lambda_{\alpha\beta\varkappa\varkappa'}(x-x')[u_{\alpha\varkappa}(x) - u_{\alpha\varkappa'}(x')][u_{\beta\varkappa}(x) - u_{\beta\varkappa'}(x')]$$

$$= (1/2)\lambda^{(s)}_{\alpha\beta\varkappa\varkappa'} (x-x')[u^{(s)}_{\alpha\varkappa}(x) - u^{(s)}_{\alpha\varkappa'}(x')][u^{(s)}_{\beta\varkappa}(x) - u^{(s)}_{\beta\varkappa'}(x')]$$

$$+ (1/2)\lambda^{(\ell)}_{\alpha\beta\varkappa\varkappa'}(x-x')[u^{(\ell)}_{\alpha\varkappa}(x) - u^{(\ell)}_{\alpha\varkappa'}(x')][u^{(\ell)}_{\beta\varkappa}(x) - u^{(\ell)}_{\beta\varkappa'}(x')].$$

$$(5.1.9)$$

We assume that in the region of longwave fluctuations, the effective size ℓ_k of the unit cell is of the order of the flucuation cor-
relation radius $\langle r_c \rangle$. Consequently, $u^{(s,\ell)}_{\alpha\varkappa}(x)$ will be a slowly
varying function of variable $\varkappa$, and this allows us to average (5.1.8)
over the cell configurations. In this case, the fluctuating part
of the Lagrangian (5.1.8) in the Fourier representation can be written
as follows:

$$\mathbb{L}^{fluct} = (1/2)(q^2 - \tau_\alpha)S^i_{\alpha 0}(q)S^i_{\alpha 0}(-q)$$

$$-(1/8)\Gamma_{1\alpha}S^i_{\alpha 0}(q_1)S^i_{\alpha 0}(q_2)S^k_{\alpha 0}(q_3)S^k_{\alpha 0}(q_4)$$
$$q_1+q_2+q_3+q_4=0$$
$$-(1/8)\Gamma_{4\alpha\beta}S^i_{\alpha 0}(q_1)S^i_{\alpha 0}(q_2)S^k_{\beta 0}(q_3)S^k_{\beta 0}(q_4)$$
$$q_1+q_2+q_3+q_4=0$$
$$-(1/2)\langle\lambda^{(s)}_{\alpha\beta\nu\nu'}\rangle_{\rho(\varkappa)}u^{(s)}_{\alpha\nu}(q)u^{(s)}_{\beta\nu'}(-q)$$

$$-(1/2\sqrt{2})\langle\Gamma_{2\alpha\nu'}\rangle_{\rho(\varkappa)} u^{(s)}_{\alpha\nu}(q_1)S^i_{\gamma 0}(q_2)S^i_{\gamma 0}(q_3)$$
$$q_1+q_2+q_3=0$$

$$-g^{(s)}_{\alpha\nu}u^{(s)}_{\alpha\nu}(q_1)\hat{\psi}^+_\sigma(p_1)\hat{\psi}_\sigma(p_2) .$$
$$q_1+p_1+p_2=0$$

$$(5.1.10)$$

Next, carrying out the standard operation of integration with respect to $u_{\alpha\gamma}^{(s)}(q)$ and $u_{\beta\gamma'}^{(s)}(q)$, we obtain (Sect. 2.2)

$$
\begin{aligned}
\mathbb{L}^{fluct}(\mathbf{q}) = \; & (1/2)(q^2-\tau_\alpha)S^i_{\alpha o}(\mathbf{q})S^i_{\alpha o}(-\mathbf{q}) \\
& -(1/8)\Gamma_{1\alpha}S^i_{\alpha o}(\mathbf{q}_1)S^i_{\alpha o}(\mathbf{q}_2)S^k_{\alpha o}(\mathbf{q}_3)S^k_{\alpha o}(\mathbf{q}_4) \\
& \qquad \mathbf{q}_1+\mathbf{q}_2+\mathbf{q}_3+\mathbf{q}_4=0 \\
& -(1/8)\Gamma_{4\alpha\beta}S^i_{\alpha o}(\mathbf{q}_1)S^i_{\alpha o}(\mathbf{q}_2)S^k_{\beta o}(\mathbf{q}_3)S^k_{\beta o}(\mathbf{q}_4) \\
& \qquad \mathbf{q}_1+\mathbf{q}_2+\mathbf{q}_3+\mathbf{q}_4=0 \\
& -(1/8)a_1^{(s)}(\Theta_k)S^i_{\gamma o}(\mathbf{q}_1)S^i_{\gamma o}(\mathbf{q}_2)S^k_{\delta o}(\mathbf{q}_3)S^k_{\delta o}(\mathbf{q}_4) \\
& \qquad \mathbf{q}_1+\mathbf{q}_2+\mathbf{q}_3+\mathbf{q}_4=0 \\
& -(1/8)a_2^{(s)}(\Theta_k)S^i_{\gamma o}(\mathbf{q}_1)S^i_{\gamma o}(\mathbf{q}_2)S^k_{\delta o}(\mathbf{q}_3)S^k_{\delta o}(\mathbf{q}_4) \\
& \qquad \mathbf{q}_1+\mathbf{q}_2+\mathbf{q}_3+\mathbf{q}_4=0 \\
& -(1/4)b_1^{(s)}(\Theta_k)S^i_{\gamma o}(\mathbf{q}_1)S^i_{\gamma o}(\mathbf{q}_2)\hat{\psi}^+_\sigma(\mathbf{p}_1)\hat{\psi}_\sigma(\mathbf{p}_2) \\
& \qquad \mathbf{q}_1+\mathbf{q}_2+\mathbf{p}_1+\mathbf{p}_2=0 \\
& -(1/4)b_2^{(s)}(\Theta_k)S^i_{\gamma o}(\mathbf{q}_1)S^i_{\gamma o}(\mathbf{q}_2)\hat{\psi}^+_\sigma(\mathbf{p}_1)\hat{\psi}_\sigma(\mathbf{p}_2) \\
& \qquad \mathbf{q}_1+\mathbf{q}_2+\mathbf{p}_1+\mathbf{p}_2 = 0 \\
& -(1/2)c_1^{(s)}(\Theta_k)\hat{\psi}^+_\sigma(\mathbf{p}_1)\hat{\psi}_\sigma(\mathbf{p}_2)\hat{\psi}^+_{\sigma'}(\mathbf{p}_3)\hat{\psi}_{\sigma'}(\mathbf{p}_4) \\
& \qquad \mathbf{p}_1+\mathbf{p}_2+\mathbf{p}_3+\mathbf{p}_4=0 \\
& -(1/2)c_2^{(s)}(\Theta_k)\hat{\psi}^+_\sigma(\mathbf{p}_1)\hat{\psi}_\sigma(\mathbf{p}_2)\hat{\psi}^+_{\sigma'}(\mathbf{p}_3)\hat{\psi}_{\sigma'}(\mathbf{p}_4), \\
& \qquad \mathbf{p}_1+\mathbf{p}_2+\mathbf{p}_3+\mathbf{p}_4=0 \quad .
\end{aligned}
\tag{5.1.11}
$$

Here, $a_{1,2}^{(s)}(\Theta_k)$, $b_{1,2}^{(s)}(\Theta_k)$, $c_{1,2}^{(s)}(\Theta_k)$ are complex functions of the angles between the directions of wave vectors of a longwave virtual phonon and the crystallographic axes. For a weakly anisotropic lattice, we can write the following final expression for the overall Lagrangian of the system:

$$
\begin{aligned}
\mathbb{L} = \; & (1/2)(\nabla S_{\alpha o})^2 - (1/2)\tau_\alpha S^2_{\alpha o} - (1/8)\Gamma_{1\alpha}S^4_{\alpha o} \\
& -(1/8)\Gamma_{4\alpha\beta}S^2_{\alpha o}S^2_{\beta o} + (1/2)\,\hat{\psi}^+_\sigma\hat{\pi}\hat{\psi}_\sigma - \mu^{d(f)}\,\hat{\psi}^+_\uparrow\,\hat{\psi}^+_\downarrow\psi_\uparrow\psi_\downarrow
\end{aligned}
$$

$$-(1/8)a_1^{(s)}S^2_{\gamma o}S^2_{\delta o} - (1/8)a_2^{(s)}\tilde{S}^2_{\gamma o}\tilde{S}^2_{\delta o} - (1/4)b_1^{(s)}S^2_{\gamma o}\,\hat{\psi}^+_\sigma\hat{\psi}_\sigma$$

$$-(1/4)b_{2s}\tilde{S}^2_{\gamma o}\,\hat{\tilde{\psi}}^+_\sigma\hat{\tilde{\psi}}_\sigma - (1/2)c_1^{(s)}\,\hat{\psi}^+_\sigma\hat{\psi}_\sigma\hat{\psi}^+_{\sigma'}\hat{\psi}_{\sigma'}$$

$$-(1/2)c_2^{(s)}\,\hat{\tilde{\psi}}^+_\sigma\hat{\tilde{\psi}}_\sigma\hat{\tilde{\psi}}^+_{\sigma'}\hat{\tilde{\psi}}_{\sigma'} + \mathbb{L}_{ph} - (1/2\sqrt{2})\nu\nabla_\alpha S_{\alpha o}\,\hat{\psi}^+_\sigma\hat{\psi}_\sigma\,, \qquad (5.1.12)$$

$$\tau_{\alpha o}\to\tau_o\,, \quad \Gamma_{1\alpha}\to\Gamma_{4\alpha\beta} + a_1^{(s)}\to\Gamma_o\,.$$

The symbol ~ denotes the fields interacting with longwave bulk deformations.

5.2 Electron-Phonon Interaction Parameter for the High-Temperature Superconducting Phase

It is known from the experimental data (Sect. 4.1) that for a phase transition to the superconducting state, bulk exchange striction is not observed. Hence we shall disregard it in the following analysis. The main problem that must be solved while determining the effective electron-phonon interaction parameter concerns the magnitude of the Coulomb interaction between conduction electrons during phase transition to the superconducting state taking fluctuations into account. While considering the effect of spin fluctuations on the superconducting transition temperature [5.11], it was shown that fluctuations must inevitably lower the transition temperature, since the effective Coulomb repulsion must increase. However, while paramagnons were taken into consideration in [5.11], longwave thermodynamic fluctuations were not. We shall show that when the latter are included a suppression of effective Coulomb repulsion is achieved. Indeed,

$$\mathbb{L} = (1/2)(q^2-\tau)S_{\alpha o}(\mathbf{q})S_{\alpha o}(-\mathbf{q}) - (1/8)\Gamma_{1o}\sum_{\mathbf{q}_1+\mathbf{q}_2+\mathbf{q}_3+\mathbf{q}_4=0}S_{\alpha o}(\mathbf{q}_1)S_{\alpha o}(\mathbf{q}_2)S_{\beta o}(\mathbf{q}_3)S_{\beta o}(\mathbf{q}_4)$$

$$-(1/4)b_{1o}^{(s)}\sum_{\mathbf{q}_1+\mathbf{q}_2+\mathbf{p}_1+\mathbf{p}_2=0}S_{\alpha o}(\mathbf{q}_1)S_{\alpha o}(\mathbf{q}_2)\hat{\psi}_\sigma^{+}(\mathbf{q}_1)\hat{\psi}_\sigma(\mathbf{p}_2) + (1/2)\,\varepsilon_\sigma(\mathbf{p})\psi_\sigma^{+}(\mathbf{p})\hat{\psi}_\sigma(-\mathbf{p})$$

$$-(1/2)g_{ph}\sum_{\mathbf{q}+\mathbf{p}_1+\mathbf{p}_2=0}\varphi(\mathbf{q})\hat{\psi}^+_\sigma(\mathbf{p}_1)\hat{\psi}_\sigma(\mathbf{p}_2) - \mu_{os}^{d(f)}\sum_{\mathbf{p}_1+\mathbf{p}_2+\mathbf{p}_3+\mathbf{p}_4=0}\hat{\psi}_\uparrow^{+}(\mathbf{p}_1)\,\hat{\psi}_\downarrow^{+}(\mathbf{p}_2)\hat{\psi}_\uparrow(\mathbf{p}_3)\hat{\psi}_\downarrow(\mathbf{p}_4)$$

$$-(1/2\sqrt{2})\nu\sum_{\mathbf{q}+\mathbf{p}_1+\mathbf{p}_2=0}\mathbf{q}_\alpha S_{\alpha o}(\mathbf{q})\,\hat{\psi}^+_\sigma(\mathbf{p}_1)\hat{\psi}_\sigma(\mathbf{p}_2) + \mathbb{L}_{ph}(\mathbf{q})\,,$$

$$\mu_{os}^{d(f)} = \mu_s^{d(f)} + (1/2)c_1^{(s)}\,, \quad c_1^{(s)} < 0\,. \qquad (5.2.1)$$

Using the standard renormalization procedure, we can show that

$$\tilde{\mu}_s^{d(f)} = \mu_{os}^{d(f)} - [4nb_{1o}^{(s)2}/(4-n)\,\Gamma_{10}]\,\{[1+(n+8)\Gamma_{1o}x]^{\frac{4-n}{n+8}} - 1\} \ . \tag{5.2.2}$$

Since $n \leq 3$ for a strong magnetic anisotropy, the effective Coulomb repulsion is suppressed by longwave magnetic fluctuations for $x \to \infty$

$$x = \frac{2}{\varepsilon(\Lambda^2)^{\varepsilon/2}} \left\{ \left[\frac{\Lambda^2}{\lambda^2(\tau)}\right]^{\varepsilon/2} - 1 \right\} , \quad \varepsilon \to 0 \ . \tag{5.2.3}$$

Let us consider the electron-phonon interaction parameter in the superconducting phase. It can be shown [5.7] that

$$|\tilde{\lambda}_{e-ph}| = \ln^{-1}\left(\frac{2}{\pi}\,\frac{\langle\omega_D\rangle}{T_N}\right) , \tag{5.2.4}$$

$$\tilde{\lambda}_{e-ph} = \lambda_{e-ph} - \tilde{\mu}_s^{d(f)} , \quad \gamma = e^C , \quad C = 0.577 ,$$

where $\tilde{\mu}^{d(f)}$ is the renormalized value of the Coulomb interaction parameter, $\langle\omega_D\rangle$ is the average Debye energy, $T_N \simeq J_{eff}$, $J_{eff} = J+A$, J is the exchange interaction parameter in the magnetic subsystem, and A is the axial anisotropy constant. In this case, the maximum value of the superconducting gap is

$$\Delta_{max} = \frac{\pi}{\gamma} T_N , \quad T_N \to T_c \ . \tag{5.2.5}$$

Let us now determine the energy of conduction electrons near the Fermi surface above the phase transition point ω_λ (Fig. 5.2). For

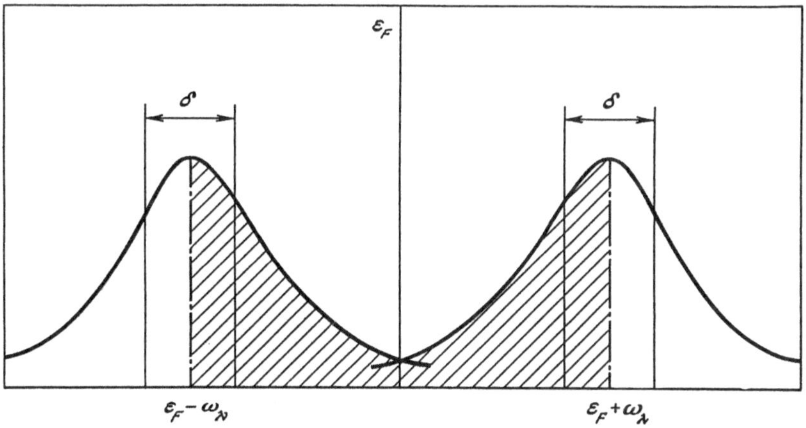

Fig. 5.2. Interaction of collective $d(f)$-electrons within $2\omega_\lambda$ of the Fermi surface

this purpose we must write the effective vertex corresponding to a four-particle interaction of electrons $T(0, \psi/T)$. The solutions for $T(0, \psi/T)$ must be joined above and below the phase transition point. Thus, we can write the RG equation for the parameter $\tilde{\lambda}_{e-ph}$ in the vicinity of the phase transition point:

$$\frac{d|\tilde{\lambda}_{e-ph}|}{dy} = |\tilde{\lambda}_{e-ph}|^2 + O(|\lambda_{e-ph}|^k) ,$$

$$k \geq 3 , \quad y = \ln \frac{\langle \omega_D \rangle}{\omega_\lambda} . \tag{5.2.6}$$

Figure 5.3 is the graphic representation of (5.2.6). Its solution has the form

$$|\tilde{\lambda}_{e-ph}| = |\tilde{\lambda}_{e-pho}| \Big/ \left[1 + |\tilde{\lambda}_{e-pho}| \ln \frac{\langle\omega_D\rangle}{\omega_\lambda}\right], \tag{5.2.7}$$

whence

$$\omega_\lambda = \langle\omega_D\rangle \exp\{(|\tilde{\lambda}_{e-ph}| - |\tilde{\lambda}_{e-pho}|)/|\tilde{\lambda}_{e-ph}||\tilde{\lambda}_{e-pho}|\} , \tag{5.2.8}$$

$|\tilde{\lambda}_{e-pho}|$ being the effective electron-phonon interaction parameter above the transition point.

Hence, the phase transition to the superconducting state is accompanied by an abrupt variation of the electron-phonon interaction parameter $|\tilde{\lambda}_{e-pho}|\,|_{T_c+0} \to |\tilde{\lambda}_{e-ph}|\,|_{T_c-0}$ and of the electron energy near the Fermi surface $\omega_\lambda \to \langle\omega_D\rangle$.[1] This means that the phase transition under consideration is a first-order phase transition, which is in accord with the results of the phenomenological theory (Chap. 4). Since the expression for $|\tilde{\lambda}_{e-ph}|$ contains the parameter T_N below the transition point (see 5.2.4), the effective electron-phonon interaction in the superconducting phase can be enhanced by the exchange interaction in the magnetic subsystem since the value of Néel tem-

---

[1]Lattice instability associated with the structural transformation is observed in superconductors with the highest transition temperatures T_c (compounds of the type A-15 and C-15), as well as in superconductors with a B-2 structure. However, the structural transformation temperature T_s is much higher than T_c and a correlation between the two parameters has not been established. In this case, there is no structural transformation in spite of the fact that the parameter λ_{e-ph} changes. This is so because the change in λ_{e-ph} is associated with a suppression of the Coulomb repulsion of the conduction electrons by fluctuations [5.4-6].

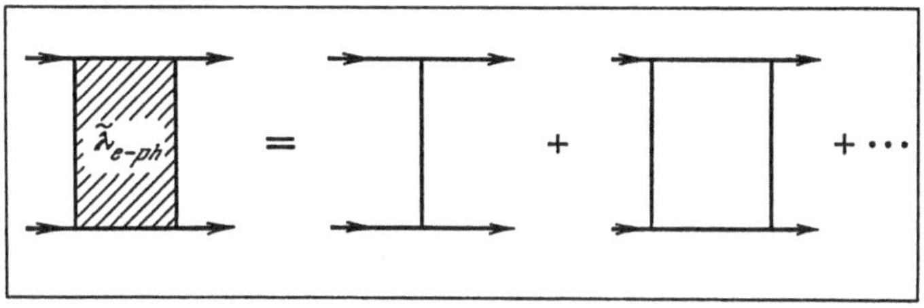

Fig. 5.3. Graphic equation for the electron-phonon interaction strength λ_{e-ph} below the first-order transition point to the superconducting state

perature may be several hundred Kelvin. This is just a manifestation of the exchange enhancement effect.

In systems with a more complex antiferromagnetic order, the most strongly fluctuating magnetic component cannot ensure the required topology of the Fermi surface, and an effective electron-phonon interaction is sufficient to cause a superconducting transition. This can be seen from (5.2.2). For $n \geq 4$, the magnetic fluctuations can no longer suppress the Coulomb repulsion of the conduction electrons. Hence the system will undergo a phase transition to the magnetic state. It follows from (5.2.2) that the strongest variation of the Coulomb repulsion parameter for the conduction electrons with temperature may be observed at $n = 1$. As $\tau \to 0$, $x \to \infty$ and the Coulomb repulsion parameter decreases and may even assume negative values which can be attributed to a strong attraction between the conduction electrons near the antiferromagnetic instability point. This leads to the possibility of a phase transition to the superconducting state for $T_c \to T_N$.

Hence, one of the main conditions for the emergence of a high-temperature superconducting phase in rare-earth metal compounds is the existence of an antiferromagnetic instability and the proximity of the Néel temperature and the superconducting transition temperatures, since the Néel temperature may be of the order of 100 K. Fluctuations of the antiferromagnetism vector cause longwave bulk deformation in the system (Sects. 5.1,2 and 3.3). Moreover, (5.2.2) was obtained by disregarding the anisotropy of the crystal lattice. However, real, experimentally investigated crystals are anisotropic [5.12,13]. A consideration of the interaction of the magnetic order parameter fluctuations with longwave bulk deformations, and the anisotropy of the crystal lattice may lead to a first-order phase transition to the antiferromagnetic state. This transition will occur before the transition to the superconducting phase. It was shown in [5.14] that antiferromagnetism does not significantly affect the suppression of superconductivity and hence the conditions for the emergence of a high-temperature superconducting phase persist in the system. This phase may emerge at a temperature lower than the Néel temperature, $T_c < T_N$. The condition $T_c < T_N$ may also be realized in systems with a multicomponent antiferromagnetic order parameter, when different components turn successively into "condensate" [5.15].

Thus, the fluctuation theory of magnetic superconductors predicted a basically new class of superconducting compounds in which the temperature of phase transition to the magnetically ordered state exceeds the superconducting transition temperature ($T_c < T_N$). A superconducting phase with a transition temperature $T_c < T_N$ was first observed in the compounds $RE(Rh_xIr_{1-x})_4B_4$, RE = Dy, Ho, and later in the compounds $RERh_2Si_2$, RE = La [5.16-18]. The superconducting transition temperature T_c in these compounds was found to be of the order of 1 K, and T_N varied between 2 and 50 K. This leads to the conclusion that although (5.2.4) represents the condition for the emergence of superconductivity in magnetic superconductors and was instrumental in predicting a new class of superconducting compounds with $T_c < T_N$, it cannot be considered to be the sufficient condition for the emergence of a high-temperature superconducting phase in magnetic superconductors, since there exist antiferromagnetic systems with low values of Néel temperature as well ($T_N \simeq 1$ K) [5.4,5]. In order to find the sufficient condition, we must set a lower limit on the value of T_N. Such a limit can be obtained by calculating the correction to the velocity of sound in the superconducting state near the phase transition point:

$$\left|\frac{s - s_n}{s_n}\right| = \left|\nu(\varepsilon_F) \frac{|\Delta_0|^2}{\langle\omega_D\rangle} \left\{ 1 + \ln\left[\left(\frac{2\gamma}{\pi}\right)^2 \left(\frac{\langle\omega_D\rangle^2 - \mu_\varkappa^2}{T^2 - \mu_\varkappa^2}\right)\right]^{\frac{1}{2}} \right\} \cdot \int_{C_\infty^+} d\omega \tilde{\lambda}_{e-ph}(\omega) \rho^{ph}(\omega) \right| . \quad (5.2.9)$$

Here, s_n is the velocity of sound in the paramagnetic (normal) phase, $\nu(\varepsilon_F)$ is the density of states on the Fermi surface, $\rho^{ph}(\omega)$ is the phonon density of states, $\tilde{\lambda}_{e-ph}(\omega)$ is the matrix element of the electron-phonon interaction as a function of frequency ω ,

$$\mu_\varkappa = \frac{p_f^2}{2\mu_e} \frac{\chi^2}{\chi\chi_{os} + \chi_{os}^2 + \chi_{os}^2 (p_f^2/p_F^2)} ; \quad (5.2.10)$$

χ, χ_{os} are the paramagnetic susceptibility and the Pauli susceptibility of collective s^-, d(f)-electrons (Chap. 1) [5.14], p_F is the Fermi momentum, $p_f = 2\pi/\langle r_c\rangle$, $\langle r_c\rangle = \langle r_{c0}\rangle$, $\tau^{-\nu}$ is the exchange correlation radius in the system of collective electron spins $\nu = \frac{1}{2}$, and

$$\langle r_{c0}\rangle = \left\langle\left(\frac{\int d\mathbf{x}\, x^2\, J(\mathbf{x})}{\int d\mathbf{x}\, J(\mathbf{x})} \right)\right\rangle^{\frac{1}{2}} . \qquad (5.2.11)$$

Since $p_f/p_F = \tau^{\nu}$, $\chi/\chi_{os} \approx \tau^{-2\nu}$, we have

$$T_N > \varepsilon_F (a/\langle r_{c0}\rangle)^2 . \qquad (5.2.12)$$

The parameter $(a/\langle r_{c0}\rangle)^2$ can be evaluated by using the exchange potential model $J(\mathbf{x}) = J_0 (r/a)^n e^{-r/a}$, $n \geq 1$. This gives $(a/\langle r_{c0}\rangle)^2 \simeq 10^{-2}$, and hence the estimate $T_N \to T_c \gtrsim 10^2$ K, i.e., the conditions (5.2.4) and (5.2.12) are the necessary and sufficient conditions for the emergence of a high-temperature superconducting phase in magnetic superconductors.

If the system first undergoes a phase transition to the antiferromagnetic state, the conditions for the emergence of a high-temperature superconducting phase deteriorate, since a system of antiferromagnetically ordered magnetic moments contains spin waves which weaken the electron-phonon interaction. The effect of spin waves on the electron-phonon interaction in the phase of coexistence of superconductivity and anti-ferromagnetic ordering was considered in [5.14]. If the activation energy in the spin wave spectrum is low ($\Delta\omega/\langle\omega_D\rangle \ll 1$), the conditions for the emergence of a high-temperature superconducting phase with $T_c < T_N$ are preserved [5.14].

It should be noted that the exchange enhancement of electron-phonon interaction due to spin fluctuations in high-temperature superconductors in the presence of a magnetic instability near the superconducting transition point was recently investigated by **Kim** [5.19].

6. Solitons in the Theory of Superconductivity of Magnetic Systems

6.1 Generalized Order Parameter in Magnetic Superconductors: Equations for the Superconducting Phase

In the preceding chapters, we have considered the phase transitions in superconducting compounds of rare-earth metals from the paramagnetic phase to the superconducting state. It was assumed that the superconducting state is homogeneous. However, since magnetic fluctuations exist in the superconducting phase (Chap. 1), they must affect the structure of the order parameter. Moreover, if we consider the phase in which superconductivity and magnetic order coexist, then in accordance with Chap. 4 it has a longwave sinusoidally modulated magnetic structure. Such a structure was later observed experimentally [6.1], and the wavelength of the structure was found to be $\lambda \sim 100\text{-}200$ Å. Hence, it can be concluded that the superconducting state is not necessarily homogeneous in magnetic superconductors due to an interaction of the superconducting electrons with the magnetic fluctuations and the nonuniform magnetic structure.

This chapter is devoted to a study of the inhomogeneous states which may appear in magnetic superconductors during phase transitions. It is well known [6.2] that the phase transition from the paramagnetic to superconducting phase in some compounds of rare-earth metals (Chap. 4) is a first-order transition, close to a second-order one. The superconducting order parameter changes rapidly, and this leads to a suppression of magnetic fluctuations. With decreasing temperature, the order parameter increases. As the $|\Delta_0(T)|$ curve attains its peak, the effective temperature $\tau_\perp(\tau_{\parallel})$ (4.2.3) decreases and the fluctuations of the magnetic subsystem beging to increase. This results in the emergence of an inhomogeneous superconducting state and to a change in the effective order parameter [6.3].

In order to write the effective Hamiltonian of the system in this case, we must introduce the interaction of the order parameter with the magnetic subsystem. The magnetic subsystem in the superconducting phase is determined by the spins of collective s, d(f)-electrons (Chap. 1) and is not ordered. Such a spin system is invariant to the transformations of the three-dimensional rotation group SO(3) in the spin space, which is isomorphic to the SU(2) group (Sect. 4.4). The complete symmetry group of the superconducting phase is defined as the product of the space group of the system and the SU(2) group, i.e., $G = G_p \times SU(2)$. The interaction of the order parameter with the magnetic subsystem is determined by introducing an effective field whose potential is transformed in both coordinate and spin spaces upon a transformation of the symmetry group G, i.e., the potential must be a tensor of rank two: $A_\nu^\alpha(x)$. Its relation with the order parameter can be expressed through the covariant derivative

$$D_\nu = \nabla_\nu - (e/c) g_1^{\#} f^{\alpha\delta\gamma} A_\nu^\delta \ . \tag{6.1.1}$$

But in this case, the potential $A_\nu^\alpha(x)$ will play the role of the gauge field potential for SU(2)-symmetry, i.e., it will coincide with the Yang-Mills field potential [6.4]. In the above formula, $\nu = 0,1,2,3$; $\alpha, \gamma = 0,1,2,3$; $\delta = 1,2,3$; $g_i^{\#}$ is the interaction parameter [6.3], and $f^{\alpha\delta\gamma}$ is the structural tensor whose explicit form is chosen from the symmetry considerations and depends on the phase in which the effective Hamiltonian is constructed. Procedding from the general considerations [6.5], we can write the effective Hamiltonian in the form

$$\begin{aligned} \mathbb{H}^{eff} &= \int d\mathbf{x}\ \{(1/2)b_1(\tau)\mathrm{Tr}(\hat{\Psi}^+\hat{\Psi}) + (1/2)\mathrm{Tr}[(\hat{D}_\nu\hat{\Psi})^+(\hat{D}_\nu\hat{\Psi})] \\ &\quad + b_2(\tau)[\mathrm{Tr}(\hat{\Psi}^+\hat{\Psi})^2]\ \ln[\mathrm{Tr}(\hat{\Psi}^+\hat{\Psi})/b_3(\tau)] - (1/4)F_{\mu\nu}^\alpha\} \\ F_{\mu\nu}^\alpha &= \nabla_\nu A_\mu^\alpha - \nabla_\mu A_\nu^\alpha + g e^{\alpha\beta\gamma} A_\mu^\beta A_\nu^\gamma \ . \end{aligned} \tag{6.1.2}$$

The order parameter has the following general form:

$$\Psi^{\alpha\beta} = \Psi_o^{\alpha\beta}\, e^{i\varphi^{\alpha\beta}} . \tag{6.1.3}$$

In this case, the equations of motion of the system can be obtained by varying the functional (6.1.2) in variables $\hat{\Psi}_o^{\alpha\beta}$, $\varphi^{\alpha\beta}$ and $A_\nu^\alpha(x)$ and will have a complex form. We have considered these equations as applied to specific phases, indicated on the phase diagram of the superconducting compounds $Er_{1-x}Ho_xRh_4B_4$ of rare-earth metals (Fig. 4.5). Let us begin with the superconducting state which is assumed to be homogeneous. The phenomenological and microscopic theories of the emergence of such a state are described in Chaps. 4 and 5 and in [6.2,6-10]. The physical mechanism behind the emergence of the superconducting phase can be explained as follows: the metastable inhomogeneous magnetically ordered state is violated and superconductivity emerges. In order to describe this process, we must take for the order parameter $\hat{\Psi}$ a quantity which effectively includes a magnetic component Ψ^{α}, $\alpha = 1,2,3$, as well as a superconducting component Ψ^{0}. Hence the order parameter can be presented as a 4-vector:

$$\Psi^{\alpha} = \Psi_{0}^{\alpha} e^{i\varphi^{\alpha}} , \quad \alpha = 1,2,3,4 , \tag{6.1.4}$$

or

$$\boldsymbol{\Psi} = \boldsymbol{\Psi}_{0} e^{i\varphi} . \tag{6.1.5}$$

In this case, the effective Hamiltonian of the system in the superconducting state near the point of transition to the paramagnetic phase is written in the form

$$\mathbb{H}^{eff} = \int d\mathbf{x} \, \{(1/2)b_1(\tau)|\boldsymbol{\Psi}|^2 + (1/2)\{(D_{\nu}\boldsymbol{\Psi})^*(D_{\nu}\boldsymbol{\Psi})\}$$

$$+ b_2(\tau)|\boldsymbol{\Psi}|^4 \ln[|\boldsymbol{\Psi}|^2/b_3(\tau)] - (1/4)(F_{\mu\nu}^{\alpha})^2\} . \tag{6.1.6}$$

Substituting (6.1.4) into (6.1.6), we obtain

$$\mathbb{H}^{eff} = \int d\mathbf{x}\Big\{\frac{1}{2}b_1(\tau)\sum_{\alpha}\Psi_{0}^{\alpha^2} + \frac{1}{2}\Big[\sum_{\alpha}\{(\nabla_{\nu}\Psi_{0}^{\alpha})^2 + (\nabla_{\nu}\varphi^{\alpha})^2\Psi_{0}^{\alpha^2}\}$$

$$- \frac{e}{c} g_1^{\#} \sum_{\alpha,\delta,\gamma} f^{\alpha\delta\gamma}(\nabla_{\nu}\Psi_{0}^{\alpha})\Psi_{0}^{\gamma}A_{\nu}^{\delta}[\exp\{i(\varphi^{\alpha}-\varphi^{\gamma})\}$$

$$+ \exp\{-i(\varphi^{\alpha}-\varphi^{\gamma})\}] - i\frac{e}{c} g_1^{\#} \sum_{\alpha,\delta,\gamma} f^{\alpha\delta\gamma}(\nabla_{\nu}\varphi^{\alpha})\Psi_{0}^{\alpha}\Psi_{0}^{\gamma}A_{\nu}^{\delta}$$

$$\times [\exp\{i(\varphi^{\alpha}-\varphi^{\gamma})\} - \exp\{-i(\varphi^{\alpha}-\varphi^{\gamma})\}]$$

$$+ \left(\frac{e}{c}\right)^2 g_1^{\#^2} \sum_{\alpha,\delta,\gamma,\delta',\gamma'} f^{\alpha\delta\gamma}f^{\alpha\delta'\gamma'}A_{\nu}^{\delta}A_{\nu}^{\delta'}\Psi_{0}^{\gamma}\Psi_{0}^{\gamma'}$$

$$\times \exp\{i(\varphi^{\gamma}-\varphi^{\gamma'})\}\Big] + b_2(\tau)\sum_{\alpha,\beta}\Psi_{0}^{\alpha^2}\Psi_{0}^{\beta^2}$$

$$\times \ln\left[\frac{1}{b_3(\tau)}\sum\Psi_{0}^{\alpha^2}\right] - \frac{1}{4}\sum_{\alpha}(F_{\mu\nu}^{\alpha})^2\Big\} . \tag{6.1.7}$$

Varying the functional (6.1.7) in the variables $\nabla_{\nu}\Psi_{0}^{\alpha}$, Ψ_{0}^{α}, $\nabla_{\nu}\varphi^{\alpha}$, φ^{α}, $\nabla_{\mu}A_{\nu}^{\alpha}$, A_{ν}^{α}, we obtain

$$\nabla^2_\nu \boldsymbol{\Psi}_o - b_1(\tau)\boldsymbol{\Psi}_o - \{(\nabla_\mu\varphi)^2\boldsymbol{\Psi}_o\}$$

$$-4b_2(\tau)\boldsymbol{\Psi}_o|\boldsymbol{\Psi}_o|^2 \ln[|\boldsymbol{\Psi}_o|^2/b_3(\tau)] - 2b_2(\tau)\boldsymbol{\Psi}_o|\boldsymbol{\Psi}_o|^2$$

$$-(e/2c)g_1^{\#}(e^{i\varphi}\nabla_\mu[\mathbf{A}_\mu\boldsymbol{\Psi}^*] + e^{-i\varphi}\nabla_\mu[\mathbf{A}_\mu\boldsymbol{\Psi}])$$
$$-(e/2c)g_1^{\#}(e^{i\varphi}[\mathbf{A}_\mu(\nabla_\mu\boldsymbol{\Psi})^*] + e^{-i\varphi}[\mathbf{A}_\mu(\nabla_\mu\boldsymbol{\Psi})]$$

$$-(1/2)(e/c)^2 g_1^{\#^2}\ (e^{i\varphi}[\mathbf{A}_\mu[\boldsymbol{\Psi}^*\mathbf{A}_\mu]] + e^{-i\varphi}[\mathbf{A}_\mu[\boldsymbol{\Psi}\mathbf{A}_\mu]]) = 0\ , \qquad (6.1.8)$$

$$(\nabla^2_\nu\varphi)\Psi_o + 2(\nabla_\mu\varphi)(\nabla_\mu\boldsymbol{\Psi}_o) - (e/2c)g_1^{\#}\{(ie^{i\varphi})\nabla_\mu[\mathbf{A}_\mu\boldsymbol{\Psi}^*]$$

$$+(ie^{i\varphi})^*\nabla_\mu[\mathbf{A}_\mu\boldsymbol{\Psi}]\} - (1/2)(e/c)^2 g_1^{\#^2}\ (ie^{i\varphi})[\mathbf{A}_\mu[\boldsymbol{\Psi}^*\mathbf{A}_\mu]]$$

$$+(ie^{i\varphi})^*[A_\mu[\boldsymbol{\Psi}\mathbf{A}_\mu]]\} = 0\ , \qquad (6.1.9)$$

$$\nabla_\mu\mathbf{F}_{\mu\nu} - g[\mathbf{A}_\mu\mathbf{F}_{\mu\nu}] - (e/2c)g_1^{\#}\ [\boldsymbol{\Psi}\nabla_\nu\boldsymbol{\Psi}*] + [\boldsymbol{\Psi}*\nabla_\nu\boldsymbol{\Psi}]$$

$$+(1/2)(e/c)^2 g_1^{\#^2}([\Psi[\mathbf{A}_\nu\boldsymbol{\Psi}^*]] + [\Psi^*[\mathbf{A}_\nu\boldsymbol{\Psi}]]) = 0\ , \qquad (6.1.10)$$

$$\nabla^2 = \sum_{\mu=0}^{4} \frac{\partial^2}{\partial x_\mu^2}\ , \qquad [\mathbf{A}_\nu\boldsymbol{\Psi}]^\alpha = f^{\alpha\delta\gamma}\mathbf{A}_\nu^\delta\Psi^\gamma,$$

$$[\mathbf{A}_\mu\mathbf{F}_{\mu\nu}]^\alpha = e^{\alpha\beta\gamma}\mathbf{A}_\mu^\beta\mathbf{F}_{\mu\nu}^\gamma\ , \qquad [\boldsymbol{\Psi}\nabla_\nu\boldsymbol{\Psi}^*]^\alpha = f^{\alpha\beta\gamma}(\nabla_\nu\boldsymbol{\Psi}^\beta)^*\Psi^\gamma\ .$$

Equations (6.1.8-10) can be written in a more compact form as follows:

$$\{D_\mu D_\mu^* \ -b_1(\tau) - 4b_2(\tau)|\boldsymbol{\Psi}|^2 \ln[|\boldsymbol{\Psi}|^2/b_3(\tau)]$$

$$- 2b_2(\tau)|\Psi\ |^2\}\boldsymbol{\Psi} = 0\ , \qquad (6.1.11)$$

$$\nabla_\mu\mathbf{F}_{\mu\nu} - g[\mathbf{A}_\mu\mathbf{F}_{\mu\nu}] + (e/2c)g_1^{\#}\{[\boldsymbol{\Psi}\nabla_\nu\ \boldsymbol{\Psi}^*] + [\boldsymbol{\Psi}^*\nabla_\nu\boldsymbol{\Psi}]\}$$

$$+(1/2)(e/c)^2 g_1^{\#^2}([\boldsymbol{\Psi}[\mathbf{A}_\nu\boldsymbol{\Psi}^*]] + [\boldsymbol{\Psi}^*[\mathbf{A}_\nu\boldsymbol{\Psi}]]) = 0\ . \qquad (6.1.12)$$

The effective Hamiltonian (6.1.7) and the equations of motion (6.1.8-10,12) are invariant relative to the SU(2) group transformations, and are also relativistically invariant in Minkowski space as well as in Euclidean space.

6.2 Soliton Mechanism for Emergence of Superconductivity Accompanying a Phase Transition from Superconducting to Paramagnetic States

Let us consider the behavior of a system (say, $Er_{1-x}Ho_xRh_4B_4$) near the line of phase transitions from a paramagnetic to a superconducting state. In the vicinity of such a phase transition point, the field $\mathbf{S}_\perp$ is known to fluctuate most strongly, and stimulates the fluctuations of $\mathbf{S}_\parallel$ and Δ [6.2,6]. In this case, $\mathbf{S}_\parallel$ and Δ , being fluctuating quantities, strive to turn into a condensate. However, states with $\Delta_o \neq 0$ and $s_{\parallel o} = 0$ will be thermodynamically stable, while the state with $s_{\parallel o} \neq 0$ and an arbitrary Δ_o will be unstable. The topological manifold corresponding to the order parameter of $\mathbf{S}_\parallel$ has the following form: $R = S^2 \times S^1$ [6.11-13], $\pi_1(R) = Z$, $\pi_2(R)=Z$. This means that the condensation of $S_\parallel$ may lead to the emergence of point singularities (topological solitons). The nontrivial nature of the second homotopic group $\pi_2(R)$ means that if the point singularity is surrounded by a second-order surface, then for a fixed phase of the order parameter of $\mathbf{S}_\parallel$, the vector $\mathbf{S}_{\parallel 0}$ circumventing the given surface maps it onto the entire sphere S^2. Consequently, the degree of mapping is nonzero. Such a singularity cannot be eliminated topologically, since the given surface cannot be reduced to a single point on the sphere S^2. However, the point singularity is unstable because of the effect of nontopological perturbations (for example, fluctuations of $\mathbf{S}_\perp$ and Δ). The destruction of solitons leads to an enhancement of fluctuations and to the appearance of the superconducting component Δ .

The theory of gauge fields [6.4] allows us to write the dynamic equation of the superconducting phase near the line of phase transitions to the superconducting state (6.1.11,12). Let us analyze these equations for the case when a phase transition from the superconducting to the paramagnetic phase is accompanied by the appearance of unstable topological solitons of "magnetic monopole" type [6.13]. Topologically, such a soliton is defined by the mapping f: $S^2 \to S^2$. The order parameter will have the form $\Psi^\alpha = \Psi^\alpha_o$. In this case, (6.1.11,12) are simplified to

$$(D_\mu^2 - b_1(\tau) - 4b_2(\tau)|\mathbf{\Psi}_o|^2 \ln[|\mathbf{\Psi}_o|^2/b_3(\tau)] - 2b_2(\tau)'|\mathbf{\Psi}_o|^2)\mathbf{\Psi}_o = 0,$$

$$\nabla_\mu \mathbf{F}_{\mu\nu} - g[\mathbf{A}_\mu \mathbf{F}_{\mu\nu}] - (e/c)g_1^\#[\mathbf{\Psi}_o \nabla_\nu \mathbf{\Psi}_o] + (e/c)^2 g_1^{\#^2}[\mathbf{\Psi}_o[\mathbf{A}_\nu \mathbf{\Psi}_o]] = 0. \quad (6.2.1)$$

We shall look for a solution for $\mathbf{\Psi}_o$ in the form

$$\Psi_o^\alpha = x_\alpha u(\xi)/\xi , \quad \xi = (x_\mu x_\mu)^{\frac{1}{2}} . \tag{6.2.2}$$

For the case of Minkowski space, $x_o = i\,\beta t$ (in Euclidean space, $x_o = \beta t$). The form of the solution for A_ν^α can be obtained by determining the current induced by the field of the order parameter Ψ_o:

$$J_\nu^\alpha = f^{\beta\alpha\gamma}(\nabla_\nu \Psi_o^\beta)\Psi_o^\gamma = -f^{\nu\alpha\gamma} x_\gamma \frac{u^2(\xi)}{\xi^2} . \tag{6.2.3}$$

This gives

$$A_\nu^\alpha = f^{\nu\alpha\gamma} x_\gamma a(\xi) . \tag{6.2.4}$$

In this case, the tensor $f^{\nu\alpha\gamma}$ is the T'Hooft tensor [6.14] (α =1,2,3; ν,γ = 0,1,2,3) :

$$f^{\nu\alpha\gamma} \equiv \eta_{\nu\gamma}^\alpha ,$$

$$\eta_{\nu\gamma}^\alpha = -e_{\alpha\nu\gamma} + \delta_{o\nu}\delta_{\alpha\gamma} - \delta_{o\gamma}\delta_{\alpha\nu} . \tag{6.2.5}$$

Substituting the values of J_ν^α and, A_ν^α into (6.2.1), we obtain the following equations for the functions $u(\xi)$, $a(\xi)$:

$$\begin{aligned} &u'' + (n-1)\xi^{-1}u' - (n-1)\xi^{-2}u - b_1(\tau)u - 4b_2(\tau)u^3 \ln[u^2/b_3(\tau)] \\ &\quad -2b_2(\tau)u^3 + 2(n-1)(e/c)g_1^{\#}au - (n-1)(e/c)^2 g_1^{\#} a^2 u = 0, \\ &a'' + (n+1)\xi^{-1}a' + 3(n-2)ga^2 - (n-2)g^2\xi^2 a^2 \\ &\quad +(e/c)g_1^{\#^2}\xi^{-2}u^2 - (e/c)^2 g_1^{\#^2} au^2 = 0 , \end{aligned} \tag{6.2.6}$$

where n is the dimensionality of Ψ_o . It should be noted that for n = 3 (6.2.6) describes a sationary bulk magnetic monopole $\xi \to r$:

$$\begin{aligned} &u'' + 2r^{-1}u' - 2r^{-2}u - b_1(\tau)u - 4b_2(\tau)u^3 \ln \frac{u^2}{b_3(\tau)} \\ &\quad -2b_2(\tau)u^3 + 4(e/c)g_1^{\#}au - 2(e/c)^2 g_1^{\#^2} r^2 a^2 u = 0 , \\ &a''' + 4r^{-1}a' + 3ga^2 - g^2r^2a^3 + (e/c)g_1^{\#}r^{-2}u^2 - (e/c)^2 g_1^{\#^2} u^2 a = 0. \end{aligned} \tag{6.2.7}$$

The boundary conditions for the current can be written in the form

$$J_\nu^\alpha(r\to 0) \to 0 , \quad J_\nu^\alpha(r\to\infty) \to 0 , \quad \nu, \alpha = 1,2,3. \tag{6.2.8}$$

In this case, we can easily write down the asymptotic forms for the functions u(r) and a(r):

(1) $r \to 0,\ u \to r\ ,\quad a \to \text{const}\ ,$

$$(2)\ r \to \infty,\ a \to 1/gr^2\ ,\quad u \to \pm\sqrt{b_1(\tau)/2b_2(\tau)}. \qquad (6.2.9)$$

In the four-dimensional case, we have obtained a soliton which is defined by the following equations:

$$u'' + 3\xi^{-1}u' - 3\xi^{-2}u - b_1(\tau)u - 4b_2(\tau)u^3 \ln \frac{u^2}{b_3(\tau)}$$

$$- 2b_2(\tau)u^3 + 6(e/c)g_1^{\#}au - 3(e/c)^2 g_1^{\#2}\ a^2u = 0\ ,$$

$$a'' + 5\xi^{-1}a' + 6ga^2 - 2g^2\xi^2a^3 + (e/c)g_1^{\#}\xi^{-2}u^2 - (e/c)^2g_1^{\#2}u^2a = 0,$$

$$A_\nu^\alpha(\mathbf{x}) = \lim_{|\xi^2|\to\infty} \eta_{\nu\gamma}^\alpha\, x_\gamma a(\xi,\ell,m)\ ,$$

$$a(\xi,\ell,m) = \frac{1}{g\xi^2}\left\{1 + -\left(\frac{2\ell^2}{m^2-\ell^2}\right)^{\frac{1}{2}} \tilde{cn}\left[\left(\frac{2}{m^2-\ell^2}\right)^{\frac{1}{2}} \ln\left|\frac{\xi}{\xi_o}\right|,\ m,\ell\right]\right\}\ .$$

$$(6.2.10)$$

Here, $\tilde{cn}\ f(\xi)$ is the generalized elliptical Jacobi function, and ℓ, m are constants of integration.

By analyzing (6.2.10), it can be concluded that for $x_o \to i0$ ($x_o \to 0$ in the Euclidean space), we have a stationary magnetic monopole which is destroyed with time. For $x_o \to i\infty$ ($x_o^E \to \infty$), the local magnetic moment becomes equal to zero and a homogeneous superconducting state appears in the system [6.2,6]. The current state will be caused by new solitons. Indeed, the solution for $A_\nu^\alpha(x)$ has the form

$$A_\nu^\alpha(\mathbf{x}) = \lim_{|\xi^2|\to\infty} \eta_{\nu\gamma}^\alpha\, x_\gamma \frac{1}{g\xi^2}\left\{1 + \left(-\frac{2\ell^2}{m^2-\ell^2}\right)^{\frac{1}{2}} \tilde{cn}\left[\left(\frac{2}{m^2-\ell^2}\right)^{\frac{1}{2}} \ln\left|\frac{\xi}{\xi_o}\right|,\ m,\ell\right]\right\}$$

$$(6.2.11)$$

where the first soliton corresponds to the values $m^2 = 1$, $\ell^2 = -1$, and the second soliton to $\ell^2 = 0$, $m^2 = 1$. Subsequent variation of m and ℓ allows us to obtain the solutions corresponding to other solitons.

Thus, we have considered the mechanism of the emergence of superconductivity upon the destruction of an unstable soliton near the

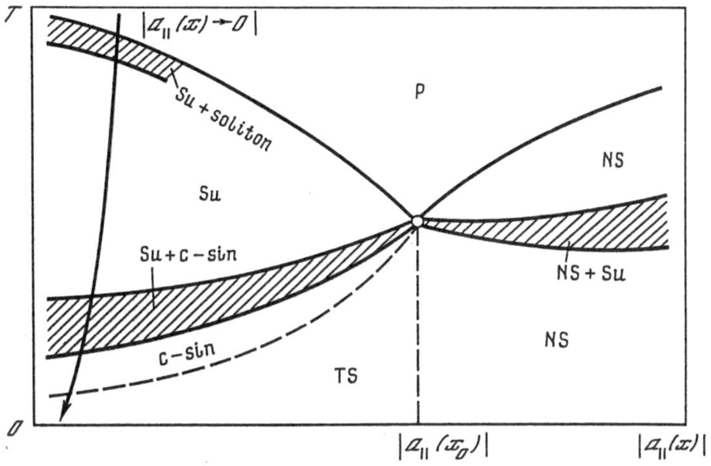

Fig. 6.1. Phase diagram of superconducting compounds of rare-earth metals containing a region in which an unstable topological magnetic soliton may emerge. The *arrow* indicates the sequence of phase transitions

phase transition from the superconducting to paramagnetic phase. This mechanism can also be used for considering the transition from the phase of coexistence of superconductivity and magnetism into a superconducting phase. However, such a transition is possible only if the latter phase is homogeneous. It can be seen from the phase diagram of the compounds $Er_{1-x}Ho_xRh_4B_4$ (Fig. 6.1) that the emergence of superconductivity can be effective only in the region of a very low anisotropy of the longitudinal magnetic subsystem $|a_{\parallel}(x)|$, since a large anisotropy hinders the formation of point singularities in the system [6.13]. In other words, the superconductivity can appear effectively only away from the point of intersection of the phase transition lines. It is also worth noting that the appearance of an unstable soliton must stabilize the superconducting state. Hence the phase of coexistence of superconductivity and magnetism will most probably be unstable, which is in accord with results of [6.2]. The resulting expression for the current J_0^{α} (6.2.3) describes the process of compensation of the magnetic moment upon a destruction of an unstable magnetic soliton. Indeed, J_0^{α} is spin-oriented, hence its integral over a sphere is nonzero. The quantity J_0^{α} can be interpreted as a current whose field compensates the magnetic moment.

6.3 Solitons in an Inhomogeneous Superconducting State and Topology

In this section, we shall consider the possible structure of the inhomogeneous superconducting state. It was shown earlier (Chap. 1,

Sects. 4.4,5.2,6.1) that magnetic fluctuations exist in the superconducting phase of a magnetic superconductor. We shall show that in the absence of an external magnetic field the magnetic fluctuatons lead to the emergence of vortex structures in the superconducting state. Mathematically, this is associated with a variation of the topological degeneracy space R corresponding to the order parameter. Indeed, in a normal superconductor $R = S^1$ while in the present case $R = SU(2)[SU(2)/Z_2] \times S^1$. Moreover, the emergence of a superconducting vortex is associated with the fact that $\int_\Gamma d\mathbf{x}_\nu \mathbf{A}_\nu^\alpha \neq 0$ (Sects. 6.1,2) [6.7,8]. In this case, a current appears in the superconducting phase and the superconducting structure will have a wave vector $\mathbf{q}$ due to the interaction of superconducting electrons with the spin-density fluctuations (Chap. 4). This means that the free energy functional contains an invariant which is linear in covariant derivatives:

$$[\hat{\Psi}(D_\nu^o\hat{\Psi})^+ - \hat{\Psi}^+(D_\nu^o\hat{\Psi})] \ .$$

Let us begin with a consideration of the equilibrium order parameter of the superconducting state when it is not homogeneous. The thermodynamic potential of the superconducting phase can be written in the form

$$F = \int d\mathbf{x} \left\{\frac{1}{2}b_1(\tau)\mathrm{Tr}(\Psi^+\Psi) + b_{5\nu}(\tau)\mathrm{Tr}(\hat{\Psi}\nabla_\nu\hat{\Psi}^+ - \hat{\Psi}^+\nabla_\nu\hat{\Psi}) + \frac{1}{2}\mathrm{Tr}(\nabla_\nu\hat{\Psi}^+\nabla_\nu\hat{\Psi}) + b_2(\tau)\mathrm{Tr}(\hat{\Psi}^+\hat{\Psi})^2] \ln\left[\frac{1}{b_3(\tau)}\mathrm{Tr}(\hat{\Psi}^+\hat{\Psi})\right]\right\}, \tag{6.3.1}$$

where $\tau = 1 - (T/T_o)$, T_o is the temperature of transition to the corresponding inhomogeneous superconducting state. The second term in (6.3.1) is associated with the emergence of a superconducting current as a result of a variation of the symmetry of the superconducting phase (Sect. 4.4). We shall seek the order parameter in the form

$$\hat{\Psi}(\mathbf{x}) = u(\mathbf{x})\,\hat{\tilde{R}}\,\hat{a} \ , \tag{6.3.2}$$

where $\hat{\tilde{R}} = e^{i\mathbf{qx}}\hat{R}$, $\hat{a} = \begin{Vmatrix}0\\1\end{Vmatrix}$. Here $\hat{R}$ is a unitary and unimodular matrix of the $SU(2)[SU(2)/Z_2]$ group. In the equilibrium state, $\hat{R}_o = \hat{1}$. Substituting (6.3.1) into (6.3.2), we can determine the wave vector of the superconducting structure from the thermodynamic potential minimum

$$q_\nu^o = 2ib_{5\nu}(\tau) - (1/2)iq_\nu(x) ,$$

$$q_\nu(\mathbf{x}) = \mathrm{Tr}\{\hat{a}\hat{a}^+[(\nabla_\nu\hat{R}^+)\hat{R} - \hat{R}^+(\nabla_\nu\hat{R})]\} \to 0.$$

Let us now consider the singularities of the field of the order parameter $\hat{\Psi}(x)$, which may appear in a given superconducting state. There may be two types of such singularities, viz., point singularities and line singularities. We begin with a consideration of point singularities. The expression for the free energy can be written in this case in the form

$$F = \int d\mathbf{x} \left\{ \frac{1}{2} b_1(\tau)u^2 + b_{\nu 5}(\tau)u^2 \right.$$
$$\times \mathrm{Tr}\left\{\hat{a}\hat{a}^+[(\nabla_\nu\hat{\tilde{R}}^+)\hat{\tilde{R}} - \hat{\tilde{R}}^+(\nabla_\nu\hat{\tilde{R}})]\right\} + \frac{1}{2}(\nabla_\nu u)^2$$
$$\left. + \frac{1}{2} u^2\mathrm{Tr}[\hat{a}\hat{a}^+(\nabla_\nu\hat{\tilde{R}}^+)(\nabla_\nu\hat{\tilde{R}})] + b_2(\tau)u^4 \ln \frac{u^2}{b_3(\tau)} \right\} . \qquad (6.3.3)$$

Suppose that $u(\mathbf{x}) \to u_\infty$ as $|\mathbf{x}| \to \infty$. In this case, the expression (6.3.3) simplifies to

$$F_\infty = (1/2)u_\infty^2 \int d\mathbf{x}\{2b_{\nu 5}(\tau)\mathrm{Tr}\{\hat{a}\hat{a}^+[(\nabla_\nu\hat{\tilde{R}}^+)\hat{\tilde{R}}$$
$$-\hat{\tilde{R}}^+(\nabla_\nu\hat{\tilde{R}})]\} + \mathrm{Tr}[\hat{a}\hat{a}^+(\nabla_\nu\hat{\tilde{R}}^+)(\nabla_\nu\hat{\tilde{R}})\} . \qquad (6.3.4)$$

Substituting $\hat{\tilde{R}} = e^{i\mathbf{q}^o\mathbf{x}}\hat{R}$ into (6.3.4), we obtain

$$F_\infty = (1/2)u_\infty^2 \int d\mathbf{x}\ \mathrm{Tr}[\hat{a}\hat{a}^+(\nabla_\nu\hat{R}^+)(\nabla_\nu\hat{R})] . \qquad (6.3.5)$$

In the general case, the matrix $\hat{R}$ has the form

$$\hat{R} = \prod_{\nu=1}^{3} \hat{T}_{x'_\nu} \exp[i\int_{x_{\nu o}}^{x_\nu} dx'_\nu\ \hat{O}(x'_\nu)] . \qquad (6.3.6)$$

The matrices $\hat{O}(x'_\nu)$ are hermitian, and $\hat{T}_{x'_\nu}$ is a time-ordering operator.

Let us consider a point singularity. In this case, the matrix $\hat{R}$ has the form

$$\hat{R} = e^{i\alpha\hat{\tau}_3}\, e^{i\beta\hat{\tau}_1}\, e^{i\gamma\hat{\tau}_3} , \qquad (6.3.7)$$

where τ_1, τ_3 are Pauli matrices. This expression can be transformed into

$$\hat{R} = \cos\beta e^{i(\alpha+\gamma)\tau_3} + i \sin\beta e^{i(\alpha-\gamma)\hat{\tau}_3\hat{\tau}_1} \qquad (6.3.8)$$

giving

$$\mathrm{Tr}[\hat{a}\hat{a} + (\nabla_\nu \hat{R}^+)(\nabla_\nu \hat{R})]$$

$$= (\nabla_\gamma \alpha)^2 + (\nabla_\gamma \beta)^2 + (\nabla_\nu \gamma)^2 + 2(\nabla_\nu \alpha \nabla_\nu \beta) \cos 2\beta \ . \qquad (6.3.9)$$

Introducing new variables $\alpha = -\gamma = \Phi/2$, $\beta = \Theta$, we obtain the following expression for the free energy of a point singularity for $|\mathbf{x}| \to \infty$:

$$F_\infty = (1/2)u_\infty^2 \int dx[(\nabla_\nu \Theta)^2 + (\nabla_\nu \Phi)^2 \sin^2\Theta] \ . \qquad (6.3.10)$$

If we have a point singularity, the functions Θ and Φ depend on ϑ and φ . Solving the equation for the extremals, we obtain

$$\Phi = n\varphi + \varphi_0 \ , \ \Theta = 2 \arctan[\tan(\vartheta/2)]^n \ . \qquad (6.3.11)$$

For n = 1, we obtain a "starfish"-type point singularity. The equation for the function u(r) is obtained by minimizing the functional

$$F = \int d\mathbf{x} \{(1/2)\tilde{b}_1(\tau)u^2 + (1/2)(\nabla_\nu u)^2$$

$$+(1/2)[(\nabla_\nu \Theta)^2 + (\nabla_\nu \Phi)^2 \sin^2\Theta] + b_2(\tau)u^4 \ln[u^2/b_3(\tau)]\} \ ,$$

$$\tilde{b}_1(\tau) = b_1(\tau) + 4 \sum_\mu b_5^2 \ \mu(\tau) \qquad (6.3.12)$$

and has the form

$$u'' + \frac{2}{r} u' - \frac{2}{r^2} u - \tilde{b}_1(\tau)u - 4b_2(\tau)u^3 \ln\left[\frac{u^2}{b_3(\tau)}\right] - 2b_2(\tau)u^3 = 0. \qquad (6.3.13)$$

This equation has the following asymtotics:

$$r \to 0 \ , \quad u \to r \ ;$$
$$r \to \infty, \ u \to u_\infty = [\tilde{b}_1(\tau)/2\tilde{b}_2(\tau)]^{\frac{1}{2}} + \delta(\tau) \ . \qquad (6.3.14)$$

When the gauge field is introduced, the expression for the free energy assumes the form

$$F = \int d\mathbf{x} \left\{\frac{1}{2} b_1(\tau)\mathrm{Tr}(\hat{\Psi}^+\hat{\Psi}) + b_{\nu 5}(\tau)\right.$$

$$\times \ \mathrm{Tr}[\hat{\Psi}(\hat{D}^{o}_{\nu}\hat{\Psi})^{+} - \hat{\Psi}^{+}(\hat{D}^{o}_{\nu}\hat{\Psi})] + \frac{1}{2}\,\mathrm{Tr}[(\hat{D}^{0}_{\nu}\hat{\Psi})^{+}(\hat{D}^{0}_{\nu}\hat{\Psi})]$$

$$+ b_2(\tau)[\mathrm{Tr}(\hat{\Psi}^{+}\Psi)^2]\ \ln\left[\frac{1}{b_3(\tau)}\,\mathrm{Tr}(\hat{\Psi}^{+}\hat{\Psi})\right] + \frac{1}{4}F^2_{\mu\nu}\Big\}. \qquad (6.3.15)$$

Here, $F^{\alpha}_{\mu\nu}$ is the "Yang-Mills field" tensor [6.7] ;

$$\hat{D}^{o}_{\nu} = \nabla_{\nu} - (i/2)\tilde{g}\hat{\tau}^{\alpha}\,A^{\alpha}_{\nu}\ ; \qquad (6.3.16)$$

$$\hat{\tau}^1 = \begin{Vmatrix} 0 & 1 \\ 1 & 0 \end{Vmatrix},\quad \hat{\tau}^2 = \begin{Vmatrix} 0 & -i \\ i & 0 \end{Vmatrix},\quad \hat{\tau}^3 = \begin{Vmatrix} 1 & 0 \\ 0 & -1 \end{Vmatrix}. \qquad (6.3.17)$$

The fields $\hat{\Psi}(\mathbf{x})$, $A^{\alpha}_{\nu}(\mathbf{x})$ are defined by

$$\hat{\Psi}(\mathbf{x}) = u(r)e^{iq^{o}\mathbf{x}} \begin{Vmatrix} -e^{-i\varphi}\sin\Theta \\ \cos\Theta \end{Vmatrix},$$

$$A^{\alpha}_{\nu}(\mathbf{x}) = e_{\nu\alpha\beta}x_{\beta}a(r)\ . \qquad (6.3.18)$$

The current induced by the fields $\hat{\Psi}(\mathbf{x})$ and $A^{\alpha}_{\nu}(\mathbf{x})$ will have the form

$$J^{\alpha}_{\nu} = -(i/4)\tilde{g}\ \mathrm{Tr}[(\nabla_{\nu}\Psi^{+}_{o})\hat{\tau}^{\alpha}\hat{\Psi}_{o}$$

$$- \hat{\Psi}^{+}_{o}\hat{\tau}^{\alpha}(\nabla_{\nu}\hat{\Psi}_{o})] + (1/4)\tilde{g}^2A^{\alpha}_{\nu}\ \mathrm{Tr}(\hat{\Psi}^{+}_{o}\hat{\Psi}_{o})\ ,$$

$$\hat{\Psi}_{o} = e^{-i\mathbf{q}^{o}\mathbf{x}}\hat{\Psi}\ . \qquad (6.3.19)$$

Taking (6.3.18) into consideration, we obtain

$$J^{\alpha}_{\nu} = (1/4)\tilde{g}^2u^2[a + (2/gr^2)]e_{\nu\alpha\beta}\ x_{\beta}\ . \qquad (6.3.20)$$

Equations for the functions u(r) and a(r) can be easily obtained:

$$u'' + (2/r)u' - b_1(\tau) - 2b_2(\tau)u^3 - 4b_2(\tau)u^3\ \ln[u^2/b_3(\tau)]$$

$$- (\tilde{g}^2r^2/2)[a + (2/\tilde{g}r^2)]^2u = 0\ ,$$

$$a'' + (4/r)a' + 3ga^2 - g^2r^2a^3 - (1/4)\tilde{g}^2u^2[a + (2/\tilde{g}r^2)] = 0. \qquad (6.3.21)$$

Taking into account the boundary conditions for the current J^{α}_{γ} (6.2.8), we obtain the following asymptotic forms of the solutions:

1) $r \to 0$, $u \to r$, $a \to \mathrm{const}$;
2) $r \to \infty$, $u \to u_{\infty}$,

$$a \to f(r/r_o)e^{-r/r_o} + a_{pi}(r/r_o)\ , \quad r_o = 2/\tilde{g}\, u_\infty\ ,$$

$$f(x) = 1/x^2, \quad a_{pi}(x) \to 0\ , \ x \to \infty \qquad (6.3.22)$$

(a_{pi} is the particular solution of the inhomogeneous equation for $a(r)$, due to the existence of the term $-u^2g/2r^2$ in it).

We have considered here a point singularity in the form of a starfish. Such singularities might, in principle, appear in the inhomogeneous superconducting state, but are found to be unstable. Indeed, if we compute the second homotopy group π_2 on the manifold $R = SO(3)\times S^1$, it turns out that $\pi_2(R) = 0$. Consequently, any point singularity on the manifold R may be contracted to a point (Fig. 6.2).

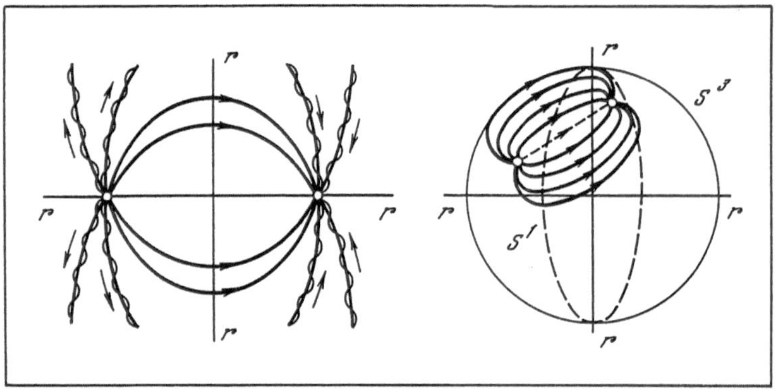

Fig. 6.2. A topologically unstable soliton in an inhomogeneous superconducting state

Let us now go over to a consideration of line singularities in the superconducting phase. As before, (6.3.2) is valid for the field $\hat{\Psi}$. The matrix $\hat{R}$ is defined by the expression

$$\hat{R} = \exp[i\xi(\boldsymbol{\tau}\mathbf{n})] \qquad (6.3.23)$$

($\mathbf{n}$ is a unit vector oriented along an isolated axis of the local coordinate system).

The field $A^\alpha_\nu(\mathbf{x})$ will be sought in the form

$$A^\alpha_\nu(\mathbf{x}) = a(\rho)e_{\alpha\beta\gamma}n^\beta\nabla_\nu n^\gamma\ , \qquad (6.3.24)$$

where ρ is the distance from the linear singularity axis.

The "Yang-Mills field" tensor will have the form

$$\mathbf{F}_{\mu\nu} = (\nabla_\nu a)[\mathbf{n}\nabla_\mu \mathbf{n}] - (\nabla_\mu a)[\mathbf{n}\nabla_\nu \mathbf{n}]$$

$$- 2[\nabla_\nu \mathbf{n}\nabla_\nu \mathbf{n}]a + \tilde{g}a^2\mathbf{n}(\mathbf{n}|\nabla_\mu \mathbf{n}\nabla_\nu \mathbf{n}]) . \qquad (6.3.25)$$

Using (6.3.23,24), we can write for the current tensor J^α_ν

$$J^\alpha_\nu = (1/2)\tilde{g}u^2[-(\nabla_\nu \xi)n^\alpha - \sin\xi \cos\xi \ (\nabla_\nu n^\alpha)$$

$$+ \sin^2\xi \ e_{\alpha\beta\gamma} n^\beta \nabla_\nu \ n^\gamma] + (1/4)\tilde{g}^2 u^2 a e_{\alpha\beta\gamma} n^\beta \nabla_\nu n^\gamma . \qquad (6.3.26)$$

Substituting (6.3.26) into (6.3.15) for the free energy, we obtain the functional

$$F = \int d\mathbf{x} \ \{(1/2)\tilde{b}_1(\tau)u^2 + b_2(\tau)u^4 \ \ln[u^2/b_3(\tau)]$$

$$+ (1/2)(\nabla_\nu u)^2 + (1/2)[(\nabla_\nu \xi)^2 + (\nabla_\nu n)^2 \sin^2\xi]$$

$$+ (1/2)u^2(\nabla_\nu \mathbf{n})^2(\tilde{g}a \sin^2\xi + (1/4)\tilde{g}^2 a^2) + (1/4)F^2_{\mu\nu} . \qquad (6.3.27)$$

If the vector $\mathbf{n}$ lies in a plane perpendicular to the linear singularity axis, the first two terms in the expression for the current component J^z_ν vanish. Moreover, the first two terms do not contribute to the free energy (6.3.27) of the system, and hence we shall disregard them below. The function ξ is a function of distance ρ from the singularity axis. In order to determine the functions $n(x)$ and $\xi(\rho)$, we must introduce the boundary conditions for the current:

$$J^\alpha_\nu(0) \to 0 , \quad J^\alpha_\nu(|\mathbf{x}| \to \infty) \to 0 . \qquad (6.3.28)$$

In this case, $n = (\cos\Phi , \sin\Phi , 0)$, $\Phi = n\varphi$.

The equation for the function $\xi(\rho)$ will have the form

$$\Delta\xi - (n^2/2\rho^2)\sin 2\xi = 0 . \qquad (6.3.29)$$

Solving this equation, we get

$$\sin \xi = \mathrm{sech}[n \ \ln(\rho/\rho_o)] . \qquad (6.3.30)$$

Thus, the solution of (6.3.29) is a soliton of size $L \sim \rho_o$. The equations for the functions $u(\rho)$ and $a(\rho)$ can then be easily written as follows:

$$u'' + (1/\rho)u' - b_1(\tau)u - 4b_2(\tau)u^3 \ln[u^2/b_3(\tau)]$$
$$- 2b_2(\tau)u^3 - (2n^2/\rho^2)u \operatorname{sech}^2[n \ln(\rho/\rho_o)]$$
$$- (\tilde{g}n^2/\rho^2)ua \operatorname{sech}^2[n \ln(\rho/\rho_o)] - (1/4)(\tilde{g}^2n^2/\rho^2)ua^2 = 0 ,$$
$$a'' - (1/\rho)a' - (1/4)\tilde{g}^2u^2\{(a + (2/\tilde{g})\operatorname{sech}^2[n \ln(\rho/\rho_o)]\} = 0. \tag{6.3.31}$$

The asymptotics of the solution have the form

$$1)\ \rho \to 0 ,\quad u \to \rho^s ,\ s \approx 1,\quad a \to \rho^2 ,\quad J_\nu^\alpha \to 0,\ \alpha = z ;$$
$$2)\ \rho \to \infty ,\quad u \to u_\infty ,$$
$$a \to (\rho/\tilde{\rho}_o)^2 f(\rho/\tilde{\rho}_o)e^{-\rho/\rho_o} + a_{pi}(\rho/\rho_o) ,\quad \tilde{\rho}_o = 2/\tilde{g}u_\infty ,$$
$$a \to x^{\frac{1}{2}} e^{-x} \left\{1 + \sum_{k=1}^{n} \frac{2^k}{k!x^k} \prod_{\ell=1}^{k} (\tfrac{1}{2} - \ell)(\tfrac{1}{2} + \ell) + (R_n \to 0)\right\} ,\quad J_\nu^\alpha \to 0 ,\ \alpha = z. \tag{6.3.32}$$

An analysis of (6.3.30,31) shows that two types of linear singularities exist in the inhomogeneous superconducting phase. The first type corresponds to an integral value of n, while the second type corresponds to half-integral n. For n = 1, the flux quantum of such a vortex has the form

$$K_o = \frac{\pi hc}{4e} \frac{\rho_o}{R_o} , \tag{6.3.33}$$

where R_o is a quantity which is a multiple of the vortex size: $\tilde{L}_o \sim R_o$, $1/4 \lesssim \rho_o/R_o \lesssim 1/2$. In this case, $K_o = \frac{\pi hc}{4e}\frac{L}{L_o} \simeq 10^{-7}$ Mx. Consequently, a quantum of nonlinear vortex flux is of the same order as a quantum of linear vortex in an ordinary type II superconductor. The problem of the stability of linear singularities in the superconducting phase is solved by computing the first homotopic group on the manifold $R = SO(3) \times S^1$, which is nontrivial: $\pi_1(R) = Z_2 + Z$. However, this is not a sufficient condition, and in order to find out which of the n = (m, m/2) correspond to stable linear singularities, we must investigate the stability of the solution of (6.3.31)

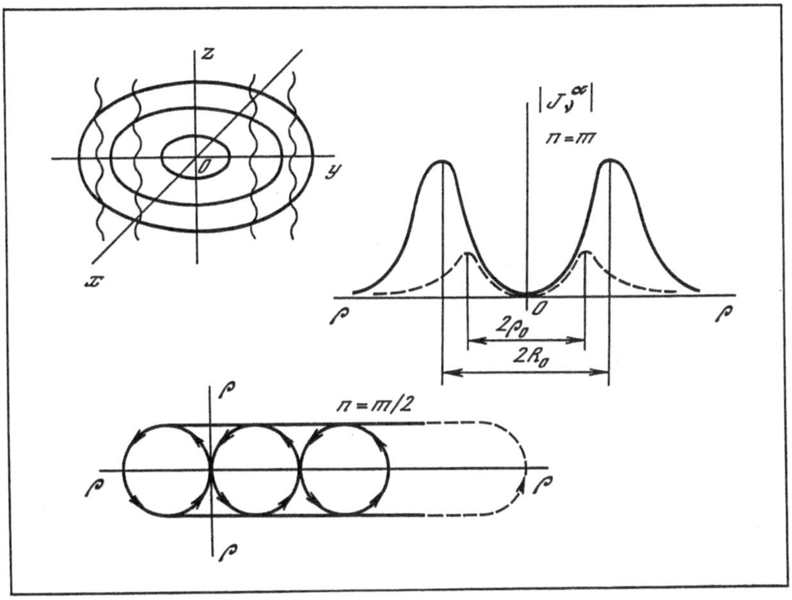

Fig. 6.3. Vortices with n = (1, 1/2) in an inhomogeneous superconducting state

to self-similar perturbations. Figure 6.3 shows stable vortices corresponding to $n = (1, \frac{1}{2})$.

6.4 Solitons in the Phase of Coexistence of Superconductivity and Magnetism

While considering the vortex structure of the phase in which superconductivity and magnetism coexist, we shall proceed from the fact that such a phase is thermodynamically stable. On the phase diagram of the compounds $Er_{1-x}Ho_xRh_4B_4$ [6.2,6], the coexistence phase Su + c-sin may be stable. In other words, the magnetic structure of such a phase is a spin-density wave modulated along the preferred z axis of the crystal. The expression for the free energy of such a state can be presented in the form

$$F = \int d\mathbf{x} \left\{ \frac{1}{2} b_1(\tau) \mathrm{Tr}(\hat{\Psi}^+\hat{\Psi}) + b_{\nu 5}(\tau) \right.$$

$$\mathrm{Tr}[\hat{\Psi}(\hat{D}^o_\nu\hat{\Psi})^+ - \Psi^+(\hat{D}^o_\nu\hat{\Psi})] + \frac{1}{2} \mathrm{Tr}[(\hat{D}^o_\nu\hat{\Psi})^+(\hat{D}^o_\nu\hat{\Psi})]$$

$$+ b_2(\tau)[\mathrm{Tr}(\hat{\Psi}^+\hat{\Psi}) \ln\left[\frac{1}{b_3(\tau)} \mathrm{Tr}(\hat{\Psi}^+\hat{\Psi})\right]$$

$$+ \frac{1}{2} a_1(\tau)(\mathbf{S}_{\parallel} \mathbf{S}^*_{\parallel}) + \frac{1}{2} c_1(\tau)\mathbf{S}^z_{\parallel} \mathbf{S}^{z*}_{\parallel}$$

$$+ a_{\nu 5}(\tau)[(\mathbf{S}^{*}_{\|}, D_\nu \mathbf{S}_{\|})^{*} - (\mathbf{S}^{*}_{\|}, D_\nu \mathbf{S}_{\|})] + \frac{1}{2}(D^{*}_{\nu} \mathbf{S}^{*}_{\|}, D_\nu \mathbf{S}_{\|})$$

$$+ a_2(\tau)(\mathbf{S}_{\|}\mathbf{S}^{*}_{\|})^2 \ln\left[\frac{1}{a_3(\tau)}(\mathbf{S}_{\|}\mathbf{S}^{*}_{\|})\right]$$

$$+ A_{\delta \|}(\tau)(\mathbf{S}_{\|}\mathbf{S}^{*}_{\|})\mathrm{Tr}(\hat{\Psi}^{+}\hat{\Psi}) + \frac{1}{4} F^2_{\mu\nu} \Big\} ,$$

$$\tau = 1 - (T/T_o) . \tag{6.4.1}$$

Here, T_o is the transition temperature to the coexistence phase Su + c-sin. The covariant derivatives $\hat{D}^o_\nu$ and D_ν are defined by the formulas

$$\hat{D}^o_\nu = \hat{\tau}_o \nabla_\nu - (i/2)\tilde{g}_1 \tau^\alpha \mathbf{A}^\alpha_\nu , \quad D_\nu = \nabla_\nu - \tilde{g}_2 e^{\alpha\beta\gamma} \mathbf{A}^\beta_\nu . \tag{6.4.2}$$

It follows from (6.4.1) that in this phase, the superconducting wave structure defined by the order parameter $\hat{\Psi} = \hat{\Psi}_o \exp(i\mathbf{q}_\delta \mathbf{x})$, coexists with a spin-density wave $\mathbf{S}_{\|} = \mathbf{S}_{\| o} \exp(i\mathbf{q}_{\|} \mathbf{x})$, where $q^o_{\delta\nu} = 2ib_{\nu 5}(\tau)$, $q^o_{\|\nu} = 2ia_{\nu 5}(\tau)$. Since, in accordance with experimental results, the phase of coexistence of superconductivity and magnetism has a structure with a fixed wavelength $\lambda \simeq 100$ Å [6.1,11,12,15], we can put $q^o_{\delta\nu} = q^o_{\|\nu}$. In this case, (6.4.1) can be reduced to

$$F = \int d\mathbf{x} \Big\{\frac{1}{2} b_1(\tau)\mathrm{Tr}(\hat{\Psi}^{+}_o \hat{\Psi}_o) + \frac{1}{2} \mathrm{Tr}[(D^o_\nu \hat{\Psi}_o)^{+} (D^o_\nu \hat{\Psi}_o)]$$

$$+ b_2(\tau)[\mathrm{Tr}(\hat{\Psi}^{+}_o \hat{\Psi}_o)^2] \ln\left[\frac{1}{b_3(\tau)} \mathrm{Tr}(\hat{\Psi}^{+}_o \hat{\Psi}_o)\right]$$

$$+ \frac{1}{2} \tilde{a}_1(\tau) S^2_{\| o} + \frac{1}{2}(D^{*}_\nu \mathbf{S}^{*}_{\| o}, D_\nu \mathbf{S}_{\| o}) + a_2(\tau) S^4_{\| o} \ln\left[\frac{S^2_{\| o}}{a_3(\tau)}\right]$$

$$+ A_{\delta \|}(\tau) S^2_{\| o} \mathrm{Tr}(\hat{\Psi}^{+}_o \hat{\Psi}_o) + \frac{1}{4} F^2_{\mu\nu}\Big\} ,$$

$$\tilde{b}_1(\tau) = b_1(\tau) + 4\sum_\nu b^2_{\nu 5}(\tau) ,$$

$$\tilde{a}_1(\tau) = a_1(\tau) + c_1(\tau) + 4\sum_\nu a^2_{\nu 5}(\tau) . \tag{6.4.3}$$

Since the magnetic subsystem has a strong anisotropy, we shall henceforth assume that $\mathbf{S}_{\|} = (0, 0, s^z_{\|})$ or $\mathbf{S}_{\|} = s_{\|}(\mathbf{x})\mathbf{n}^3_s$. The order parameter is written in the form

$$\hat{\Psi}_o = u(\mathbf{x})\hat{R}(\mathbf{x})\hat{a} , \quad \hat{R}(\mathbf{x}) = \exp[i\xi(\boldsymbol{\tau}\mathbf{n}_\delta)] .$$

Let us consider a linear singularity in which the potential of the "electromagnetic" field has the form

$$A^{\alpha}(\mathbf{x}) = a(\rho)e^{\alpha\beta\gamma}n^{\beta}_{\delta}\nabla_{\nu}n^{\gamma} ,$$

$$\mathbf{n}_{\delta} = (\cos\Phi , \sin\Phi , 0) , \quad \Phi = n\varphi . \tag{6.4.4}$$

In this case,

$$F^2_{\mu\nu} = \{ (\nabla_{\nu} a)(\nabla_{\mu}\mathbf{n}) - (\nabla_{\mu} a)(\nabla_{\nu}\mathbf{n}) \}^2 + 4a^2[\nabla_{\mu} n\nabla_{\nu}\mathbf{n}]$$
$$-4ga^3(\mathbf{n}[\nabla_{\mu}\mathbf{n}\nabla_{\nu}\mathbf{n}])^2 + g^2a^4(\mathbf{n}[\nabla_{\mu}\mathbf{n}\nabla_{\nu}\mathbf{n}])^2 . \tag{6.4.5}$$

The expression for the superconducting current J^{α}_{ν} is identical to (6.3.26). The final expression for the free energy is

$$F = \int d\mathbf{x} \Big\{ (1/2)\tilde{b}_1(\tau)u^2 + b_2(\tau)u^4 \ln[u^2/b_3(\tau)]$$
$$+ (1/2)(\nabla_{\nu} u)^2 + (1/2)u^2[(\nabla_{\nu}\xi)^2 + (\nabla_{\nu}\mathbf{n})^2\sin^2\xi]$$
$$+ (1/2)u^2(\nabla_{\nu} n)^2(\tilde{g}_1 a \sin^2\xi + \tilde{g}_1^2 a^2/4)$$
$$+ (1/2)\tilde{a}_1(\tau)s_{||}^2 + (1/2)(\nabla_{\nu}\mathbf{s}_{||})^2 + a_2(\tau)s_{||}^4 \ln[s_{||}^2/a_3(\tau)]$$
$$+ A_{\delta ||}(\tau)s_{||}^2 u^2 + (1/4)F^2_{\mu\nu} \Big\}. \tag{6.4.6}$$

The functions u, $s_{||}$, a, and $\sin\xi$ are functions of distance ρ from the linear singularity axis (see also 6.3.30). Minimizing the functional (6.4.6) in variables u, $s_{||}$ and a, we obtain

$$u'' + (1/\rho)u' - b_1(\tau)u - 4b_2(\tau)u^3\ln[u^2/b_3(\tau)]$$
$$- 2b_2(\tau)u^3 - 2A_{\delta ||}(\tau)us_{||}^2 - (2n^2/\rho^2)u \,\mathrm{sech}^2[n \ln(\rho/\rho_o)]$$
$$- (\tilde{g}_1 n^2/\rho^2)ua \,\mathrm{sech}^2[n \ln(\rho/\rho_o)] - (\tilde{g}_1^2 n^2/4\rho^2)ua^2 = 0 ,$$
$$s_{||}'' + (1/\rho)s_{||}' - a_1(\tau)s_{||} - 4a_2(\tau)s_{||}^3 \ln[s_{||}^2 /a_3(\tau)]$$
$$- 2a_2(\tau)s_{||}^3 - 2A_{\delta ||}(\tau)s_{||}u^2 = 0 ,$$
$$a'' - (1/\rho)a' - (\tilde{g}_1 u^2/4)\{a + (2/\tilde{g}_1)\mathrm{sech}^2[n \ln(\rho/\rho_o)]\} = 0.$$
$$\tag{6.4.7}$$

The boundary conditions for the current J^{α}_{ν} coincide with the conditions (6.3.28). This gives

$$1)\ \rho \to 0,\quad u \to \rho^s,\ s \simeq 1,\quad s_{\parallel} \to \rho,\quad a \to \rho^2,\quad J^{\alpha}_{\nu} \to 0,\ \alpha = z;$$

$$2)\ \rho \to \infty,\quad u \to u_{\infty},\quad s_{\parallel} \to s_{\infty},$$

$$a \to x^{\frac{1}{2}} e^{-x} \left\{ 1 + \sum_{k=1}^{n} \frac{2^k}{k!\,x^k} \prod_{\ell=1}^{k} \left(\frac{1}{2} - \ell\right)\left(\frac{1}{2} + \ell\right) + (R_n \to 0) \right\}$$

$$x = \rho/\tilde{\rho}_0,\quad \tilde{\rho}_0 = 2/\tilde{g}_1 u_{\infty},\quad J^{\alpha}_{\nu} \to 0,\quad \alpha = z. \tag{6.4.8}$$

It follows from (6.4.8) that as far as the symmetry of the superconducting component is concerned, this singularity does not differ in any way from a linear singularity of the inhomogeneous superconducting phase (6.3.32). This is because the magnetic component does not interact with the field A^{α}_{ν} (6.4.6). If such an interaction existed, it would cause a strong distortion of the magnetic structure. Hence the thermodynamic stability of the vortex structure in the phase of coexistence of superconductivity and magnetism can be ensured only when the interaction between the magnetic and superconducting subsystems is weak, as is indeed the case in the above computations.

Let us now consider a point singularity in the coexistence phase. In this case, $\mathbf{s}_{\parallel} = s_{\parallel}(r)\mathbf{n}_s$, and the fields $\Psi^{\alpha\beta}_0(\mathbf{x})$ and $A^{\alpha}_{\nu}(\mathbf{x})$ have the form described in Sect. 6.3. We can then write the following expression for the free energy functional:

$$\begin{aligned} F = \int d\mathbf{x} \Big\{ & (1/2)\tilde{b}_1(\tau)u^2 + (1/2)(\nabla_{\nu} u)^2 + (1/2)u^2(\nabla_{\nu}\mathbf{n}_{\delta})^2 \\ & + b_2(\tau)u^4 \ln[u^2/b_3(\tau)] + \tilde{g}_1 a u^2 + (1/4)\tilde{g}_1^2 u^2 a^2 r^2 \\ & + (1/2)\tilde{a}_1(\tau)s_{\parallel}^2 + (1/2)(\nabla_{\nu} s_{\parallel})^2 + (1/2)s_{\parallel}^2(\nabla_{\nu}\mathbf{n}_s)^2 \\ & + a_2(\tau)s_{\parallel}^4 \ln(s_{\parallel}^2/a_3(\tau)) - 2\tilde{g}_2 s_{\parallel}^2 a^2 + \tilde{g}_2^2 r^2 s_{\parallel}^2 a^2 \\ & - B_{\delta\parallel}(\tau)s_{\parallel}^2 u^2 + (1/4)F_{\mu\nu}^2 \Big\}. \end{aligned}$$

$$u = u(r),\quad s_{\parallel} = s(r),\quad a = a(r). \tag{6.4.9}$$

The expression for the current J^{α}_{ν} is

$$\begin{aligned} J^{\alpha}_{\nu} = & -(i/4)\tilde{g}_1 \operatorname{Tr}[(\nabla_{\nu}\hat{\Psi}_0)^{+}\hat{\tau}^{\alpha}\hat{\Psi}_0 - \hat{\Psi}_0^{+}\hat{\tau}^{\alpha}(\nabla_{\nu}\hat{\Psi}_0)] \\ & + (1/4)\tilde{g}_1^2 A^{\alpha}_{\nu} \operatorname{Tr}(\hat{\Psi}^{+}\hat{\Psi}) - \tilde{g}_2[\mathbf{s}_{\parallel 0}\nabla_{\nu}\mathbf{s}_{\parallel 0}]^{\alpha} + \tilde{g}_2^2[\mathbf{s}_{\parallel 0}[\mathbf{A}_{\nu}\mathbf{s}_{\parallel 0}]]^{\alpha}. \end{aligned} \tag{6.4.10}$$

Considering that $\Phi_o = u(r) \begin{Vmatrix} e^{-i\varphi} \sin\theta \\ \cos\theta \end{Vmatrix}$ and $A_\nu^\alpha = a(r) e_{\nu\alpha\beta} x_\beta$, we obtain

$$J_\nu^\alpha = (1/4)\tilde{g}_1^2 u^2 [a + (2/\tilde{g}_1 r^2)] e_{\nu\alpha\beta} x_\beta + \tilde{g}_2^2 s_{||}^2 [a - (1/\tilde{g}_2 r^2)] e_{\nu\alpha\beta} x_\beta \,. \quad (6.4.11)$$

We can now write the expressions for the scalar functions u, $s_{||}$ and a :

$$u'' + (2/r)u' - \tilde{b}_1(\tau)u - 2b_2(\tau)u^3 - 4b_2(\tau)u^3 \ln[u^2/b_3(\tau)] - (\tilde{g}_1^2 r^2/2)[a + (2/\tilde{g}_1 r^2)]^2 u - 2B_{\delta ||}(\tau) s_{||}^2 u = 0 \,,$$

$$s_{||}'' + (2/r)s_{||}' - (2/r^2)s_{||} - a_1(\tau)s_{||} - 2a_2(\tau)s_{||}^3 - 4a_2(\tau)s_{||}^3 \ln[s_{||}^2/a_3(\tau)] + 4\tilde{g}_2 s_{||} a - 2\tilde{g}_2^2 r^2 a^2 s_{||} - 2B_{\delta ||} s_{||} u^2 = 0 \,,$$

$$a'' + (4/r)a' + 3ga^2 - g^2 r^2 a^3 - (1/4)\tilde{g}_1^2 u^2 (a + 2/\tilde{g}_1 r^2) - \tilde{g}_2^2 s_{||}^2 [a - (1/\tilde{g}_2^2 r^2)] = 0 \quad (6.4.12)$$

which have the asymptotic forms

1) $r \to 0$, $u \to r$, $s_{||} \to r$, $a \to \text{const}$;

2) $r \to \infty$, $u \to u_\infty$, $s_{||} \to s_{||\infty}$,

$$a \to f(r/r_o) e^{-r/r_o} + a_{pi}(u_\infty, s_{||\infty}, r) \,,$$

$$r_o = [(4/\tilde{g}_1^2 u_\infty^2) + (1/\tilde{g}_2^2 s_{||\infty}^2)]^{\frac{1}{2}} \,. \quad (6.4.13)$$

Suppose that $\tilde{g}_1 = -g$, $\tilde{g}_2 = g$. In this case, we obtain for $r \to \infty$

$$a(r) = f(r/r_o) e^{-r/r_o} + 2/gr^2 + 1/gr^2, \quad f(x) = 1/x^2 \,.$$

The "Yang-Mills field" tensor $F_{\mu\nu}^\alpha$ is a gauge invariant of the function $A_\nu^\alpha = (2/gr^2) e_{\nu\alpha\mu} x_\mu$. Hence we can write $a \to f(r/r_o) e^{-r/r_o} + 1/gr^2 \to 1/gr^2$. But this means that as $r \to \infty$, the field is determined by the magnetic component, which indicates the supression of superconductivity in the region $r \to \infty$. The superconducting component is nonzero inside the singularity, but since the spatial configuration corresponding to it is topologically unstable (Fig. 6.2), the singularity under consideration will also be unstable. As a result, we obtain an ordinary magnetic monopole of starfish-type. Such a monopole is topologically stable, but it can exist only with small magnetic anisotropies $|a_{\parallel}| \to 0$. Hence, in view of the above statements, we cannot expect the appearance of a stable point singularity in the coexistence phase of superconductivity and magnetism.

7. Spin Glasses Based on Rare-Earth Metal Compounds

7.1 Phase Transitions in Heavy Rare-Earth Metals

The phase diagram of heavy rare-earth metals and their alloys, constructed in [7.1] on the basis of the experimental data (Fig. 3.5f), has a larger number of physical states than the diagrams obtained theoretically in Chap. 3 (Fig. 3.5a-e). The experimental diagram in the temperature-anisotropy coordinates shows the trajectories of different systems during phase transitions. Trajectories 3-5 correspond to phase transitions in alloys of Er-Tm type, and also in the pure rare-earth metal Tm (space group $P6_3/mmc - D^4_{6h}$).[1] The phase transitions in this class of materials have the following special features.

From the paramagnetic phase, the system is transformed to a state with a spin-density wave modulated along the preferred axis z of the crystal (sinusoidal c-sin modulation) with a wave vector q. As the temperature decreases, the magnetic structure period $2\pi/q$ remains practically unchanged, but the spin distribution in the wave changes, i.e., the system modulation changes. At a certain temperature $T < T_{c-sin}$ (T_{c-sin} is the temperature of transition to the c-sin state), a phase transition takes place to a state in which all magnetic moments in the spin-density wave have nearby values and are oriented along the axis z. As before, the axial anisotropy is quite large, i.e., $|a_{\|}| \lesssim J$. However, a part (n_1) of spins is oriented upwards, and a part (n_2) is oriented downwards, $n_1 \neq n_2$. Such a configuration is termed a "ferrimagnetic antiphase domain". This means that the system

---

[1] As a matter of fact, the complete group of complex magnetic systems is a combination of the space and spin groups: $G = G_p \times G_s$, where G_s is defined on the topological degeneracy space R of the given magnetic structure. Since the phase lamination effect is observed in these systems (Chap. 3), a phase transition to each state leads to the emergence of a magnetic structure which has its own corresponding order parameter degeneracy space $R_i \subset R$. However, this means that as a result of phase lamination, the phase transitions take place in subgroups G_i of the group G, where $G_i \subset G$.

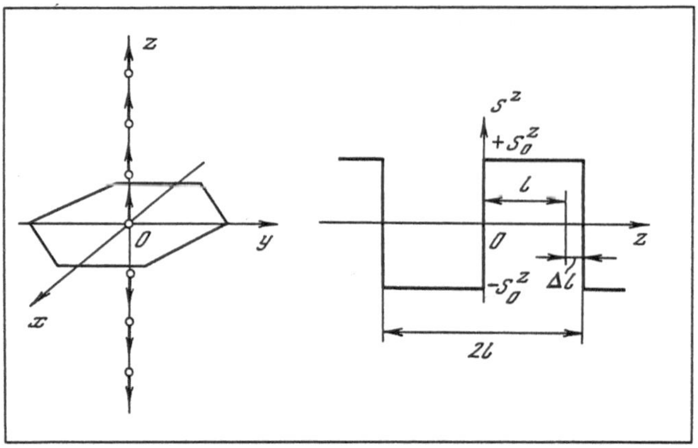

Fig. 7.1. Magnetic structure of thulium (Tm) in the lowest ordered state

has a magnetic moment in the ground state. Let us try to explain the phase transitions in these compounds. For this purpose, we again use the phase lamination mechanism, which was proposed by us and used successfully. Let us first determine the order parameter of the magnetic structure, i.e., the ferrimagnetic antiphase domain. The spin-density distribution function for this structure is presented in Fig. 7.1, and can be described by

$$S_0^z(z) = S_0^z f(z) , \tag{7.1.1}$$

where S_0^z is the value of the z-component of the magnetic moment at a lattice site. The function f(z) can be expanded into a Fourier series. In this case, we can write

$$f(z) = a_0 + \sum_{n=1}^{\infty} [a_n \cos(n\pi z/\ell) + b_n \sin(n\pi z/\ell)] . \tag{7.1.2}$$

Here, $2\ell = 2\pi/q$ is the structure period. The Fourier coefficients a_0, a_n, and b_n are given by

$$a_0 = 1, \quad a_n = (-1)^n \frac{\ell}{\pi n \Delta\ell} \sin \frac{n\pi\Delta\ell}{\ell} ,$$

$$b_n = (\ell/\pi n\Delta\ell)\, [1 - (-1)^n \cos(n\pi\Delta\ell/\ell)] , \tag{7.1.3}$$

where $\Delta\ell$ is the "detuning" of the structure half-period, which ensures a nonzero magnetic moment of the system (Fig. 7.1). With the help of elementary transformations, we can transform (7.1.2) as follows:

$$f(z) = 1 + \sum_{n=1}^{\infty} [f_n^{0+} \exp\{i(q_n z + \varphi_{n+})\} + i f_n^{0-} \exp\{-i(q_n z + \varphi_{n-})\}] , \tag{7.1.4}$$

where

$$q_n = \pi n/\ell \ , \quad f_n^{0+} = f_n^{0-} = (\ell/\sqrt{2}\ \pi n \Delta\ell)[1 - (-1)^n \cos(\pi n \Delta\ell/\ell)]^{\frac{1}{2}} \ ,$$

$$\varphi_{n+} = -\arctan\{ [1 - (-1)^n \cos(n\pi\Delta\ell/\ell)]/(-1)^n \sin(n\pi\Delta\ell/\ell)\} \ ,$$

$$\varphi_{n-} = \arctan\{ (-1)^n \sin(n\pi\Delta\ell/\ell)/[1 - (-1)^n \cos(n\pi\Delta\ell/\ell)]\} \ .$$

It follows from (7.1.4) that the magnetic structure (i.e., ferrimagnetic antiphase domain) can be presented as a superposition of N+1 spin-density waves with a periodicity along the preferred z-axis (hexagonal axis) of the crystal. In this case, in order to write down the expression for the free energy of the system in the paramagnetic region, we must introduce N+1 complex vectors S_0, S_n, $1 \le n \le N$. These vectors are defined by the expressions

$$\mathbf{S}_0 = \mathbf{S}_0 \exp(i2\pi z/a) \ , \quad \mathbf{S}_n = \mathbf{S}_n^+ + i\mathbf{S}_n^- \ . \qquad (7.1.5)$$

The vectors $\mathbf{S}_n^+$ and $\mathbf{S}_n^-$ have the form

$$\mathbf{S}_n^+(z) = \mathbf{S}_{no}^+ \operatorname{Re}\{\exp[i(q_n z + \varphi_{n+})]\} \ ,$$

$$\mathbf{S}_n^-(z) = \mathbf{S}_{no}^- \operatorname{Im}\{\exp[i(q_n z + \varphi_{n-})]\}. \qquad (7.1.6)$$

The number of spin-density waves N must have an upper limit. Indeed, the minimum length of the structure wave is $2\pi/a$ (ferromagnet). In this case, $N = 2\ell/a$. After this we proceed in the same way as in the preceding chapters while considering phase transitions in complex magnetic structures.

We can write the second- and fourth-order invariants in the following form:

$$(1) \quad S_0^2, \ S_0^z \ , \ (\mathbf{S}_n\mathbf{S}_n^*) \ , \ S_n^z S_n^{z*} \ ,$$

$$(2) \quad S_0^4 \ , \ (\mathbf{S}_n\mathbf{S}_n^*)^2 \ , \ S_0^2(\mathbf{S}_n\mathbf{S}_n^*) \ , \ (\mathbf{S}_0\mathbf{S}_n)(\mathbf{S}_0^*\mathbf{S}_n^*) \ ,$$

$$(\mathbf{S}_\ell\mathbf{S}_\ell^*)(\mathbf{S}_m\mathbf{S}_m^*) \ , [(\mathbf{S}_\ell\mathbf{S}_m)(\mathbf{S}_\ell^*\mathbf{S}_m^*) + (\mathbf{S}_\ell\mathbf{S}_m^*)(\mathbf{S}_\ell^*\mathbf{S}_m)] \ . \qquad (7.1.7)$$

In this case, the phase transitions in such a system can be constructed in the same way as in Chaps. 3 and 4. The only difference is that N+1 stationary states having different spin-wave modulations along the z-axis may appear in the system. Phase transitions to these states must be first-order. Experimentally, this is manifested in the form of additional peaks on the temperature dependence of magnetic susceptibility χ_{zz} and heat capacity C, associated with new phase transitions.

Thulium is the most thoroughly investigated among the rare-earth materials which undergo a phase transition to the ordered state along trajectory 5 on the experimental phase diagram (Fig. 3.5f) [7.2]. Its magnetic structure in the ordered ground state is presented in Fig. 7.1 and is a ferrimagnetic antiphase domain in which four spins are oriented upwards along the z-axis, and three spins are oriented downwards. Thus the structure's period is equal to seven atomic spacings. Néel temperature (temperature of transition to the c-sin state) of thulium is 58 K, and the axial magnetic anisotropy continues to be quite large, $|a_{\parallel}| \lesssim J$. Starting from a temperature of $\simeq$ 42 K, the strongest variations in the sinusoidal modulation start taking place, and at T_c = 25 K a ferrimagnetic antiphase domain structure is established in the system, which can be tentatively denoted as ++++--- (the plus sign denotes upward spin, and the minus sign, downward spin).

In this section, we shall consider the possible phase transitions induced in Tm by temperature variation. The expression for the free energy of Tm can be constructed on the basis of the invariants (7.1.7). Since the anisotropy of the system is large, we can write the following expression for the free energy [7.3]:

$$F = \frac{1}{2}\tau_o s_o^2 + \frac{1}{2}\sum_{i=1}^{N}\tau_i(s_i^{+2} + s_i^{-2}) + \frac{1}{8}\Gamma_o s_o^4$$
$$+ \frac{1}{8}\sum_{i=1}^{N}\Gamma_i(s_i^{+2} + s_i^{-2})^2 + \frac{1}{4}\sum_{i=1}^{N}\Gamma_{oi}s_o^2(s_i^{+2} + s_i^{-2})$$
$$+ \frac{1}{8}\sum_{i,j=1}^{N}\Gamma_{ij}(s_i^{+2} + s_i^{-2})(s_j^{+2} + s_j^{-2}) . \qquad (7.1.8)$$

The initial conditions for the amplitudes Γ_o, Γ_i, Γ_{oi}, Γ_{ij} and temperatures τ_o and τ_i are chosen from the following considerations. In the absence of an external magnetic field, macroscopic magnetic moment appears in the ordered state only at a temperature T = 32 K [7.4]. There is no magnetic moment in the temperature interval 32-38 K, and the system has a complex modulated structure. Taking this into consideration, we introduce the following initial conditions:

$$\tau_{oo}<\tau_{1o}<\ldots<\tau_{No},\ \tau_{oo} = \tau - |a_{\parallel o}|,\quad \tau_{io} = \tau - |a_{\parallel i}|,$$
$$\Gamma_{1o} < \Gamma_{2o}<\ldots<\Gamma_{No},\Gamma_{o1o} = \Gamma_{o2o} = \ldots = \Gamma_{oNo}, \tag{7.1.9}$$
$$\Gamma_{ijo} = \Gamma_{i'j'o},\quad 1\le i,j,i',j' \le N .$$

It follows from these conditions that as we move from the paramagnetic phase, the ferromagnetic component S_o is the one that fluctuates most strongly, while the intrinsic fluctuations S_i of the fields are weaker. In this case, we can write the RG equations in the form

$$-\Gamma_o' = (n_o + 8)\Gamma_o^2,\quad -\Gamma_{oi}' = (n_o + 2)\Gamma_o\Gamma_i,$$
$$-\Gamma_i' = n_o\Gamma_{oi}^2,\quad -\Gamma_{ij}' = n_o\Gamma_{oi}\Gamma_{oj},$$
$$-\tau_o' = (n_o + 2)\tau_o\Gamma_o,\quad -\tau_i' = n_o\tau_{oi}\Gamma_{oi},\quad n_o = 1, \tag{7.1.10}$$

where
$$x = \frac{1}{(\varepsilon/2)(\Lambda^2)^{\varepsilon/2}}\left\{\left(\frac{\Lambda^2}{\max[\lambda^2(\tau_o),s_{io}^2]}\right)^{\varepsilon/2} - 1\right\},\quad \varepsilon\to 0. \tag{7.1.11}$$

The solutions of (7.1.10) have the form

$$\Gamma_o = \frac{\Gamma_{oo}}{1 + 9\Gamma_{oo}x},\quad \Gamma_{oi} = \frac{\Gamma_{oio}}{(1 + 9\Gamma_{oo}x)^{1/3}},$$
$$\Gamma_i = \Gamma_{io} - \frac{\Gamma_{oio}^2}{3\Gamma_{oo}}[(1 + 9\Gamma_{oo}x)^{1/3} - 1],$$
$$\Gamma_{ij} = \Gamma_{ijo} - \frac{\Gamma_{ijo}\Gamma_{oio}}{3\Gamma_{oo}}[(1 + 9\Gamma_{oo}x)^{1/3} - 1],$$
$$\tau_o = \frac{\tau_{oo}}{(1+9\Gamma_{oo}x)^{1/3}},\quad \tau_i = \tau_o - \frac{\tau_{oo}\Gamma_{oio}}{3\Gamma_{oo}}[(1 + 9\Gamma_{oo}x)^{1/3} - 1]. \tag{7.1.12}$$

It follows from (7.1.9-12) that initially, the first harmonic turns into a condensate and magnetic order is observed in the system. This corresponds to a sinusoidal spin-density wave with a periodicity along the preferred z (c-sin) axis of the crystal. In this case, a first-order phase transition takes place. The jump in the order parameter and the transition temperature are given by

$$s_{1o}^2 = [(1 - \varepsilon/2)/(1 + \varepsilon x_{1s}/2)]^{2/\varepsilon},$$

$$\tau_{1s} = 2b(x^*_{1s})s^2_{1o} , \quad b(x^*_{1s}) = (1/8)\Gamma^2_{o1}(x^*_{1s}) ,$$

$$x^*_{1s} = (1/9\,\Gamma_{oo})\ \{[(3\Gamma_{oo}\Gamma_{1o}/\Gamma^2_{o1o}) + 1]^3 - 1\} . \qquad (7.1.13)$$

The temperature increments for the ferromagnetic component and other harmonics accompanying a first-order phase transition are given by

$$\Delta\tau_o = (1/4)\Gamma_{o1}(x^o_{1s})s^2_{1o} ,$$

$$\Delta\tau_i = (1/4)\Gamma_{i1}(s^o_{1s})s^2_{1o} , \quad 2 \le i \le N , \qquad (7.1.14)$$

where x^o_{1s} is the value of the variable in the RG method at the transition point. For $\Gamma_{oi}(x^o_{1s}) < \Gamma_{i1}(x^o_{1s})$, the inequality between temperatures $\tau_{oo} < \tau_{2o} < \ldots < \tau_{No}$ is preserved. This means that the second harmonic will be the next to form a condensate; this will be followed by the third harmonic, and so on. The phase transitions in this case are first-order transitions. The ferromagnetic component, which is the last to form a condensate, undergoes a second-order phase transition. Thus, N+1 stationary states appear on the phase diagram of rare-earth metals and their alloys. Since the structure period of Tm in the magnetically ordered ground state is 7c/2 (c is the height of the hexagonal cell), the number of harmonics must be N = 6. The phase diagram of heavy rare-earth metals and their alloys taking the above analysis into consideration is presented in Fig. 7.2. It follows from this diagram that phase transitions in complex magnetic structures may have a multi-step nature. It is interesting to note

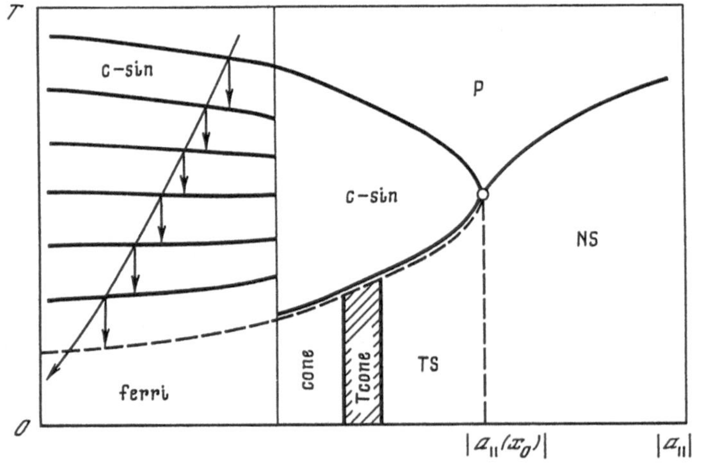

Fig. 7.2. Theoretical phase diagram of heavy rare-earth metals and their alloys

that, as in the alloys Er-Dy, Er-Tb, Er-Ho considered earlier (Chap. 3), Tm also experiences a phase lamination upon phase transitions from paramagnetic phase to the ordered state. Indeed, following a phase transition to the c-sin state, the degeneracy space (manifold) corresponding to the order parameter of S_1 has the form $R_1 = S^2 \times S^1$ (Sect. 3.2). When the second harmonic forms a condensate, the degeneracy space changes and assumes the form $R_2 = S^2 \times S^1 \times S^1$. Subsequent spatial sequences can be presented as follows:

$$R_3 = S^2 \times S^1 \times S^1 \times S^1, \ \ldots, \ R_N = S^2 \times S^1 \times S^1 \times \ldots \times S^1 .$$

The effective contour on the surface can be presented as

$$S^1 \times S^1 \times S^1 \times \ldots \times S^1 , \quad \gamma^{N-1} .$$

Let us now compute the first homotopy groups π_1 on the spaces $R_1, \ldots, R_N$. These groups are equal to

$$\pi_1(R_1) = Z, \ \pi_1(R_2) = Z + Z , \ \ldots, \ \pi_1(R_N) = Z + Z + \ldots + Z. \tag{7.1.15}$$

Since the complete vector **S** of the order parameter describes the effective contour γ^{N-1} on the manifold $S^1 \times S^1 \times S^1 \times \ldots \times S^1$, which is then mapped onto the complex plane (v, iw), the group Z+Z+...+Z can be interpreted as the effective group of integrals Z^{N-1}, $\pi_1(\gamma^{N-1}) = Z^{N-1}$, where $Z^{N-1} \subset Z + \ldots + Z$. The spaces $R_1, \ldots, R_N$ correspond to c-sin type structures, which do not contain a ferromagnetic component. The addition of the ferromagnetic component does not change any of these spaces and in this case, $R_{N+1} = R_N$. Proceeding from the above considerations, we can write the following relations which are satisfied for phase transitions in ferrimagnetic antiphase domain structures:

$$R_1 \subset R_2 \subset \ldots \subset R_N \subseteq R_{N+1} ,$$

$$\pi_1(R_1) \subset \pi_1(R_2) \subset \pi_1(R_3) \subset \ldots \subset \pi_1(R_N) \subseteq \pi_1(R_{N+1}) ,$$

$$R_1 \xleftarrow[\gamma_o^1]{\underbrace{S^1 \times S^1}} R_2 \xleftarrow[\underline{S^1 \times S^1}]{\underbrace{S^1 \times S^1}_{\gamma_o^2}} R_3 \xleftarrow{\underbrace{S^1 \times S^1}_{\gamma_o^3}} \ldots \xleftarrow{\underbrace{S^1 \times S^1}_{\gamma_o^{N-1}}} R_N + R_{N+1} . \tag{7.1.16}$$

Here the symbol $\xleftarrow{\gamma_o^i}$ indicates that upon a transition from one magnetically ordered phase into another, a layer $S^1 \times S^1$ is removed from the manifold R_{i+1}. In the complex plane (v, iw), this corresponds to "cutting" a contour γ_o^i from the contour γ_i^{i+1} which is consequently transformed into the contour $\gamma^i(\gamma^{i+1} \xrightarrow{\gamma_o^i} \gamma^i)$. For N = 6, (7.1.16) describes phase lamination during phase transitions in thulium.

Before concluding this section, we note that the second homotopy group π_2 is nontrivial on any of the manifolds $R_1,\ldots,R_N,R_{N+1}$: $\pi_2(R_i) = Z$. This means that in all phases of the type of materials under consideration (Er-Tm alloys, Tm), point singularities of the vector field of the order parameter may exist in principle. However, a strong axial magnetic anisotropy prevents the emergence of such singularities. Hence only linear singularities of the vector fields can be observed in these substances.

7.2 Modern Spin Glass Models

During the past decade, there has been a considerable rise in the interest regarding amorphous magnetically ordered systems [7.5-8]. These include disordered ferro- and antiferromagnets, spin glasses, and the so-called metal glasses. The first two types of materials possess the following properties. Atoms are randomly distributed in space, and hence a short-range order exists in the crystal structure. In disordered ferro- and antiferromagnets, the spin system represents a short-range magnetic order, although finite regions with steady-state ferro- or antiferromagnetic order (clusters) may be observed against the background of the short-range magnetic order. Outside the clusters, the spins are distributed randomly and have a uniform distribution on the average. Another distinguishing feature of the disordered ferromagnets is that they possess a macroscopic spontaneous magnetic moment, while antiferromagnets do not possess such a moment. In a spin glass, all atomic spins are randomly oriented in space and hence it does not possess a macroscopic magnetic moment. Metal glasses also have a short-range crystal order. However, they differ from disordered magnets and spin glasses primarily in that their magnetic structure may have a short-range as well as a long-range order.

Theoretical studies of disordered systems and spin glasses are quite common [7.9-17]. On the other hand, metal glasses have been discovered more recently [7.5]. These are state-of-the-art materials and are widely used in industry. Hence the spurt in activity concerning disordered magnetic system analysis should be associated with the appearance of metal glasses [7.5,8].

Metal glasses include two types of alloys, viz., the compounds of transition or noble metals Ni, Co, Fe, Mn, Zr, La, Pa, Pt (80%) with metalloids B, C, Si, P (20%), and alloys of rare-earth and transition metals $GdFe_2$, $TbFe_2$, $GdCo_2$, $TbCo_2$, $DyCo_{3,4}$. A remarkable property of the second type of materials is that their structure shows a complex long-range magnetic order [7.8]. Hence, it can be expected that they will possess anomalous properties during phase transitions to the ordered state. The fluctuation theory can be used successfully for disordered magnetic systems as well. We shall confine ourselves to a phenomenological approach since a microscopic analysis is required only for calculating the parameters of the effective Hamiltonian. Hence we shall consider thermodynamic fluctuations in a spin subsystem, which are especially significant near the phase transitions.

In this section, we shall use the idea of phase lamination and the exchange enhancement effect to develop the method of formation and the theory of disordered magnetic systems of a new type. The new class of materials is based on rare-earth metals, their alloys, and compounds [7.1]. Disordered magnetic systems are frequently investigated with the help of Heisenberg's model. The exchange integral J_{ij} is a random function of the lattice site number i. Moreover, the function $J(R_{ij})$, $R_{ij} = R_i - R_j$, defined at a fixed lattice site i may be a complex oscillatory function in space.

Relativistic interactions are taken into account in the same way as for ordered magnets [7.12], the only difference being that the direction of the magnetic anisotropy axis of the local spin changes randomly from one site to another. Such models have been, and are still being, used for calculating the static and dynamic properties of magnetically disordered systems [7.12]. The main computational technique employed for this purpose is the self-consistent field method (the coherent potential method is one of its variations). The distribution functions of exchange and relativistic parameters are either assumed to be given by the conditions of the problem, or determined from the corresponding self-consistent equations. Such an approach yields results for disordered ferro- and antiferromagnets, which are found to be in good agreement with the experimental data [7.12].

However, the situation is more complicated for the case of spin glasses. It was mentioned above that by spin glass we frequently mean a system with random atomic distribution and spin orientation in space. Such a state usually appears during annealing of disordered magnets. In the general case, the annealing of a magnet may lead to such a state. Therefore, it is not a phase state of matter, i.e., it does not possess an intrinsic parameter whose variation may result in a spontaneous violation of symmetry and in a phase transition to a new magnetic (or paramagnetic) state. Hence, by the term "spin glass" we mean a phase that does not correspond to the phase diagram, and in which a disordered magnetic system exists for some definite values of a certain external parameter. The impurity concentration x is one such parameter. Disordered magnets are usually obtained by introducing nonmagnetic impurities or a different type of magnetic impurity (for example, an antiferromagnetic material into a ferromagnet or a ferromagnetic material into an antiferromagnet) into a magnetically ordered crystal.

Thus, we can obtain the following aggregate of states of a system. Suppose that we have a disordered ferromagnet in the region $0 < x < x_1$. In this state, the system has a nonzero magnetic moment. Moreover, we can isolate in this system macroscopic regions with ferromagnetic or nearly ferromagnetic order against the background of the disorder. Antiferromagnetic cells do not form macroscopic clusters. For $x_2 < x < 1$, we obtain a disordered antiferromagnet. The system does not have a magnetic moment, and a large number of randomly distributed antiferromagnetic cells are formed in the system, thus resulting in the possibility of antiferromagnetic cluster formation. A phase called the spin glass is formed in the system in the concentration range $x_1 < x < x_2$. In this state, the material does not have a magnetic moment, and its magnetic structure is a particular mixture of ferro- and antiferromagnetic orders. Ferro- and antiferromagnetic clusters may be formed in such a system. Spin glasses possess the following characteristic properties. At a certain temperature T_c, the static magnetic susceptibility has a sharp peak (cusp) [7.7], while the temperature dependence of the susceptibility of an annealed sample is smooth. The heat capacity at low temperatures is proportional to temperature: $C \approx \alpha_T$. Finally, the spectrum of spin waves has no (or very low) dispersion.

The static properties of a spin glass can be explained by using the cluster model with an exchange Hamiltonian [7.13]

$$\mathbb{H} = -\sum_{i,j=1}^{N\ N} J_{ij}\mathbf{S}_i\mathbf{S}_j$$

which can be written in the equivalent form:

$$\mathbb{H} = \sum_{i,j=1}^{N_{sp}} \sum_{\nu,\lambda=1}^{N_{cl}} J_{i\nu j\lambda} S_{i\nu} S_{j\lambda} \qquad (7.2.1)$$

if we assume that our system has N_{cl} clusters with N_{sp} spins in each cluster: $N = N_{cl}N_{sp}$. Dividing the sum (7.2.1) into terms $\nu = \lambda$ and $\nu \neq \lambda$, we obtain

$$\mathbb{H} = -\sum_{\nu<\lambda} J_{\nu\lambda} \mathbf{S}_\nu \mathbf{S}_\lambda - \sum_{\nu} \sum_{i<j} J^o_{ij} \mathbf{S}_{i\nu} \mathbf{S}_{j\lambda} , \qquad (7.2.2)$$

where J^o_{ij} is the exchange interaction between clusters and $J_{\nu\lambda}$ is the random intracluster exchange interaction between nearest neighbors of ferromagnetic type.

In the same way as in the **Edwards-Anderson** theory of spin glasses [7.9], the distribution function of exchange integrals in a cluster is defined by the relation

$$P(J_{\nu\lambda}) = (1/\sqrt{2\pi}J)\exp[-(J_{\nu\lambda}-J_1)^2/2J^2] , \quad J > 0 , \qquad (7.2.3)$$

i.e., the distribution is Gaussian. Using the self-consistent (mean) field theory, we can calculate the free energy inside a cluster and hence obtain the equations for the three parameters M, q, and m, $M = \langle\langle S^2_\nu\rangle\rangle_{(\Gamma_v)}$. Here, $\langle \ldots \rangle_{(\Gamma_v)}$ indicates averaging over configurations, $q = \langle\langle \mathbf{S}\rangle\rangle^2_{(\Gamma_v)}$ is the spin glass order parameter introduced by Edwards and Anderson, and $m = \langle\langle S_\nu\rangle\rangle_c$ is the ordinary ferromagnetic order parameter which is equal to zero in a spin glass and in a paramagnet. Thus, the free energy in a cluster is given by

$$F(q,M) = -T \{(\beta^2 J^2/4)(q^2 - M^2) + (1/\sqrt{2\pi}) \cdot \int_{-\infty}^{\infty} d\xi \, e^{-\xi^2/2} \, \mathrm{Tr}[\exp(-\beta \mathbb{H}^{eff})] \} , \qquad (7.2.4)$$

where

$$\mathbb{H}^{eff} = \sum_{i<j} J^o_{ij} \mathbf{S}_{i\nu} \mathbf{S}_{j\nu} + \bar{J}\sqrt{q} \, \mathbf{S}_\nu \xi + (1/2)\beta \, \bar{J}^2 (M - q) S^2_\nu , \qquad (7.2.5)$$

$\bar{J} = z^{\frac{1}{2}} J$; $\beta = 1/T$, and z is the number of nearest neighbors in the cluster. The equations for the parameters M and q will have the form

$$M = (1/\sqrt{2\pi}) \int_{-\infty}^{\infty} d\xi \, e^{-\xi^2} \mathrm{Tr}[(S^2_\nu \, e^{-\beta \mathbb{H}^{eff}})/Z] , \qquad (7.2.6)$$

$$M-q = (1/\sqrt{2\pi}) \int_{-\infty}^{\infty} d\xi e^{-\xi^2} \xi(1/\beta Jq^{\frac{1}{2}}) \mathrm{Tr}[(\mathbf{S}_\nu e^{-\beta \mathbb{H}^{eff}})/Z] , \quad (7.2.7)$$

where $Z = \mathrm{Tr}[\exp(-\beta\mathbb{H}^{eff})]$.

The static magnetic susceptibility $\chi(T)$ is defined by the expression

$$\chi(T) = \chi_o(T)/[1 - \bar{J}_1\chi_o(T)] , \quad (7.2.8)$$

where $\chi_o(T) = \beta(M-q)^2\mu_B^2$ for $J_1 \neq 0$, $\bar{J}_1 = zJ_1$. The heat capacity C_m is defined by the relation

$$C_m = C_m^{inter} + C_m^{intra} , \quad (7.2.9)$$

where the intra- and intercluster contributions are defined by the expressions

$$C_m^{inter} = k \frac{d}{dT}[(\bar{J}^2/2T)(q^2 - M^2)] , \quad (7.2.10)$$

$$C_m^{intra} = k \frac{d}{dT} \int_{-\infty}^{\infty} d\xi\, e^{-\xi^2/2}\, \mathrm{Tr}\left\{\left[- \sum_{i<j} J_{ij}^o \mathbf{S}_{i\nu}\mathbf{S}_{j\nu}\, e^{-\beta\mathbb{H}^{eff}}\right]/\sqrt{2\pi} Z \right\}, \quad (7.2.11)$$

k being the Boltzmann constant.

By solving (7.2.6,7), we can construct the dependences M(T), q(T), $\chi(T)$, and $C_m(T)$. These dependences are shown in Fig. 7.3, and have all the properties inherent in spin glasses. The transition temperature T_C is defined by the relation

$$T_C = JM(T_C) . \quad (7.2.12)$$

It can be seen from Fig. 7.3a that the heat capacity C_m has a peak at $T > T_C$. this contradiction is due to the fact that the nearest neighbor approximation takes into consideration the exchange interaction within a cluster. Normal behavior is restored if we take into account the interaction with the next neighbors as well, assuming this interaction to be antiferromagnetic (of opposite sign), Fig. 7.3b.

This theory (7.2.6-11) is useful for describing the static characteristics of spin glasses obtained by introducing magnetic impurities in a nonmagnetic matrix (or vice versa). Among the materials of this type are the spin glasses AuFe and CuMn, which have been thoroughly investigated experimentally [7.14,15]. The results of this theory compare well with the corresponding experimental data.

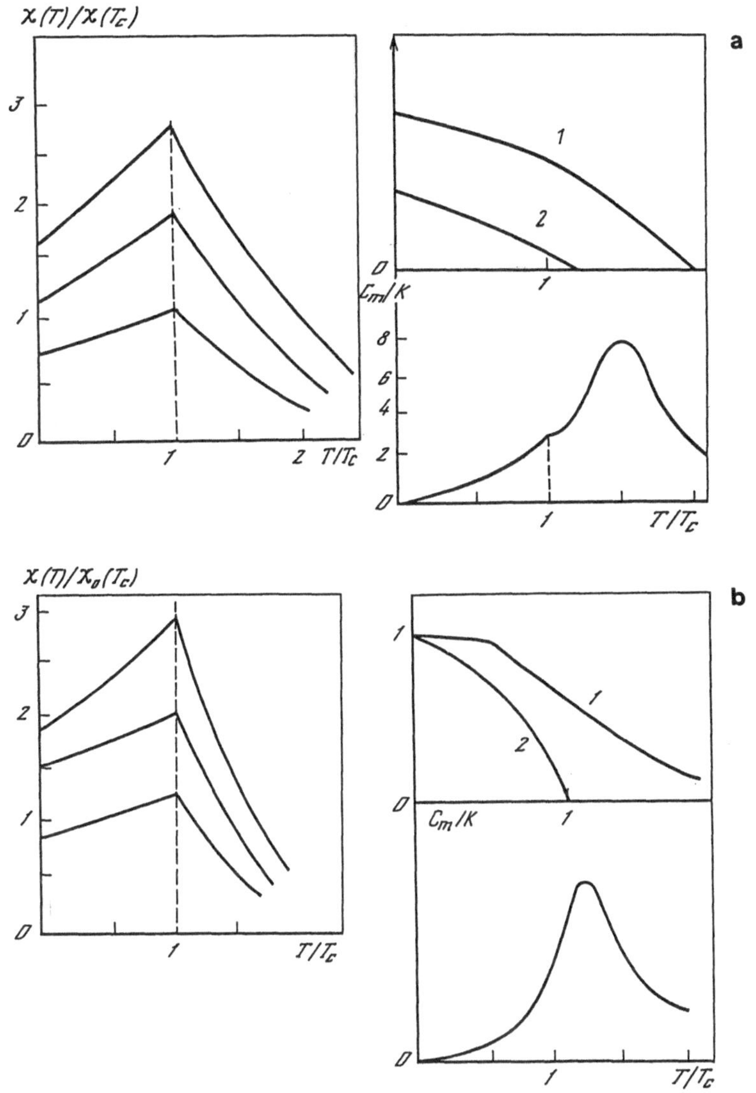

Fig. 7.3. Temperature dependences of the magnetic susceptibility order parameter q, M, and heat capacity C_m of an "impurity" spin glass, calculated on the basis of Heisenberg's cluster model using the approximation of nearest-neighbor interaction in the cluster (**a**), and by taking into account the interaction with the second coordination sphere in the cluster (**b**); *curve 1* – $M(T)/M(0)$; *curve 2* – $q(T)/q(0)$

In order to explain the properties of spin glasses, the renormalized group approach was also used in addition to the mean field approximation [7.16,17]. However, this approach allows us to describe $\chi(T)$ only in the vicinity of T_C, while the $C_m(T)$ dependence cannot be explained on account of the fact that the $C_m(T)$ peak is smoothed at $T \to T_C$. Hence the renormalized group method is less appropriate here than the mean field theory.

Thus, the cluster model provides a better interpretation of the static properties of spin glasses than the self-consistent field theory. However, this approach is unsuitable for describing the spin waves in a spin glass. Investigations of the spin-wave dynamics were carried out [7.18] by using a model which assumes that on the average, all atoms and spins are distributed in space uniformly and isotropically. In this case, the energy of a spin system is found to be invariant relative to the rotation group transformations in the spin space SO(3). This means that the ground state of a spin glass is multiply degenerate. The spin-wave spectrum will then contain [7.18] three acoustic branches with frequencies

$$\omega_{1,2,3} = c_s k \ . \tag{7.2.13}$$

In such a spin glass, the relativistic interactions are anomalously weak. Consequently, local spin-flipping may occur in the system and hence the spin waves are strongly attenuated. In other words, the modes (7.2.13) are transformed into diffusion modes [7.18-20].

Hence, on the basis of the above discussion, it can be concluded that unlike in the case of disordered magnets, a unified theory for describing the static and dynamic magnetic properties of impurity spin glasses has not been constructed so far. This is a serious drawback in the investigations of spin glasses.

7.3 Peculiarities of Vector Fields in Heavy Rare-Earth Metals

In this section, we shall consider a new class of materials based on rare-earth metal compounds with magnetic structures of NS, c-sin, TS, or cone-type. The phase diagram of rare-earth metals and their alloys (Fig. 7.2) shows a region of temperature and magnetic anisotropy $|a_{\|}|$ values, in which the phase transitions are of multi-step type (Tm and Er-Tm alloys). Let us assume that before going over to the ground state, these materials must pass through a number of ordered phases, i.e., they must experience the phase lamination effect. These phases may include states with anomalous electromagnetic properties.

It was shown in [7.21; see also 7.22] that the existence of linear defects in liquid and magnetically ordered crystals leads to the formation of special lines on which the vector field has a singularity (disclination) (Sects. 6.3,4). It is possible to construct new materials by radiative introduction of crystal lattice defects, dislocations, and plastic disclinations into a magnetically ordered crystal. The introduction of a large number of linear defects in a crystal with any of the above-mentioned magnetic structures causes a strong distortion of the magnetic order. The system becomes disordered and its magnetic structure corresponds to that of a spin glass.

Heavy rare-earth metals and their alloys have a strong magnetic anisotropy. this means that at low temperatures, the spin glasses considered above may have a short-as well as long-range order, and a nondegenerate ground state. Consequently, their properties may differ significantly from those of normal impurity spin glasses.

Thus, let us study the properties of vector fields in complex magnetic structures such as NS (normal spiral), c-sin (longitudinal sinusoidal modulation), TS (tilted spiral), and cone (conic spiral) into which linear defects have been introduced.

1. **Normal spiral**. The order parameter in this case has the form (Sect. 3.2)

$$\boldsymbol{\gamma}_{\perp} = \frac{\mathbf{s}_{\perp}^{+} + i\mathbf{s}_{\perp}^{-}}{|\mathbf{s}_{\perp}^{+} + i\mathbf{s}_{\perp}^{-}|} \; , \; \boldsymbol{\gamma}_{\perp}(\mathbf{x}) = \boldsymbol{\gamma}_{\perp o}\, e^{i\mathbf{q}_{\perp}\mathbf{x}} \; , \qquad (7.3.1)$$

or

$$\boldsymbol{\gamma}_{\perp}(\mathbf{x}) = \boldsymbol{\gamma}_{\perp o} \cos q_{\perp}(\mathbf{n}_{\perp}\mathbf{x}) + [\mathbf{n}_{\perp}\boldsymbol{\gamma}_{\perp o}] \sin q_{\perp}(\mathbf{n}_{\perp}\mathbf{x}),$$

where $\mathbf{n}_{\perp}$ is a unit vector oriented along the spiral axis. The degeneracy space corresponding to the order parameter is the rotational group R = SO(3). The fundamental group on the manifold R is $\pi_1(SO(3)) = Z_2$, while the second homotopy group is trivial, i.e., $\pi_2(SO(3)) = 0$. This means that there are no point singularities in a normal spiral. It is known from algebraic topology that $SO(3) = SU(2)/Z_2$ [7.23,24]. In order to classify linear singularities, we must factorize the group SU(2) in the quaternion group Q [7.22,23]. In this case, the required manifold will have the form

$$\tilde{R} = SU(2)/Q = S^3/Q \; . \qquad (7.3.2)$$

The expression for the free energy of the system can be written as follows:

$$F = \int dx \left\{ \frac{1}{2}\delta_\perp(\boldsymbol{\gamma}_\perp \boldsymbol{\gamma}^*_\perp) + \frac{1}{2} b_\perp(\gamma_{\perp x}\gamma^*_{\perp x} + \gamma_{\perp y}\gamma^*_{\perp y}) + \frac{1}{2} c_\perp \gamma_{\perp z}\gamma^*_{\perp z} \right.$$

$$\left. + d_{i\perp}\left(\boldsymbol{\gamma}_\perp \frac{\partial \boldsymbol{\gamma}^*_\perp}{\partial x_i} - \boldsymbol{\gamma}^*_\perp \frac{\partial \boldsymbol{\gamma}_\perp}{\partial x_i}\right) + \frac{1}{2} f_\perp \left(\frac{\partial \boldsymbol{\gamma}_\perp}{\partial x_i} \frac{\partial \boldsymbol{\gamma}^*_\perp}{\partial x_i}\right) \right\} . \quad (7.3.3)$$

Here, $b_\perp < 0$, $c_\perp > 0$ are the anisotropy constants, $f_\perp$ and $\delta_\perp$ are the exchange constants, and $\mathbf{d}_\perp \to (0, 0, d_{\perp z})$ in the absence of perturbations in the crystal lattice. Since the spin-density vector varies along the z-axis, we shall be interested in linear defects whose generalized Burgers vector $\mathbf{B} = \mathbf{b} + [\boldsymbol{\Omega}(\mathbf{x} - \mathbf{x}_0)]$ ($\boldsymbol{\Omega}$ is the Frank vector) is also directed along the same axis. These defects include:

(a) a screw dislocation whose line is parallel to the isolated z-axis of the crystal;

(b) an edge dislocation whose line is perpendicular to the z-axis;

(c) a torsional dislination with $\boldsymbol{\Omega}_{\perp z}$; and all possible combinations of these defects.

Unlike biophysical systems and cholesterine liquid crystals, the rotation (torsion) about the z-axis in the systems considered here does not lead to the formation of disclinations, since the magnetic moment is not rigidly connected to the crystal lattice during rotations of crystallographic unit cells, and the magnetic symmetry does not always determine the crystallographic symmetry.

Let us first consider the case of a screw dislocation, since its symmetry corresponds to the magnetic symmetry of an NS. If a magnetically ordered crystal contains aline defect, the following conditions are satisfied during the circumvention of its line along the contour Γ^ν:

$$\int_{\Gamma^\nu} dx_k u_{ik} = -b_i - e_{ijk}\Omega_j(x_k - x_{ko}) , \quad (7.3.4)$$

$$\int_{\Gamma^\nu} dx_k \gamma_{\perp ik} = 0 . \quad (7.3.5)$$

The condition (7.3.5) indicates the continuity of the order parameter on the contour Γ^ν. In this case, $\boldsymbol{\Omega} = 0$, and $\mathbf{b} = (0,0,b_z)$, $|\mathbf{b}| = nc/2$, where c/2 is the distance between two adjacent magnetic planes. Since $\boldsymbol{\gamma}_\perp = \boldsymbol{\gamma}_{\perp 0}\exp(iq_\perp z)$ in equilibrium, the condition (7.3.5) can be used to construct an operator of translation along the z-axis, acting on the vector $\boldsymbol{\gamma}_\perp$:

$$\hat{T}_{nc/2}\boldsymbol{\gamma}_{\perp} = \boldsymbol{\gamma}_{\perp} , \tag{7.3.6}$$

or

$$\hat{T}_{nc/2}\, \boldsymbol{\gamma}_{\perp o} = \exp(i\mathbf{q}_{\perp}\mathbf{b})\boldsymbol{\gamma}_{\perp o} .$$

Since

$$q_{\perp} = \frac{2\pi}{N_{\perp}c/2} ,$$

we obtain

$$T_{nc/2}\, \boldsymbol{\gamma}_{\perp o} = \exp(-i2\pi n/N_{\perp})\boldsymbol{\gamma}_{\perp o} . \tag{7.3.7}$$

As we circumvent the contour Γ^{ν}, the vector $\boldsymbol{\gamma}_{\perp o}$ moves along the circle S^1 described on the sphere S^3, $S^3 \subset Q$. The contour Γ^{ν} can be mapped onto an entire circle if $n = N_{\perp}/2$, i.e., if the direction of the vector $\boldsymbol{\gamma}_{\perp o}$ is reversed. All the remaining contours on S^1 can be contracted to a point, i.e., they are homotopic to zero (Fig. 7.4).

In the general case, $n = \pm \ell N_{\perp}/2$, $\ell = 2m+1$, $m = 0,1,\ldots,$ since we normally have $N_{\perp} = 8$, $n = \pm 4\ell$. As we circumvent a singular line, the contour Γ^{ν} is mapped onto a fundamental group element of a circle: $f : \Gamma^{\nu} \to \pi_1(S^1) = Z$. Since $S^1 \subset S^3 \subset Q$, we have $Z \subset Q$, but the group Q contains only two elements of the group Z, viz., e{1} and e{-1}, corresponding to $Z = \pm 1$ and $\ell = 1$.

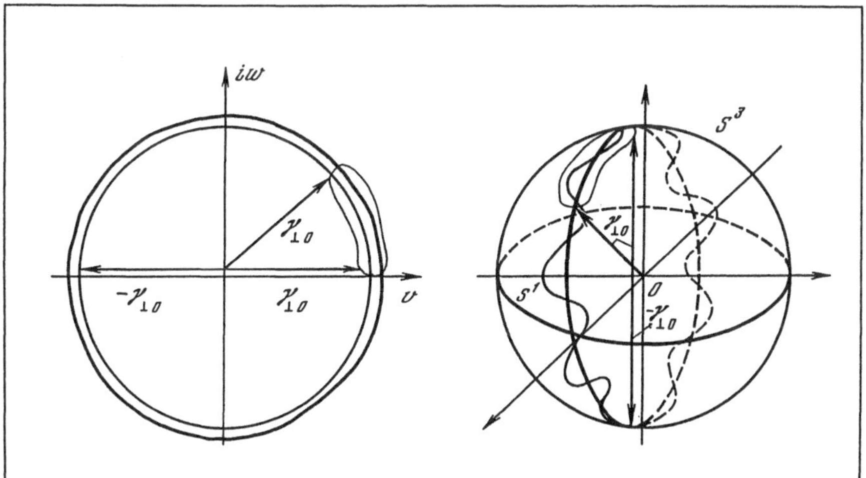

Fig. 7.4. Degeneracy space corresponding to the order parameter $\gamma_{\perp}$ for the normal spiral (NS) magnetic structure, and mapping of the contour enclosing the singular line in the magnetic structure on this space

The nature of this singularity can be found by minimizing the free energy (7.3.3). The screw dislocation does not change the magnitude of the wave vector $q_\perp$ of the helix and its orientation in space. Hence (7.3.3) can be reduced to the form

$$F = (1/2)\int dx\left\{\tilde{\delta}_\perp(\boldsymbol{\gamma}_{\perp o}\boldsymbol{\gamma}^*_{\perp o}) + \tilde{a}\gamma_{\perp oz}\gamma^*_{\perp oz} + f_\perp\left(\frac{\partial\boldsymbol{\gamma}_{\perp o}}{\partial x_i}\frac{\partial\boldsymbol{\gamma}^*_{\perp o}}{\partial x_i}\right)\right\}, \quad (7.3.8)$$

$$\tilde{\delta}_\perp = \delta_\perp + b_\perp, \quad \tilde{a}_\perp = a_\perp - b_\perp .$$

Introducing the variables

$$\gamma_{\perp ox} = (1/\sqrt{2})\cos\Theta\, e^{i\Phi}, \quad \gamma_{\perp oy} = (1/\sqrt{2})\cos\Theta\, e^{-i\Phi},$$
$$\gamma_{\perp oz} = \sin\Theta, \quad (7.3.9)$$

we obtain the functional for the corresponding variational problem:

$$F = (1/2)L\int_{\eta_{min}}^{\eta_{max}} d\eta\int_0^{2\pi} d\varphi\left\{\rho^2\tilde{a}_\perp\sin^2\Theta\right.$$
$$\left. + f_\perp[\Theta_\eta^2 + \Theta_\varphi^2 + (\Phi_\eta^2 + \Phi_\varphi^2)\cos^2\Theta]\right\}, \quad (7.3.10)$$

where ρ, φ, z are variables of the cylindrical system of coordinates, $\eta = \ln\rho$, L is the length of the singular line, and Θ_η, Θ_φ, Φ_η and Φ_φ stand for $\partial\Theta/\partial\eta$, $\partial\Theta/\partial\eta$, $\partial\Phi/\partial\eta$, and $\partial\Phi/\partial\varphi$, respectively.

The equations for the extremum have the form

$$\Theta_{\eta\eta} + \Theta_{\varphi\varphi} - (1/2)\sin 2\Theta\,(\rho^2 a_\perp/f_\perp - \Phi_\eta^2 - \Phi_\varphi^2) = 0, \quad (7.3.11)$$
$$(\Phi_{\eta\eta} + \Phi_{\varphi\varphi})\cos^2\Theta = 0. \quad (7.3.12)$$

Since we are considering the spin distribution away from the line of disclination, we assume that the departure of magnetic moments from the (x,y) plane is insignificant. In this case, $\max(\Theta, \partial\Theta/\partial\varphi, \partial\Theta/\partial\eta) \ll 1$. Hence (7.3.12) does not contain the derivatives Θ_φ and Θ_η.

We introduce the following boundary conditions at infinity:

$$\Theta(\varphi,\infty) = 0, \quad \Phi(0,\infty) = 0, \quad \Phi(2\pi,\infty) = \pi. \quad (7.3.13)$$

Thus for $\eta\to\infty$ $\Theta\to 0$, $\Phi\to\Phi(\varphi) = (\ell/2)\varphi$, we obtain

$$F = (1/4)\pi L f_\perp \ell^2 \ln(R_{max}/R_{min}). \quad (7.3.14)$$

In order to find out the stability of a singularity, we must consider the second variation $\delta^2 F$ of the functional (7.3.10). In this case, the equation for a small self-simulating perturbation $\delta\Theta$ will have the form

$$\delta\Theta_{\eta\eta} + \delta\Theta_{\varphi\varphi} - \delta\Theta\left(\rho^2 \frac{\tilde{a}_\perp}{f_\perp} - \ell^2/4\right) = 0 \ . \qquad (7.3.15)$$

The solution of this equation is given by

$$\delta\Theta(\xi,\varphi) = \sum_k C_k K_{\nu(k)}(\xi)\sin(k/2)\varphi \ . \qquad (7.3.16)$$

Here, $\xi = \rho\sqrt{\tilde{a}_\perp/f_\perp}$; $\nu^2 = (k^2 - \ell^2)/2$ and $K_\nu(\xi)$ is the Macdonald function. For $\xi \to \infty$, we obtain

$$K_\nu(\xi) = \sqrt{(\pi/2\xi)}\, e^{-\xi}[1 + O(\xi^{-1})] \ . \qquad (7.3.17)$$

For $\ell = 1$, all ν^2 are nonnegative, and hence disclinations with Frank's index ±1 are stable. This is in agreement with the topological analysis.

Hence, topological analysis in this case allows us to do away with cumbersome calculations. Figure 7.5 shows the spin disclination with Frank's index +1. Since the system has a strong crystallographic anisotropy, and the lattice is "indented" as a result of

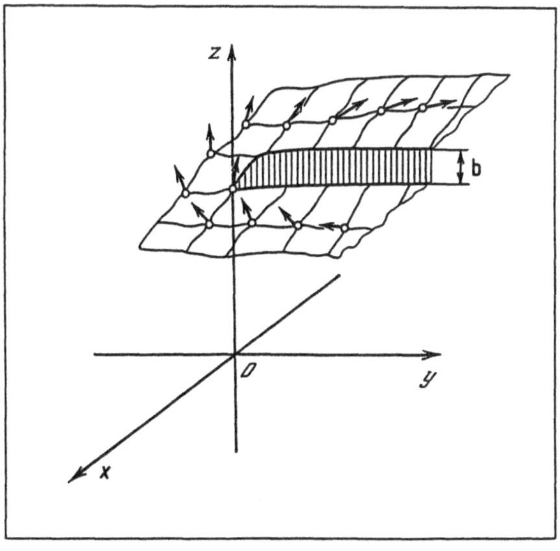

Fig. 7.5. Spin disclination with Frank's index +1 in the normal spiral magnetic structure following the introduction of screw dislocation in the crystal

the introduction of a dislocation, the magnetic moments will depart from the plane near the dislocation line. Hence the Néel domain wall, which appears at infinity, will turn into a Bloch-type domain wall near the dislocation core. This means that the magnetic symmetry of the crystal changes near the dislocation core. Indeed Fig. 7.5 shows that antiferromagnetic order is established along the faces of a cut, and the crystallographic unit cell is found to be orthorhombic. This is a very important fact which we shall use at a later stage.

Let us now consider an edge dislocation (Fig. 7.6). A special feature of such dislocations is that the distribution of magnetic moments near them cannot be considered in a single magnetic plane, as in the case of a screw dislocation, but requires an investigation of the range of variation of the order parameter which covers the entire system of planes. Hence the vector field singularities at the edge dislocations must be considered in the framework of a large-distance continuum theory. The topological classification for an edge dislocation remains the same as for a screw dislocation: $n = \pm 4\ell$, $\ell = 1$. However, a characteristic feature of an edge dislocation is that its line must be parallel or perpendicular to the order parameter $\boldsymbol{\gamma}_\perp$ ($\mathbf{L} \parallel \boldsymbol{\gamma}_\perp$, $\mathbf{L} \perp \boldsymbol{\gamma}_\perp$).

We take the order parameter in the form

$$\boldsymbol{\gamma}_\perp = \boldsymbol{\gamma}_{\perp o}^{+} \cos\psi_\perp + i\boldsymbol{\gamma}_{\perp o}^{-} \sin\psi_\perp \,, \qquad (7.3.18)$$

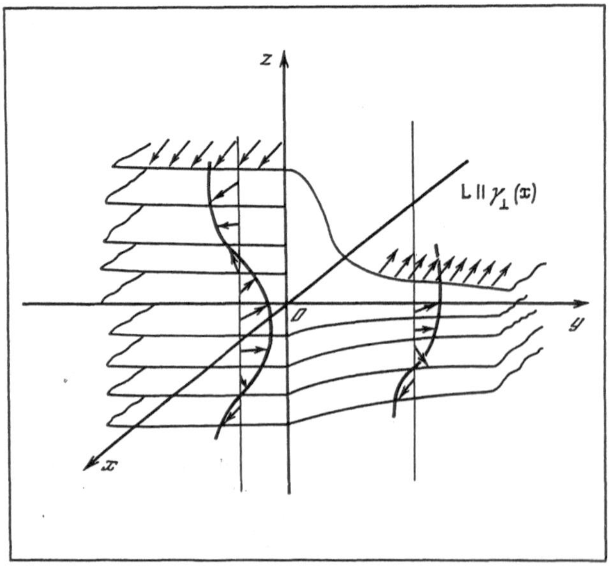

Fig. 7.6. Disclination with Frank's index +1 in the normal spiral magnetic structure in a system containing an edge dislocation

where the variable $\psi_\perp$ satisfies the condition $\nabla\psi_\perp = q_\perp \mathbf{n}_\perp$, $\mathbf{n}_\perp = [\boldsymbol{\gamma}^+_{\perp o}\boldsymbol{\gamma}^-_{\perp o}]$, $|\boldsymbol{\gamma}_\perp| = 1$, $n^2_\perp = 1$. The free energy functional can be written in the form

$$F = (1/2)f_\perp \int d\mathbf{x}\Big\{(\tilde{a}_\perp/f_\perp)\gamma_{\perp z}\gamma^*_{\perp z} + 2q_\perp \mathbf{n}_{\perp i}[|\boldsymbol{\gamma}^-_\perp|(\partial|\boldsymbol{\gamma}^+_\perp|/\partial x_i)$$

$$-|\boldsymbol{\gamma}^+_\perp|(\partial|\boldsymbol{\gamma}^-_\perp|/\partial x_i)] + [(\partial\boldsymbol{\gamma}_\perp/\partial x_i)(\partial\boldsymbol{\gamma}^*_\perp/\partial x_i)]\Big\} . \qquad (7.3.19)$$

Substituting (7.3.18) into (7.3.19) and averaging over the angular variable $\psi_\perp$, we obtain

$$\langle F\rangle_{\rho(\psi_\perp)} = (1/2)f_\perp \int d\mathbf{x}\Big\{-q^2_\perp n^2_{\perp i} + (\tilde{a}_\perp/f_\perp)(\boldsymbol{\gamma}_{\perp \mathbf{n}_\perp}\boldsymbol{\gamma}^*_{\perp \mathbf{n}_\perp})$$

$$+ (1/2)[(\partial\boldsymbol{\gamma}^+_{\perp o}/\partial x_i)^2 + (\partial\boldsymbol{\gamma}^-_{\perp o}/\partial x_i)^2]\Big\} . \qquad (7.3.20)$$

Introducing the new variable $v_{si} = -(\boldsymbol{\gamma}^-_{\perp o}\nabla_i\boldsymbol{\gamma}^+_{\perp o})$, we obtain the final expression for the free energy

$$\langle F\rangle_{\rho(\psi_\perp)} = (1/2)f_\perp \, d\mathbf{x}\Big\{(\tilde{a}_\perp/f_\perp)(\boldsymbol{\gamma}_{\perp \mathbf{n}_\perp}\boldsymbol{\gamma}^*_{\perp \mathbf{n}_\perp}$$

$$+ (1/2)\alpha_{1\perp}(\partial\mathbf{n}_\perp/\partial x_i)^2 + \alpha_{2\perp}(\mathbf{n}_\perp\mathbf{v}_s)^2 + \alpha_{3\perp}[\mathbf{n}_\perp\mathbf{v}_s]^2\Big\}. \qquad (7.3.21)$$

Since the system is isotropic in its magnetic properties, we have $\alpha_{1\perp} = \alpha_{2\perp} = \alpha_{3\perp} = 1$. The vector v_s is analogous to the superfluid velocity in He^3-A.

At the line of a large-scale defect in the form of a system of edge dislocations, the helical vector $q_\perp$ turns by an angle $\Phi_\perp$, and $v_s = 0$. Away from the defect line, the magnetic moments lie in one plane and the free energy functional has the standard form

$$\langle F\rangle_{\rho(\psi_\perp)} = (1/4)f_\perp \int d\mathbf{x}(\partial\mathbf{n}_\perp/\partial x_i)^2 . \qquad (7.3.22)$$

A plastic rotation about the (-x)-axis leads to a rotation of the vector $\mathbf{n}_\perp$ in the (z,y)-plane by an angle $\Phi_{\perp o} = \pi/4 + \lambda_o/2$; the distribution $\Phi_\perp(\varphi) = \Phi_{\perp o}(\varphi/\varphi_o)$, $\varphi_o = \pi/2 + \lambda_o$ (Fig. 7.7), $0 \le \varphi \le \pi/2 + \lambda_o$. It is interesting to note that the condition imposed on $\psi_\perp$ leads to $[\nabla\mathbf{n}_\perp] = 0$. On the one hand, this condition means that the magnetic layers of a normal spiral are equidistant; on the other hand, it allows point singularities which are not taken into consideration by the manifold (7.3.2). However, the existence of such singularities is not possible due to a high rigidity of the crystal lattice.

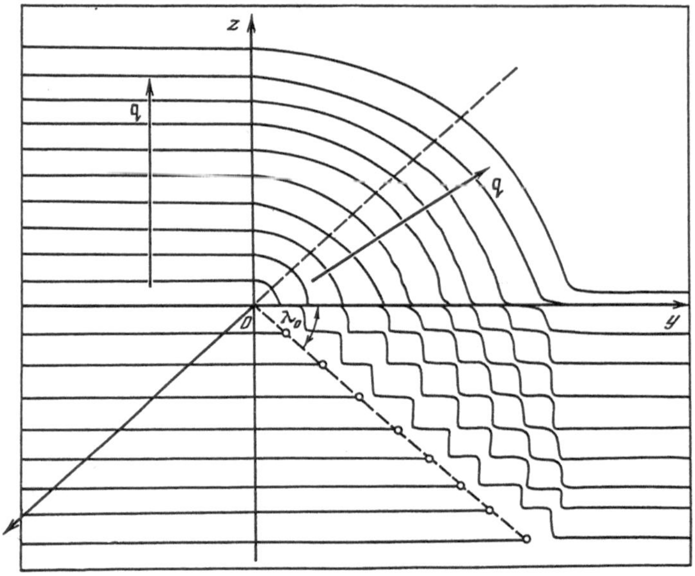

Fig. 7.7. Distortion of magnetic structure in a normal spiral following a torsional plastic disclination

Since dislocations and plastic disclinations are lattice defects, a decomposition of the singular line into two is not possible due to a high rigidity of the crystal. Hence the homotopy classes {i,-i}, {j,-j} , and {k,-k} on the manifold $\tilde{R} = S^3/Q$ are not filled.

The vector field singularities in c-sin, TS, and cone-type phases (Fig. 7.2) are constructed in the same way as for a normal spiral. Hence we shall confine ourselves to a description of the corresponding order parameter manifolds in these phases (Sects. 3.2, 4.4).

2. Sinusoidal modulation (c-sin). The order parameter in this phase has the form

$$\boldsymbol{\gamma}_{\parallel} = \boldsymbol{\gamma}^{+}_{\parallel} + i\boldsymbol{\gamma}^{-}_{\parallel} \quad , \quad \boldsymbol{\gamma}^{+}_{\parallel}(\mathbf{x}) = \mathrm{Re}\{\boldsymbol{\gamma}^{+}_{\parallel 0} \ \exp[i(\mathbf{q}_{\parallel}\mathbf{x} + \varphi_{+})]\} \ ,$$

$$\boldsymbol{\gamma}^{-}_{\parallel}(\mathbf{x}) = \mathrm{Im}\{\boldsymbol{\gamma}^{-}_{\parallel 0} \ \exp[i(\mathbf{q}_{\parallel}\mathbf{x} + \varphi_{-})]\} \ . \tag{7.3.23}$$

The degeneracy space of the order parameter is $R = A^2 \times B^1$ (in the simplest case, $A^2 = S^2$, $B^1 = S^1$). This gives $\pi_1(R) = Z$, and $\pi_2(R) = Z$. Since the state of the system with $\mathbf{q}_{\parallel}$, $\boldsymbol{\gamma}_{\parallel 0}$ is equivalent to the state with $-\mathbf{q}_{\parallel}$, $-\boldsymbol{\gamma}_{\parallel 0}$, this means that the manifold $S^2 \times S^1$ must be factorized on the center group Z_2. Hence $\tilde{R} = S^2 \times S^1/Z_2$. The free energy differs from (7.3.3) only in the substitution $\gamma_{\perp} \to \gamma_{\parallel}$ and in that $b_{\parallel} > 0$, $a_{\parallel} < 0$. Since the second homotopy group is nontrivial, i.e., $\pi_2(\tilde{R}) \neq 0$ (Sect. 4.4), point singularities may appear in the system with a low anisotropy.

3. **Tilted spiral (TS).** In this case, the order parameter is defined by two vectors $\gamma_{\parallel}$ and $\gamma_{\perp}$, $q_{\parallel} = q_{\perp}$:

$$R = SO(3) \times S^1 , \quad \pi_1(R) = Z_2 + Z , \quad \pi_2(R) = 0.$$

The factorization is carried out on the group Q, and the free energy is

$$F = F_{\perp} + F_{\parallel} + \Phi(\gamma_{\parallel}, \gamma_{\perp}) , \tag{7.3.24}$$

where $\Phi(\gamma_{\parallel}, \gamma_{\perp})$ is a functional of fourth-order invariants.

4. **Conic spiral (cone).** In this case also, the order parameter is defined by two vectors

$$\boldsymbol{\gamma}_{\perp}, \boldsymbol{\gamma}_f , \quad \boldsymbol{\gamma}_f = \boldsymbol{\gamma}_{fo} \exp[i2\pi(\mathbf{n}_{\perp}\mathbf{x})] ,$$

$$R = S^3/Q , \quad \pi_1(S^3/Q) = Q , \quad \pi_2(S^3/Q) = 0.$$

Hence the singularities in a conic spiral are the same as in a normal spiral.

7.4 Magnetic Susceptibility in the New Spin Glasses

It was shown in Sect. 7.3 that the existence of linear defects in magnetically ordered crystals leads to a strong distortion of the magnetic structure. If an anomalously large number of dislocations are introduced into the system, the magnetic structure will be completely disordered, i.e., it will be a structure of spin-glass type. If ring dislocations of radius R_d are introduced in a magnetic structure of NS type ($P6_3/mmc$ - D^4_{6h}), it is found that although such dislocations lead to the emergence of disclinations, they do not change the symmetry of the unit cell and do not introduce a singularity in the crystal lattice. Introduction of impurities in a crystal cannot lead to the formation of disclinations since they do not cause a relative displacement of atomic planes. Consequently, the introduction of ring dislocations may preserve the spatial symmetry group of the crystal D^4_{6h}. By introducing ring dislocations and averaging over their distribution, we can construct a dislocation lattice. The unit cell of this lattice has a symmetry group G_p whose elements (two second-order rotational axes C_2, one sixth-order axis C_6, six mirror planes $_6$, and a mirror plane σ_o) are contained in $P6_3/mmc-D^4_{6h}$, $G_p \subset P6_3/mmc-D^4_{6h}$ [7.25].

If we consider one atomic plane, it is found that the ring dislocations form in it a system of antiphase domains analogous to the one in thulium [7.2]. The magnetic structure period in the (x,y) plane is found to be a multiple of the mean distance $L^i = 2\langle R^i_\nu \rangle_{(D^\nu)}$ between dislocations. The spin vector $\mathbf{S}$ is transformed according to the irreducible representation $\Phi_f(\mathbf{x})\exp(i\mathbf{fx})$, where

$$\Phi_f(\mathbf{x}) = \psi_f(x,y)\exp(iq_\perp z), \quad \Psi_f(x,y) = \sum_\ell c_{fi}\varphi_{f\ell}(x,y) \; . \tag{7.4.1}$$

Here, $\varphi_{f\ell}$ are the basis functions of Ψ_f. The maximum number ℓ_{max} is determined by the period of the magnetic structure in the (x,y) plane: $\ell_{imax} = L^i/a$, $\ell_{max} = N$. In this case, $N = \ell_{x\,max}\ell_{y\,max}$. It is quite obvious that if L^i/a is finite, N will also be finite. This means that the magnetic structure of such a spin glass can be represented as a superposition of N spin-density waves with N corresponding complex vectors

$$\mathbf{S}_\ell = \mathbf{S}^+_\ell = i\mathbf{S}^-_\ell \; , \quad 1 \leq \ell \leq N \; . \tag{7.4.2}$$

Hence in such a system the phase lamination effect will occur upon phase transitions from the paramagnetic region to the ordered state. Physically, this corresponds to the emergence of anomalies in the temperature dependences of the heat capacity and magnetic susceptibility of the system (Sect. 7.1). The magnetic symmetry group of each phase can be constructed by adding to the group G_p the rotation group elements in the spin space SO(3). If ring dislocations are introduced in an analogous manner into a c-sin magnetic structure and the analysis is carried out in the same way, we obtain a phase lamination with different modulations of the magnetic structure along the axis. The topological manifolds corresponding to the order parameter in each phase have the form

$$R_1 = S^2 \times S^1, \quad R_2 = S^2 \times S^1 \times S^1 \; , \; \ldots$$

In this case, the following conditions are satisfied:

$$R_1 \subset R_2 \subset \ldots \subset R_N \; ,$$

$$\pi_1(R_1) \subset \pi_1(R_2) \subset \ldots \subset \pi_1(R_N) \tag{7.4.3}$$

and we can construct a sequence analogous to the lamination sequence obtained for heavy rare-earth metals and their alloys with multi-step phase transitions (Sect. 7.1)

$$R_1 \xleftarrow{W_1} R_2 \xleftarrow{W_2} \dots \xleftarrow{W_{N-1}} R_N \ . \tag{7.4.4}$$

Let us now consider the temperature dependence $\chi(T)$ of the magnetic susceptibility of a spin glass. In the general case, the expression for the free energy in a disordered (paramagnetic) state can be presented in the form

$$F = \sum_{i=1}^{N} \left\{ \frac{1}{2}\tau_i (s_i^{+2} + s_i^{-2}) + \frac{1}{2}a_{i\alpha}(s_{i\alpha}^{+2} + s_{i\alpha}^{-2}) \right.$$

$$\left. + \frac{1}{8}\Gamma_{1i}(s_i^{+4} + s_i^{-4}) + \frac{1}{4}\Gamma_{2i}s_i^{+2}s_i^{-2} + \frac{1}{2}\Gamma_{3i}(s_i^{+}s_i^{-})^2 \right\}$$

$$+ \sum_{i \neq j}^{N} \left\{ \frac{1}{8}\Gamma_{4ij}(s_i^{+2}s_j^{+2} + s_i^{+2}s_j^{-2} + s_i^{-2}s_j^{+2} + s_i^{-2}s_j^{-2}) \right.$$

$$\left. + \frac{1}{8}\Gamma_{5ij}[(\mathbf{s}_i^{+}\mathbf{s}_j^{+})^2 + (\mathbf{s}_i^{+}\mathbf{s}_j^{-})^2 + (\mathbf{s}_i^{-}\mathbf{s}_j^{-})^2] \right\}$$

$$\Gamma_{2i0} = \Gamma_{1i0} - 2\Gamma_{3i0} \ , \tag{7.4.5}$$

and the system is found to have a strong crystallographic anisotropy $|a_i| \lesssim J_i$. Each isolated direction can be put in correspondence with three spin-density waves (one longitudinal and two transverse). Decomposing each of them into a Fourier series, we obtain three classes of complex vectors each of which has a spin-density wave with a corresponding sinusoidal modulation:

$$s_{i_1x}^{+}(x) + is_{i_1x}^{-}(x), \quad s_{j_1y}^{+}(x) + is_{j_1y}^{-}(x), \quad s_{k_1z}^{+}(x) + is_{k_1z}^{-}(x),$$

$$s_{i_2x}^{+}(y) + is_{i_2x}^{-}(y), \quad s_{j_2y}^{+}(y) + is_{j_2y}^{-}(y), \quad s_{k_2z}^{+}(y) + is_{k_2z}^{-}(y),$$

$$s_{i_3x}^{+}(z) + is_{i_3x}(z), \quad s_{j_3y}^{+}(z) + is_{j_3y}^{-}(z), \quad s_{k_3z}^{+}(z) + is_{k_3z}^{-}(z),$$

$$i_1 = j_1 = k_1 = \ell_1, i_2 = j_2 = k_2 = \ell_2, i_3 = j_3 = k_3 = \ell_3 . \tag{7.4.6}$$

In the isotropic case, $\ell_1 = \ell_2 = \ell_3$. As a result, the expression for the free energy is considerably simplified:

$$F = \sum_{i=1}^{N} [\frac{1}{2} \tau_i(s_i^{+2} + s_i^{-2}) + \frac{1}{8} \Gamma_{1i}(s_i^{+2} + s_i^{-2})^2]$$

$$+ \sum_{i \neq j}^{N} \frac{1}{8} \Gamma_{4ij}(s_i^{+2} + s_i^{-2})(s_j^{+2} + s_j^{-2}) \, . \qquad (7.4.7)$$

Introducing the field $\varphi_i^2 = s_i^{+2} + s_i^{-2}$, we obtain

$$F = \sum_{i=1}^{N} \frac{1}{2} \tau_i \varphi_i^2 + \frac{1}{8} \Gamma_{1i} \varphi_i^4 + \sum_{i \neq j}^{N} \frac{1}{8} \Gamma_{4ij} \varphi_i^2 \varphi_j^2 \, . \qquad (7.4.8)$$

Applying the renormalization group technique, we obtain equations for the amplitudes Γ_{1i} and Γ_{ij} , as well as for the temperatures τ_i. If all the fields φ_i are found to be strongly fluctuating, these equations can be written in the form

$$- \Gamma'_{1i} = (n+8) \Gamma_{1i}^2 + n \sum_{j \neq i} \Gamma_{4ij}^2 \, ,$$

$$- \Gamma'_{4ij} = (n+2)(\Gamma_{1i} + \Gamma_{1j})\Gamma_{4ij} + 4\Gamma_{4ij}^2 + n \sum_{k \neq i,j} \Gamma_{4ik}\Gamma_{4kj} \, ,$$

$$- \tau'_i = (n+2)\tau_i \Gamma_{1i} + n \sum_{i \neq j}^{N} \Gamma_{4ij}\tau_j \, . \qquad (7.4.9)$$

The variable in the renormalization group method is defined by

$$x = \frac{2}{\varepsilon(\Lambda^2)^{\varepsilon/2}} \left\{ \left(\frac{\Lambda^2}{\max[\lambda^2(\ldots,\tau_i,\ldots),\ldots,\varphi_i^2,\ldots]} \right)^{\varepsilon/2} - 1 \right\} ,$$

$$\varepsilon \to 0 \, , \qquad (7.4.10)$$

where $\lambda^2(\ldots,\tau_i,\ldots)$ is the scaling parameter as a function of $(\ldots,\tau_i,\ldots)$ and n is the dimensionality of the field φ_i (in the present case, n = 2).

Since each harmonic has nine equivalent vectors corresponding to it, N = 9m, and we can introduce the following initial conditions:

$$1 \leq i \, , \, j \leq 9 \, , \quad \tau_{i0} = \tau_{10} \, , \quad \Gamma_{1i0} = \Gamma_{4ij0} = \Gamma_{10} \, ,$$

$$9(m'-1) + 1 \leq i \, , \quad j \leq 9m' \, , \quad \tau_{i0} = \tau_{m'0} \, , \quad \Gamma_{1i0} = \Gamma_{4ij0} = \Gamma_{m'0} \, ,$$

$$1 \leq i \leq 9, \quad 9(m'-1)+1 \leq j \leq 9m' \, , \quad \Gamma_{4ij0} = \Gamma_{41m'0} = \Gamma_{410} \, ,$$

$$9(m'-1)+1 \leq i \leq 9m' \;,\; 9(m''-1)+1 \leq j \leq 9m'' \;,\; 2 \leq m',m'' \leq m,$$

$$\Gamma_{4ij0} = \Gamma_{4m'm''0} = \Gamma_{420} \;,\; \tau_{10} < \tau_{20} < \ldots < \tau_{m0} \;. \tag{7.4.11}$$

As we pass from the paramagnetic phase to $T_C = T_0$, the first harmonic will fluctuate most strongly, and the RG equations are

$$-\Gamma_1' = 26\,\Gamma_1^2 \;,\quad -\Gamma_{4ij}' = 18\,\Gamma_{41i}\,\Gamma_{41j} \;,$$

$$-\Gamma_{1i}' = 18\Gamma_{41i}^2 \;,\quad -\tau_1' = 20\,\tau_1\Gamma_1 \;,$$

$$-\Gamma_{41i}' = 20\,\Gamma_1\Gamma_{41i} \;,\quad -\tau_1' = 18\,\tau_1\,\Gamma_{41i} \;. \tag{7.4.12}$$

These equations can be easily solved:

$$\Gamma_1 = \frac{\Gamma_{10}}{1 + 26\Gamma_{10}x} \;,\quad \Gamma_{41i} = \frac{\Gamma_{41i0}}{(1 + 26\Gamma_{10}x)^{10/13}} \;,$$

$$\Gamma_{1i} = \Gamma_{1i0} + \frac{9\,\Gamma_{41i0}^2}{7\Gamma_{10}}[(1 + 26\Gamma_{10}x)^{-7/13} - 1] \;,$$

$$\Gamma_{4ij} = \Gamma_{4ij0} + \frac{9\Gamma_{41i0}\Gamma_{41j0}}{7\Gamma_{10}}[(1 + 26\Gamma_{10}x)^{-7/13} - 1] \;,$$

$$\tau_1 = \frac{\tau_{10}}{(1 + 26\Gamma_{10}x)^{10/13}} \;,$$

$$\tau_i = \tau_{i0} + \frac{9\tau_{i0}\Gamma_{41i0}}{7\,\Gamma_{10}}[(1 + 26\,\Gamma_{10}x)^{-7/13} - 1] \;. \tag{7.4.13}$$

As $x \to \infty$, $\Gamma_1 \to 0$, $\Gamma_{41i} \to 0$, $\Gamma_{1i} \to \Gamma_{1i0} - 9\Gamma_{41i0}^2/7\Gamma_{10}$,

$$\Gamma_{4ij} \to \Gamma_{4ij0} - 9\Gamma_{41i0}\,\Gamma_{41j0}/7\,\Gamma_{10} \;,\quad \tau_1 \to \tau_{i0} - 9\tau_{10}\Gamma_{41i0}/7\Gamma_{10}.$$

If $(\Gamma_{1i0} - 9\Gamma_{41i0}/7\Gamma_{10}) > 0$, $(\Gamma_{4ij0} - 9\Gamma_{41i0}\Gamma_{41j0}/7\Gamma_{10}) > 0$,

$(\tau_{i0} - 9\tau_{10}\Gamma_{4i0}/7\Gamma_{10}) > 0$,

the system will first experience a second-order phase transition to a state with a longwave magnetic order with critical susceptibility indices $\gamma = 1 + (5/13)\varepsilon + \ldots$, and correlation radii $\nu = \frac{1}{2} + (5/26)\varepsilon + \ldots$. After a transition to the ordered phase, the temperatures of the remaining harmonics receive increments

$$\Delta\tau_i(\varphi_{10}^2 + \ldots + \varphi_{90}^2) = (1/2)\Gamma_{41i}(\ldots, \varphi_{k0}, \ldots) \sum_{k=1}^{9} \varphi_{k0}^2 . \tag{7.4.14}$$

It follows from the conditions (7.4.11) that the temperature inequality is preserved. The second harmonic is the next to form a condensate. This is followed by the third harmonic, and so on. Hence, the temperature dependence of the magnetic susceptibility acquires a number of peaks which are smoothed out as we approach the ground state because the temperature range in which corresponding phase can exist becomes narrower and approaches the region of strongly developed thermodynamic fluctuations . $\Delta T_{i,i+1} \to \max(\tau_i T_i, \tau_{i+1} T_{i+1})$ (Fig. 7.8). For the initial peaks to remain distinct, the following condition must be satisfied:

$$\varphi_{i0}^2(T_{i+1}) < (\Delta T_{i,i+1}/T_{i+1}) . \tag{7.4.15}$$

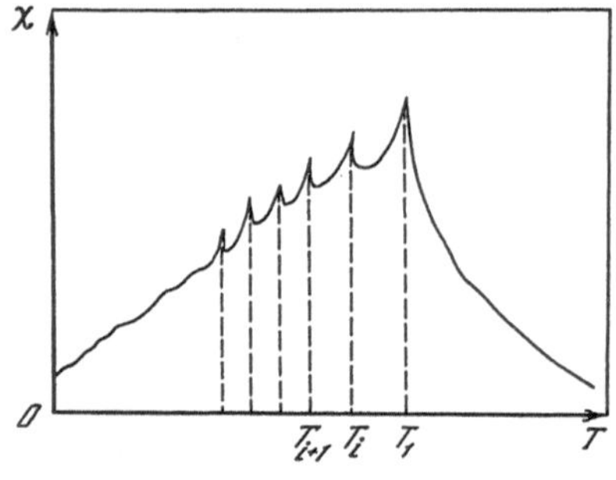

Fig. 7.8. Temperature dependence of the magnetic susceptibility of a new class of spin glasses. The peaks correspond to second-order phase transtions. The density of dislocations introduced in the crystal is $n_d \lesssim 10^7 \mathrm{cm}^{-2}$

If $\Gamma_{1i0} - 9\Gamma_{41i0}/7\Gamma_{10} < 0$, the system will first undergo a first-order phase transition to the state with the second harmonic. After this, the third harmonic will be condensed through a first-order phase transition, and so on. The first harmonic will be the last to condense, as a result of a second-order phase transition. This is reflected in the $\chi(T)$ dependence (Fig. 7.9).

The above analysis was carried out for the case when the density of introduced dislocations is $n_d \lesssim 10^7$ cm^{-2}. If the dislocation density is anomalously high, e.g., $n_d \gg 10^7$ cm^{-2}, and the dislocations are introduced and interlocked at random (as is the case for

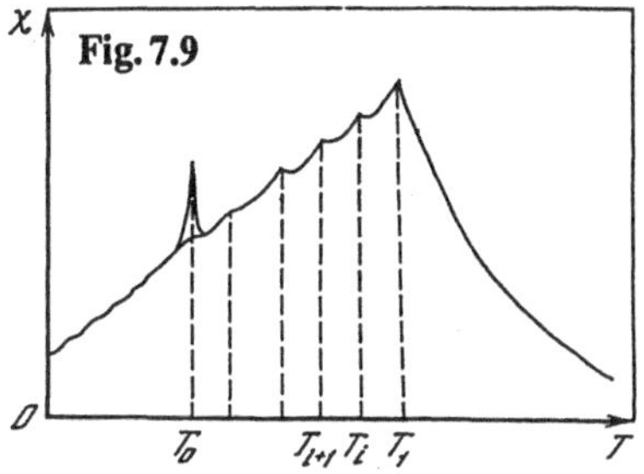

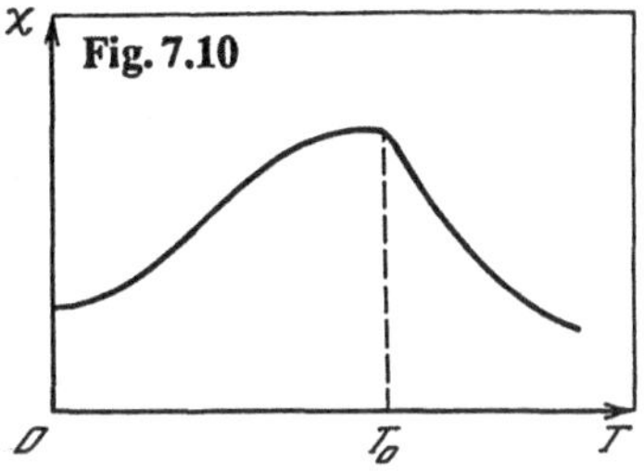

Fig. 7.9. Temperature dependence of the magnetic susceptibility of a new class of spin glasses. The peaks correspond to first-order phase transitions. The peak at T_0 may correspond to a second-order phase transition; $n_d < 10^7\text{cm}^{-2}$

Fig. 7.10. Temperature dependence of $\chi(T)$ for the case when the density of dislocations introduced in the crystal is $n_d \gg 10^7\text{cm}^{-2}$

an annealed sample), N increases sharply and may become on the order of the number of introduced dislocations. In this case, the space symmetry group coincides with the impurity spin glass group G_p = SO(3) × I (I stands for inversion). The peaks in the temperature dependence are washed out and the dependence becomes smooth (Fig. 7.10). This means that in such a spin glass, fluctuations exist at low temperatures near the ground state.

7.5 Superconducting Phases in Spin Glasses

It was shown in Chaps. 4 and 5 that a high-temperature superconducting phase may be formed in compounds of rare-earth metals. The conditions necessary for the emergence of such a phase include the existence of antiferromagnetic order and orthorhombic symmetry of the crystal lattice. It was mentioned in Sect. 7.3 that the introduction of line defects into a crystal with a complex magnetic structure leads to a change in the spatial symmetry of the system and to the appearance of antiferromagnetic order on the defect line. Hence, by introducing line defects of a certain type, e.g., dislocations with a high density $n_d \simeq 10^7 - 10^8\ \text{cm}^{-2}$, we can obtain a class of spin glasses in which antiferromagnetic order will dominate. An antiferromagnetic structure can be presented with the help of nine vectors:

$$S_{1i\alpha}(x_\alpha) = S_{1i\alpha 0}e^{i\pi x_\alpha}, \quad i, \alpha = 1,2,3. \qquad (7.5.1)$$

In this approach the superconducting component can be taken into consideration by introducing the following invariants in addition to the purely magnetic invariants (Chap. 4):

$$\Delta\Delta^* \ , \ (\Delta\Delta^*)^2 \ , \quad (\Delta\Delta^*)[(\mathbf{S}_{1i\alpha}\mathbf{S}^*_{1i\alpha}) + (\mathbf{S}_j\mathbf{S}^*_j)] \ . \tag{7.5.2}$$

In this case, the expression for the free energy is

$$\begin{aligned}
F = {} & \frac{1}{2}\sum_{1\le\ell\le 9}\tau_{1\ell}s_{0\ell}^2 + \frac{1}{2}\sum_{10\le\ell\le 9m}\tau_{2\ell}(s_\ell^{+2} + s_\ell^{-2}) + \frac{1}{2}\tau\delta|\Delta|^2 \\
& + \frac{1}{8}\sum_{1\le\ell\le 9}\Gamma_{1\ell}s_{0\ell}^4 + \frac{1}{8}\sum_{10\le\ell\le 9m}\Gamma_{2\ell}(s_\ell^{+2} + s_\ell^{-2})^2 + \frac{1}{8}\Gamma_\delta|\Delta|^4 \\
& + \frac{1}{8}\sum_{1\le\ell,\ell'\le 9}\Gamma_{4\ell\ell'}s_{0\ell}^2 s_{0\ell'}^2 + \frac{1}{4}\sum_{\substack{1\le\ell\le 9\\ 10\le\ell'\le 9m}}\Gamma_{5\ell\ell'}s_{\ell 0}^2(s_\ell^{+2}+s_\ell^{-2}) \\
& + \frac{1}{8}\sum_{10\le\ell',\ell\le 9m}\Gamma_{6\ell\ell'}(s_\ell^{+2}+s_\ell^{-2})(s_{\ell'}^{+2}+s_{\ell'}^{-2}) + \frac{1}{4}\sum_{1\le\ell\le 9}\Gamma_{7\ell}|\Delta|^2 s_{0\ell}^2 \\
& + \frac{1}{4}\sum_{10\le\ell'\le 9m}\Gamma_{8\ell}|\Delta|^2(s_\ell^{+2}+s_\ell^{-2}) \ .
\end{aligned} \tag{7.5.3}$$

The RG equations for the general case can be written

$$-\Gamma'_{1\ell} = (n_1+8)\Gamma_{1\ell}^2 + n_1\sum_{1\le\ell'\le 9}\Gamma_{4\ell\ell'}^2 + n_2\sum_{10\le\ell'\le 9m}\Gamma_{5\ell\ell'}^2 + \Gamma_{7\ell}^2 \ ,$$

$$-\Gamma'_{2\ell} = (n_2+8)\Gamma_{2\ell}^2 + n_2\sum_{10\le\ell'\le 9m}\Gamma_{6\ell\ell'}^2 + n_1\sum_{1\le\ell'\le 9}\Gamma_{5\ell\ell'}^2 + \Gamma_{8\ell}^2 \ ,$$

$$-\Gamma'_\delta = 9\Gamma_\delta^2 + n_1\sum_{1\le\ell\le 9}\Gamma_{7\ell}^2 + n_2\sum_{10\le\ell\le 9m}\Gamma_{8\ell}^2 \ ,$$

$$\begin{aligned}
-\Gamma'_{4\ell\ell'} = {} & (n_1+2)(\Gamma_{1\ell}\Gamma_{4\ell\ell'}+\Gamma_{1\ell'}\Gamma_{4\ell\ell'}) + 4\Gamma_{4\ell\ell'}^2 + n_1\sum_{1\le\ell''\le 9}\Gamma_{4\ell\ell''}\Gamma_{4\ell''\ell'} + \\
& + n_2\sum_{10\le\ell''\le 9m}\Gamma_{5\ell\ell''}\Gamma_{5\ell''\ell'} + \Gamma_{7\ell}\Gamma_{7\ell'}
\end{aligned}$$

$$\begin{aligned}
-\Gamma'_{5\ell\ell'} = {} & (n_1+2)\Gamma_{1\ell}\Gamma_{5\ell\ell'} + (n_2+2)\Gamma_{2\ell}\Gamma_{5\ell\ell'} + 4\Gamma_{\ell\ell'}^2 \\
& + n_1\sum_{1\le\ell''\le 9}\Gamma_{4\ell\ell''}\Gamma_{5\ell\ell''\ell'} + n_2\sum_{10\le\ell''\le 9m}\Gamma_{5\ell\ell''}\Gamma_{5\ell''\ell'} + \Gamma_{7\ell}\Gamma_{7\ell'};
\end{aligned}$$

$$\begin{aligned}
-\Gamma'_{6\ell\ell''} = {} & (n_2+2)(\Gamma_{2\ell}+\Gamma_{2\ell'})\Gamma_{6\ell\ell'} + 4\Gamma_{6\ell\ell'}^2 + n_1\sum_{1\le\ell'\le 9}\Gamma_{5\ell\ell''}\Gamma_{5\ell\ell''\ell'} \\
& + n_2\sum_{10\le\ell''\le 9m}\Gamma_{6\ell\ell''}\Gamma_{6\ell''\ell'} + \Gamma_{8\ell}\Gamma_{8\ell'} \ ,
\end{aligned}$$

$$-\Gamma'_{7\ell} = 3\Gamma_\delta\Gamma_{7\ell} + (n_1+2)\Gamma_{1\ell}\Gamma_{7\ell} + 4\Gamma^2_{7\ell} + n_1 \sum_{1\le\ell''\le 9} \Gamma_{4\ell\ell''}\Gamma_{7\ell''}$$

$$+ n_2 \sum_{10\le\ell''\le 9m} \Gamma_{5\ell\ell''}\Gamma_{8\ell''} ,$$

$$-\Gamma'_{8\ell} = 3\Gamma_\delta\Gamma_{8\ell} + (n_2+2)\Gamma_{2\ell}\Gamma_{8\ell} + 4\Gamma^2_{8\ell} + n_1 \sum_{1\le\ell''\le 9} \Gamma_{5\ell\ell''}\Gamma_{7\ell''}$$

$$+ n_2 \sum_{10\le\ell'\le 9m} \Gamma_{6\ell\ell'}\Gamma_{8\ell'} ,$$

$$-\tau'_{1\ell} = (n_1+2)\tau_{1\ell}\Gamma_{1\ell} + n_1 \sum_{1\le\ell'\le 9} \tau_{1\ell'}\Gamma_{4\ell\ell'}$$

$$+ n_2 \sum_{10\le\ell'\le 9m} \tau_{2\ell'}\Gamma_{5\ell\ell'} + \tau_\delta\Gamma_{7\ell} ,$$

$$-\tau'_{2\ell} = (n_2+2)\tau_{2\ell}\Gamma_{2\ell} + n_2 \sum_{10\le\ell'\le 9m} \tau_{2\ell}\Gamma_{6\ell\ell'}$$

$$+ n_1 \sum_{1\le\ell'\le 9} \tau_{1\ell'}\Gamma_{5\ell\ell'} + \tau_\delta\Gamma_{8\ell} ,$$

$$-\tau'_\delta = 3\tau_\delta\Gamma_\delta + n_1 \sum_{1\le\ell\le 9} \tau_\ell\Gamma_{7\ell} + n_2 \sum_{10\le\ell\le 9m} \tau_\ell\Gamma_{8\ell} , \qquad (7.5.4)$$

where

$$n_1 = 1, \quad n_2 = 2 ,$$

$$x = \frac{2}{\varepsilon(\Lambda^2)^{\varepsilon/2}} \left\{ \left(\frac{\Lambda^2}{\max[\lambda^2(\tau_\delta,\tau_{1\ell},\ldots),\ldots,\varphi^2_{\ell 0},\ldots,|\Delta_0|^2]} \right)^{\varepsilon/2} - 1 \right\} ,$$

$$\varepsilon \to 0 ,$$

$$\varphi^2_{\ell 0} \to s^2_{0\ell} , \quad s^{+2}_{\ell 0} + s^{-2}_{\ell 0} .$$

We introduce the initial conditions

$$1\le\ell ,\ \ell' \le 9,\ \tau_{1\ell 0} = \tau_{10} ,\ \Gamma_{1\ell 0} = \Gamma_{10} ,\ \Gamma_{4\ell\ell' 0} = \Gamma_{10} ,\ \Gamma_{7\ell 0} = \Gamma_{710} ,$$

$$9(m'-1)+1 \le \ell,\ \ell' \le 9m',\ \tau_{2\ell 0} = \tau_{m'0} ,\ \Gamma_{2\ell 0} = \Gamma_{m'0} ,$$

$$\Gamma_{6\ell\ell' 0} = \Gamma_{m'0} ,\ \Gamma_{8\ell 0} = \Gamma_{810} ,$$

$$1\le\ell\le 9 ,\ 9(m'-1)+1\le \ell' \le 9m' ,\ \Gamma_{5\ell\ell' 0} = \Gamma_{510} ,$$

$$9(m'-1)+1\le\ell \le 9m' ,\ 9(m''-1)+1 \le\ell' \le 9m'' ,$$

$$2\le m' ,\ m''\le m ,\ \Gamma_{6\ell\ell' 0} = \Gamma_{610} . \qquad (7.5.5)$$

If the antiferromagnetic component has higher energy than any of the remaining harmonics, it will fluctuate most strongly at low temperatures. Hence the initial temperature values must satisfy the condition

$$\tau_{10} \leq \min(\tau_{\delta 0}, \tau_{20}, \dots, \tau_{m0}) \ . \qquad (7.5.6)$$

In this case, (7.5.4) can be easily solved and their solution shows that if

$$\Gamma_{\delta 0} < \Gamma_{2\ell 0} < \Gamma_{6\ell\ell' 0}\ , \ \Gamma_{5\ell\ell' 0}\ , \ \Gamma_{7\ell 0}\ , \ \Gamma_{8\ell 0}\ ,$$

$$\Gamma_{\delta 0} - 9\Gamma_{7\ell 0}^2/5\Gamma_{10} > 0\ , \qquad (7.5.7)$$

the system will first undergo a second-order phase transition to the state with antiferromagnetic order. The complex antiferromagnetic order will prevent the emergence of superconductivity (Chap. 4). Hence the superconducting phase, if it appears at all, will be metastable. For

$$\Gamma_{\delta 0} - 9\Gamma_{7\ell 0}/5\Gamma_{10} < 0 \qquad (7.5.8)$$

the system may first undergo a first-order phase transition to the superconducting state, and a superconducting phase may be observed in this case (Fig. 7.11). However, the presence of dislocations and a complex antiferromagnetic order in the system may change the topology of the Fermi surface in such a way that the electron-phonon interaction is not sufficiently effective for the formation of bound electron states. The Fermi surface may have singular points, as well as holes and pockets since the dislocation potential is a complex function of coordinates [7.26,27]. This inevitably leads to a de-

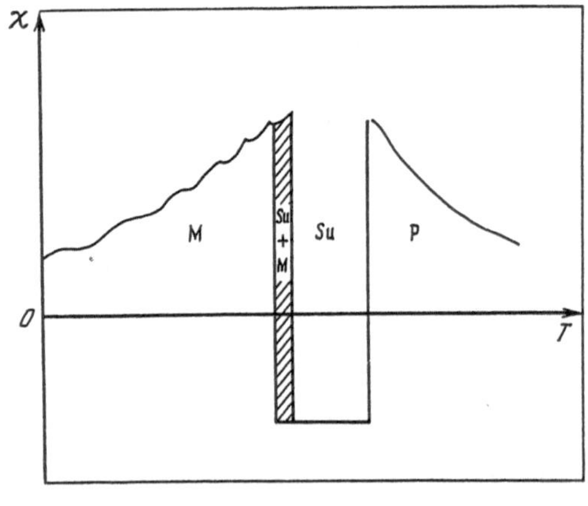

Fig. 7.11. Temperature dependence of the magnetic susceptibility of a new type of spin glass containing a superconducting phase

crease in the density of states on the Fermi surface. Hence such a superconducting phase will be a low-temperature phase $1 \lesssim T_c \lesssim 10$ K [7.26]. A lowering of temperature leads to a magnetic order and to the disappearance of superconductivity (Chap. 4). Hence reentrant superconductivity will appear in spin glasses in the same way as in rare-earth compounds considered in Chaps. 4 and 5.

7.6 Peculiarities of Spin-Wave Dynamics in Spin Glasses

At low temperatures, the magnetic structure of a spin glass near the ground state is a combination of a large number of disclinations. Hence, it seems clear at first sight that a short-range magnetic order must exist in the system. However, it was shown in the preceding sections that the magnetic structure of the ground state of a spin glass is formed as a result of superposition of a large number of spin-density waves, each of which has long-range order. The effective anisotropy of the system decreases with temperature, and at low temperatures, it is small in comparison with the exchange interaction. Hence the temperatures τ_i of spin-density waves that are not in a condensed state are nearly equal. This means that near the ground state of a spin glass, long-wave fluctuations still exist in the spin-density waves that have not yet been transformed into condensates.

A system of spin disclinations can be described as follows. Let us assume that a certain effective magnetic field corresponds to this system (Sect. 6.1). We isolate a differentially small macroscopic volume into the system of disclinations. It follows from the continuum theory of dislocations [7.27] that each such volume has, on the average, an identical distribution of disclinations. However, since the total macroscopic magnetic moment of the system is equal to zero, each such distribution must have its own orientation in space. Thus, a local system of coordinates must be attached to each differentially small volume. In this case each volume is transformed not only in the coordinate space, but also in the space of three-dimensional rotations (spin space). Consequently, the effective field potential A_ν^α of spin disclinations must be a tensor. However, in analogy with the discussion in Sect. 6.1, it can be assumed that A_ν^α is just the Yang-Mills field potential [7.28]. Spin waves are elementary excitations of the system of magnetic disclinations and are space-time oscillations of the nonequilibrium magnetization of the system. We juxtapose the magnetization of the system to the time component of the Yang-Mills field A_0^α, i.e., $m^\alpha = \chi A_0^\alpha$. In this case, the dispersion in the system may be described with the help of fields A_ν^α, where ν, α = 1, 2, 3.

The spin glass considered here has a nonzero magnetic anisotropy even at zero temperature. Hence its ground state is nondegenerate relative to the transformations of the rotation group O(3). The energy of magnetic anisotropy must depend on the rotation vector in the spin space

$$E_a = a(\mathbf{\Omega})^2 + b(\mathbf{\Omega})^4 + \dots \quad (7.6.1)$$

Using this equation, we can write the Hamiltonian for the excited state of the spin glass in the absence of an external magnetic field:

$$\mathbb{H} = \frac{1}{2}\int d\mathbf{x}\left\{\chi(A_o)^2 + \alpha_s(A_\nu)^2 + 2a(\mathbf{\Omega})^2 + \sum_{j=M}^{N}[\tau_j(\varphi_j^{+2} + \varphi_j^{-2}) + \beta(D_\nu\boldsymbol{\varphi}_j^+)^2 + \beta(D_\nu\boldsymbol{\varphi}_j^-)^2]\right\}, \quad (7.6.2)$$

where $\varphi_j^\pm$ is the fluctuating field of the spin-density waves that have not yet been condensed and $D_\nu = \boldsymbol{\nabla}_\nu - ge^{\alpha\beta\gamma}A_\nu^\beta$ is the covariant derivative. The constants χ and α_s correspond to the susceptibility and the exchange rigidity of the system [7.26].

With the help of (7.6.2), we can easily write the equations of motion for the fields $\mathbf{A}_o$, $\mathbf{A}_\nu$, $\mathbf{\Omega}$, and $\boldsymbol{\varphi}_i^\pm$ correct to nonlinear order:

$$\frac{\partial \mathbf{A}_o}{\partial t} = \{\mathbb{H}A_o\} = \left(\frac{\alpha_s}{\chi}\right)\boldsymbol{\nabla}_\nu\mathbf{A}_\nu - \left(\frac{2a}{\chi}\right)\mathbf{\Omega} - \beta g\boldsymbol{\nabla}_\nu[\boldsymbol{\varphi}_j^+\boldsymbol{\nabla}_\nu\boldsymbol{\varphi}_j^+] - \beta g\boldsymbol{\nabla}_\nu[\boldsymbol{\varphi}_j^-\boldsymbol{\nabla}_\nu\boldsymbol{\varphi}_j^-], \quad (7.6.3)$$

$$\frac{\partial\mathbf{\Omega}}{\partial t} = \{\mathbb{H}\mathbf{\Omega}\} = \mathbf{A}_o, \quad (7.6.4)$$

$$\frac{\partial \mathbf{A}_\nu}{\partial t} = \{\mathbb{H}A_\nu\} = \boldsymbol{\nabla}_\nu\mathbf{A}_o - [\mathbf{A}_\nu\mathbf{A}_o], \quad (7.6.5)$$

$$\frac{\partial\boldsymbol{\varphi}_j^\pm}{\partial t} = \{\mathbb{H}\boldsymbol{\varphi}_j^\pm\} = -\tau_j\boldsymbol{\varphi}_j^\pm + \beta\boldsymbol{\Delta}\boldsymbol{\varphi}_j^\pm - g\beta\boldsymbol{\nabla}_\nu[\mathbf{A}_\nu\boldsymbol{\varphi}_j^\pm] - g\beta[\mathbf{A}_\nu\boldsymbol{\nabla}_\nu\boldsymbol{\varphi}_j^\pm]. \quad (7.6.6)$$

Solving (7.6.3-6), we can show that the spectrum of the system has the form

$$\omega_k = [(2a/\chi) + (\alpha_s/\chi)k^2]^{1/2}, \quad \omega_{jk}^{(o)} = i(\tau_j + \beta k^2),$$

$$\omega_{jk}^{(\ell)} = \omega_{jk}^{(o)} + \ell\omega_k. \quad (7.6.7)$$

The first mode in this formula corresponds to longwave oscillations of the disclination density which is described by the Yang-Mills fields [7.28]. The modes $\omega_{jk}^{(o)}$ are diffusive and correspond to long-wave fluctuations of spin-density waves. Finally, the modes $\omega_{jk}^{(\ell)}$ are due to the nonlinear interaction of the Yang-Mills fields with fluctuations.

In conclusion, it should be observed that the theory of Yang-Mills fields as applied to spin glasses, allows us to show that point singularities (topological solitons) of the starfish magnetic monopole type [7.26] can emerge in them. However, such singularities will be unstable since they are destroyed by the magnetic fluctuations existing in the system.

By using the idea of phase lamination, we have constructed a model for spin glasses which allows us to predict a wide range of new spin glass types. These include spin glasses with a superconducting phase which, however, is a low-temperature phase ($1 \lesssim T_c \lesssim 10$ K). A notable feature of this model is that it leads to the construction of a unified theory of spin glass, which can explain the appearance and disappearance of a cusp on the temperature dependence $\chi(T)$ of the magnetic susceptibility, and also predict new modes in the spectrum of spin waves.

It is well known that there exists a class of spin glasses (Sect. 7.2) in which the phase lamination effect does not take place and the $\chi(T)$ dependence is explained within the framework of the impurity model. It can be shown that the impurity model is a special case of our model. Indeed, let us consider a spin glass obtained by introducing ring dislocations in a magnetically ordered crystal. The impurity limit corresponds to the case when the dislocation density $n_d > 10^8$ cm^{-2}, and their radius $R_d \to 0$. In this case, the space group $G_p = SO(3) \times I$, while the microscopic Hamiltonian is defined by the expression

$$\mathbb{H} = (-1/2) \sum_{\mathbf{r},\mathbf{r}'} \langle (J(\mathbf{r} - \mathbf{r}') \rangle_{(D^\nu)} \mathbf{S}_{\mathbf{r}} \mathbf{S}_{\mathbf{r}'} \quad . \tag{7.6.8}$$

The exchange integral averaged over the dislocation distribution is given by

$$\langle J(\mathbf{R}) \rangle_{(D^\nu)} = \frac{\eta^4 R_d^4 b^2 n_d}{3V} \frac{\sin(q_o R) - q_o R \cos(q_o R)}{R^3} + O(R \to \infty) \ , \tag{7.6.9}$$

where $R_d \to 0$; $\eta = k/(\lambda_o + k)$; λ_o and k are the elastic moduli, b is the modulus of the Burgers vector, and q_o is the dislocation wave vector. It follows from (7.6.9) that we have obtained an impurity model with a modified exchange interaction. Consequently, this limiting transition emphasizes the universal nature of our model.

8. High-Temperature Superconducting Ceramics

8.1 High-Temperature Superconducting Oxides

The high-temperature superconductivity of complex compounds of rare-earth metals and electromagnetic materials, viz., oxides, was predicted in Chap. 5 and in [8.1,2]. The predictions based on our theory were fully supported by **Bednorz** and **Müller** and by **Chu**'s group [8.3-6], who experimentally discovered complex superconducting oxides based on rare-earth metals, e.g., La-Ba-Cu-O ($T_c \simeq 30$ K), La-Sr-Cu-O ($T_c \simeq 40$ K), and Y-Ba-Cu-O ($T_c = 92$ K). Of these, $YBa_2Cu_3O_{7-y}$ ($T_c = 92$-98 K) has been most thoroughly investigated, and is being actively studied today [see, for example, 8.7]. Some of the recently discovered compounds are: Tl-Ba-Cu-O, $T_c = 90$ K [8.8], Bi-Al-Ca-Sr-Cu-O, $T_c = 114$ K [8.9], and Ba-K-Bi-O,[1)] $T_c \simeq 30$ K. The crystal structure of new high-temperature superconductors (HTS) is basically of perovskite type, and their unit cell has, as a rule, orthorhombic or tetragonal symmetry, which is in accord with the predictions made in Chap. 5. Figure 8.1 shows the crystal structure of the high-temperature superconducting ceramic $YBa_2Cu_3O_{7-y}$ and also of other HTS compounds while Fig. 8.2 shows the temperature dependence of its electrical resistivity. Above the superconducting transition temperature, the temperature dependence of the electrical resistivity is close to linear, a characteristic feature of metallic systems. The only difference is that the resistivity in the normal phase is three orders of magnitude higher than in good metals. This means that the effective electron-phonon interaction parameter is much stronger than for low-temperature superconductors. Experimental investigations have shown [8.10] that the system $YBa_2Cu_3O_{7-y}$, as well as the other HTS ceramics, is quite sensitive to the oxygen concentration. For example, with increasing oxygen deficiency y, this material loses its superconducting properties and the critical temperature T_c becomes equal to zero for $y = 0.5$. Moreover, the properties of the new superconductors are quite different from those of low-temperature superconductors. This difference is manifested most strikingly in the

1) See R.J. Cava, B. Batlogg, J.J. Krajewski et al.: Nature 332, 814 (1988)

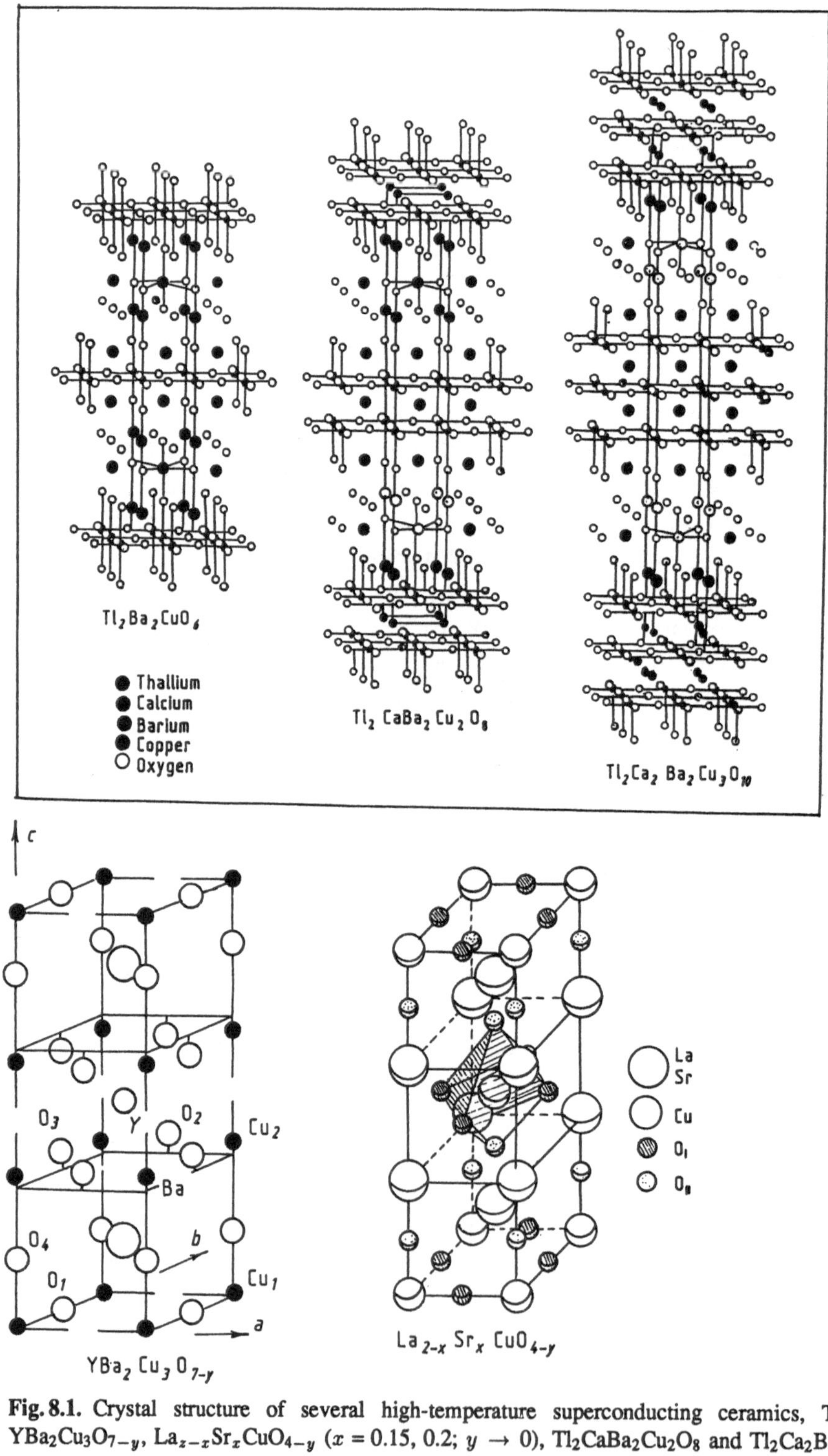

Fig. 8.1. Crystal structure of several high-temperature superconducting ceramics, $Tl_2Ba_2CuO_6$, $YBa_2Cu_3O_{7-y}$, $La_{z-x}Sr_xCuO_{4-y}$ ($x = 0.15, 0.2$; $y \to 0$), $Tl_2CaBa_2Cu_2O_8$ and $Tl_2Ca_2Ba_2Cu_3O_{10}$

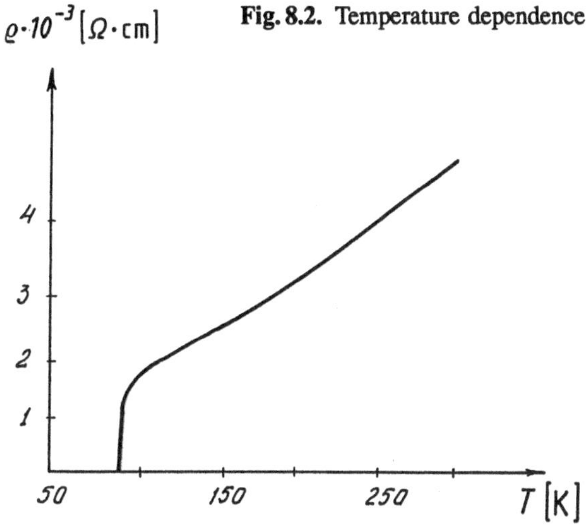

Fig. 8.2. Temperature dependence of the electrical resistivity of $YBa_2Cu_3O_{7-y}$

unusual isotope effect in HTS ceramics. It is well known from BCS theory that the criticial temperature $T_c \sim M^{-\frac{1}{2}}$, since $T_c = (2\gamma/\pi)\langle\omega_D\rangle \exp[-1/(\lambda-\mu^*)]$ (μ^* is the effective Coulomb repulsion parameter), and the mean Debye energy $\langle\omega_D\rangle \sim M^{-\frac{1}{2}}$. The latter relation is satisfied quite well for low-temperature superconductors and confirms the phonon mechanism of superconductivity. For superconductors of the new type $T_c \simeq M^{-\alpha}$, where α varies between 0.2 and 0.1 as the isotope O^{16} is replaced by O^{18} [8.11-13]. This effect cannot be explained within the framework of the BCS theory. The new superconductors also reveal strong spin and superconductivity fluctuations [8.14-18], which conforms to the predictions of our theory. The final confirmation of our predictions is the low nonexponential temperature dependence of the specific heat of the electron system in the superconducting phase. The existence of the isotope effect in HTS ceramics rules out the possibility of using a nonphonon superconductivity mechanism for explaining the properties of high-temperature superconductors. It follows from the experimental results that the BCS theory, which provides a satisfactory description of the properties for most low-temperature superconductors, is also inapplicable in this case. The theory of high-temperature superconductivity must take into consideration the electron-phonon interaction. At the same time, it must take into account the interaction facilitating the attraction between electrons which form pairs by exchanging virtual phonons. It was shown in Chap. 5 that the exchange interaction between electrons is an interaction of this type. The exchange interaction can effectively enhance the attraction between electrons through phonons only if electron spins fluctuate. Such fluctuations do exist in all superconductors. However, while these fluctuations are small in ordinary low-temperature superconductors and may not significantly affect the electron interaction, the critical temperature T_c of high-temperature superconductors will be determined by the exchange enhancement effect of electron-phonon interaction due to fluctuations. The fluctuation theory of high-temperature superconductivity predicted a new class of high-temperature superconductors and the values of critical temperature $T_c \gtrsim 100$ K. A major problem in the physics of the superconducting state at present is not only to explain the properties of new superconductors, but also to synthesize compounds with room-temperature superconductivity. In order to find methods of synthesizing such compounds, we must determine the system parameters on which the enhancement of electron-phonon interaction and the increase in T_c depend. This can be done by finding the spectrum of quasi-particles exchanged by electron pairs which form the superconducting Bose condensate.

8.2 Spectrum of Quasi-Particles Exchanged by Electrons in HTS Ceramics

It follows from the fluctuation theory of high-temperature superconductivity (Chap. 5) that in contrast to the BCS theory where the electrons exchange virtual phonons, the electrons must exchange quasi-particles which are quanta of the coupled oscillations of the crystal lattice ions with the spin fluctuations of the conduction electrons. The spectrum of such quasi-particles was first considered in [8.19].

Let us consider the Hamiltonian of a superconducting system taking into account the exchange interaction of conduction electrons:

$$\begin{aligned}
\mathbb{H} = \int d\mathbf{x} \Big\{ & \tilde{\psi}^+_{\alpha'}(\mathbf{x})\Big(\frac{\tilde{p}^2}{2\mu_e} - \hat{\mu}\Big)\hat{\psi}_{\alpha'}(\mathbf{x}) + \frac{1}{2}[\hat{\psi}^+_{\alpha'}(\mathbf{x})\hat{\Delta}_{\alpha'\gamma'}(\mathbf{x})\hat{\psi}^{+t}_{\gamma'}(\mathbf{x}) \\
& + \hat{\psi}^t_{\alpha'}\Delta^+_{\alpha'\gamma'}(\mathbf{x})\hat{\psi}_{\gamma'}(\mathbf{x})]\Big\} + \frac{1}{2}\int d\mathbf{x}\, d\mathbf{x}' \, [\hat{\psi}^+_{\alpha'}(\mathbf{x})\hat{\psi}^+_{\beta'}(\mathbf{x}') \\
& \times V(\mathbf{x} - \mathbf{x}')\hat{\psi}_{\beta'}(\mathbf{x}')\hat{\psi}_{\alpha'}(\mathbf{x}) \\
& - J(\mathbf{x} - \mathbf{x}')\hat{\psi}^+_{\alpha'}(\mathbf{x})\tau^{\alpha}_{\alpha'\gamma'}\hat{\psi}_{\gamma'}(\mathbf{x})\, \hat{\psi}^+_{\alpha''}(\mathbf{x}')\tau^{\alpha}_{\alpha''\gamma''}\hat{\psi}_{\gamma''}(\mathbf{x}')] \\
& + \int d\mathbf{x}\Big(\frac{p^2_{\gamma}}{2M_i} + \frac{1}{2}\lambda_{ik\ell m}u_{ik}u_{\ell m}\Big) + g_{ph}\int d\mathbf{x}\, u_{ii}\hat{\psi}^+_{\alpha'}(\mathbf{x})\hat{\psi}_{\alpha'}(\mathbf{x}).
\end{aligned} \tag{8.2.1}$$

In this formula, $\hat{\mu}$ is the chemical potential of the system, $V(\mathbf{x} - \mathbf{x}')$ is the Coulomb potential, $\hat{\tau}^{\alpha}$ is the Pauli matrix ($\alpha = 1,2,3$), $\hat{\Delta}_{\alpha'\gamma'}(\mathbf{x})$ and $\hat{\Delta}^{+}_{\alpha'\gamma'}(\mathbf{x})$ are the functions determining the energy gap in the spectrum of conduction electrons, and $\hat{\psi}^{+}_{\alpha'}(\mathbf{x})$ and $\hat{\psi}_{\alpha'}(\mathbf{x})$ are operators of creation and annihilation of electrons. In $\mathbf{k}$-representation, the electron operators have the following form:

$$\hat{\psi}_{\alpha'}(\mathbf{x}) = \frac{1}{\sqrt{V}} \sum_{\mathbf{k}} \varphi_{\mathbf{k}\alpha'}(\mathbf{x})\hat{C}_{\mathbf{k}\alpha} , \tag{8.2.2}$$

where $\varphi_{\mathbf{k}\alpha'}(\mathbf{x})$ is a one-electron wave function [in the simplest case, $\varphi_{\mathbf{k}\alpha}(\mathbf{x}) = \exp(i\mathbf{k}\mathbf{x})$]. It was shown in Sect. 5.2 that the smaller the exchange correlation radius $\langle r_c \rangle$, the more effective the exchange enhancement of the electron-phonon interaction. Consequently, the exchange interaction between electrons must be short-range. Thus we expand the electron magnetization operator $\hat{\Omega}^{\alpha}(\mathbf{x}') = \frac{1}{\sqrt{s}}\hat{\psi}^{+}_{\alpha'}(\mathbf{x}') \times \tau^{\alpha}_{\alpha'\gamma'}\hat{\psi}_{\gamma'}(\mathbf{x}')$ to second order in the difference $(\mathbf{x}' - \mathbf{x})_k$

$$\hat{\Omega}^{\alpha}(\mathbf{x}') = \hat{\Omega}^{\alpha}(\mathbf{x}) + \frac{\partial \hat{\Omega}^{\alpha}}{\partial x_k}(\mathbf{x}' - \mathbf{x})_k + \frac{1}{2}\frac{\partial^2 \hat{\Omega}^{\alpha}}{\partial x_k \partial x_\ell}(\mathbf{x}' - \mathbf{x})_k(\mathbf{x}' - \mathbf{x})_\ell + \dots . \tag{8.2.3}$$

Integrating with respect to $\mathbf{x}'$, the term taking into account the exchange interaction of electrons in (8.2.1), we can reduce the exchange Hamiltonian to the form

$$\mathbb{H}_{\mathrm{exch}} = -\frac{1}{2}\int d\mathbf{x}(J_0 s\,\hat{\Omega}^2 - \alpha_s \hat{A}^2_{\nu}) . \tag{8.2.4}$$

The operator $\hat{A}^{\alpha}_{\nu}(\mathbf{x})$ has the form $\hat{A}^{\alpha}_{\nu} = \frac{\partial \hat{\Omega}^{\alpha}}{\partial x_{\nu}}$, and $\alpha_s = \frac{1}{2}J_0 s \langle r_c \rangle^2$, $J_0 = \int d\mathbf{x} J(\mathbf{x})$. Averaging (8.2.4) over the quantum states, we obtain the exchange free energy. By minimizing the free energy functional, we arrive at the following expression for the electron magnetization vector $\mathbf{\Omega}$:

$$J_0 s\mathbf{\Omega} + \alpha_s \frac{\partial^2 \mathbf{\Omega}}{\partial x^2_{\nu}} = 0 . \tag{8.2.5}$$

The solution of this equation can be presented as follows:

$$\Omega^{\alpha}(\mathbf{x}) = A^{\alpha} \cos(\mathbf{k}_c\mathbf{x}) + B^{\alpha} \sin(\mathbf{k}_c\mathbf{x}) ;$$

$$\mathbf{k}_c = \sqrt{3}\,\mathbf{n}\frac{2\pi}{\langle \tau_c \rangle} ; \quad \mathbf{n} = \frac{1}{2\pi}\left(\frac{1}{\sqrt{3}} ; \frac{1}{\sqrt{3}} ; \frac{1}{\sqrt{3}}\right). \tag{8.2.6}$$

This function may make a nonzero contribution to the ground state energy of the electron system only if $\langle r_c \rangle \to \infty$. This means that for finite values of $<r_c>$, long-range order cannot appear in the system of electron spins. Consequently, the spin-density waves in the electron system with a wavelength on the order of $<r_c>$ are non-equilibrium waves. Accordingly, two types of fluctuations may appear in the spin system: (a) slowly fluctuating spin-density waves with a wavelength on the order of $<r_c>$, and (b) fluctuations of non-equilibrium magnetization with wavelengths $\lambda < <r_c>$, i.e., short-wave fluctuations (ripples) against a background of quasi-equilibrium spin-density waves. In this case, the effective Hamiltonian of the spin system can be presented in the form

$$\hat{\mathbb{H}}_s = \int d\mathbf{x}\left(\frac{\hat{m}^2}{2\chi} + \frac{1}{2}\alpha_s\hat{A}_\nu^2 - \frac{1}{2}J_0 s\hat{\Omega}^2\right) . \tag{8.2.7}$$

In this expression, $\hat{m}^2/2\chi$ is the kinetic energy operator of the spin system, $\hat{\mathbf{m}} = \chi\hat{\mathbf{\Omega}}$, and $\chi = 1/J_0 s$ is the effective paramagnetic susceptibility (or the effective "mass" of the spin system); $s = \frac{1}{2}$ is the electron spin. The interaction of the spin system with the crystal lattice can be taken into consideration by introducing a generalized momentum operator of the spin system:

$$\hat{\mathbf{M}} = \hat{\mathbf{m}} + (g\chi/M_i)\hat{p}_\nu\hat{\mathbf{A}}_\nu \tag{8.2.8}$$

where $\hat{p}_\nu$ is the phonon momentum, $g = U/J_0$, U being the electron-ion potential. Consequently, the effective spin-phonon Hamiltonian assumes the form

$$\hat{\mathbb{H}}_{s\text{-}ph} = \int d\mathbf{x}\left\{\frac{\hat{m}^2}{2} + \frac{1}{2}\alpha_s\hat{A}_\nu^2 - \frac{1}{2}J_0 s\hat{\Omega}^2 + \frac{\hat{p}_\nu^2}{2M_i} + \frac{1}{2}\lambda_i u_{k\ell}^2 + (g/M_i)\hat{p}_\nu[(\hat{\mathbf{m}}\hat{\mathbf{A}}_\nu) + (\hat{\mathbf{A}}_\nu\hat{\mathbf{m}})] + (g^2\chi/2M_i^2)\hat{p}_\nu\hat{p}_{\nu'}(\hat{\mathbf{A}}_\nu\hat{\mathbf{A}}_{\nu'})\right\} . \tag{8.2.9}$$

For the sake of simplicity, we shall consider the case of an isotropic crystal lattice (real HTS ceramics are anisotropic as in Fig. 8.1). We divide the spin variables into "slow" and "fast" variables by introducing the functions

$$\hat{\mathbf{\Omega}} = \mathbf{\Omega}_0^z + \delta\hat{\mathbf{\Omega}} , \quad \hat{A}_\nu = \Delta\mathbf{A}_\nu^z + \delta\hat{\mathbf{A}} .$$

Using the equations of motion analogous to (7.6.3-5) for the spin operators $\hat{\mathbf{m}}$, $\hat{\mathbf{A}}_\nu$, and $\hat{\mathbf{\Omega}}$ and for the phonon operators $\hat{p}_\nu$ and $\hat{u}_i$, we can show that the spectrum of spin fluctuations consists of three branches, one of which ($\omega_{s\parallel k}$) is longitudinal and the other ($\omega_{s\perp 1k}$ and $\omega_{s\perp 2k}$) are transverse:

$$\omega_{s\parallel \mathbf{k}} = J_0 s\,[(k/k_c)^2 - 1]^{\frac{1}{2}} \,, \tag{8.2.10}$$

$$\omega_{s\perp 1\mathbf{k}} = J_0 s[(k/k_c)^2 - (k/k_c) - 1]^{\frac{1}{2}} \,, \tag{8.2.11}$$

$$\omega_{s\perp 2\mathbf{k}} = J_0 s[(k/k_c)^2 + (k/k_c) - 1]^{\frac{1}{2}} \,. \tag{8.2.12}$$

Of the three spin branches, the longitudinal spin mode is linearly coupled with the phonons. Let us consider the interaction of spin fluctuations with phonons in the secondary quantization representation. We confine the analysis to the interaction of phonons with the longitudinal spin mode which is linearly coupled with one of the three spin modes. It follows from (8.2.10) that the longitudinal phonon mode is real only in the range of values $|\mathbf{k}| > 2\pi/\langle r_c\rangle$ of the wave vectors. For $|\mathbf{k}| < 2\pi/\langle r_c\rangle$ this mode is diffusive. This is because a long-range magnetic order cannot be established in the spin system at distances larger than $\langle r_c\rangle$, while for distances smaller than or on the order of $\langle r_c\rangle$, a nonequilibrium long-range order may be created by the slowly fluctuating spin-density waves. The symmetry of the spin oscillations corresponding to the longitudinal mode allows us to represent the operator $\delta\hat{\Omega}^z$ as a linear combination of the second quantization operators which obey Bose statistics:

$$\delta\hat{\Omega}^z = \frac{1}{\sqrt{V}} \sum_{\mathbf{k}} e^{i\mathbf{kx}} [\hbar/2\chi\omega_{s\mathbf{k}}]^{\frac{1}{2}} (\hat{a}^2_{\mathbf{k}} + \hat{a}^2_{-\mathbf{k}}) \,, \tag{8.2.13}$$

$$\hat{m}^z = \frac{1}{\sqrt{V}} \sum_{\mathbf{k}} e^{i\mathbf{kx}}\, i[\hbar\chi\omega_{s\mathbf{k}}/2]^{\frac{1}{2}} (\hat{a}^2_{\mathbf{k}} - \hat{a}^2_{-\mathbf{k}}) \,, \tag{8.2.14}$$

$$\delta\hat{A}^z_\nu = \frac{1}{\sqrt{V}} \sum_{\mathbf{k}} e^{i\mathbf{kx}}\, ik_\nu [\hbar/2\chi\omega_{s\mathbf{k}}]^{\frac{1}{2}} (\hat{a}^2_{\mathbf{k}} + \hat{a}^z_{-\mathbf{k}}) \,, \quad \omega_{s\mathbf{k}} \equiv \omega_{s\parallel\mathbf{k}} \,. \tag{8.2.14'}$$

The second quantization representation for phonon operators is introduced in the standard manner:

$$\hat{u}_{\nu i} = \frac{1}{\sqrt{V}} \sum_{\mathbf{k}} e^{i\mathbf{kx}}\, ik_i e_\nu\, [\hbar/2M_i\, \omega_{c\mathbf{k}\nu}]^{\frac{1}{2}} (\hat{b}_{\mathbf{k}\nu} + \hat{b}^+_{-\mathbf{k}\nu}) \,, \tag{8.2.15}$$

$$\hat{p}_\nu = \frac{1}{\sqrt{V}} \sum_{\mathbf{k}} e^{i\mathbf{kx}}\, ie_\nu\, [\hbar M_i \omega_{c\mathbf{k}\nu}/2]^{\frac{1}{2}} (\hat{b}_{\mathbf{k}\nu} - \hat{b}^+_{-\mathbf{k}\nu}) . \tag{8.2.16}$$

The frequency of the phonon mode linearly coupled with the spin mode is renormalized and has the form

$$\omega_{\mathbf{c k}3} = ck[1 + \zeta^2]^{\frac{1}{2}} . \quad (8.2.17)$$

Here, ζ is the spin and phonon subsystem coupling parameter

$$\zeta = \frac{\sqrt{3}\, g\hbar}{\pi v_0^{1/3} (J_0 s M_i)^{\frac{1}{2}}} 2\pi^2 (v_0^{1/3}/\langle r_c\rangle) . \quad (8.2.18)$$

As a result, the effective Hamiltonian taking into account the linear coupling of spin fluctuations with phonons is

$$\mathbb{H}^*_{s\text{-}ph} = \sum_{\mathbf{k},\nu=1,2} \omega_{\mathbf{c k}\nu} \hat{b}^+_{\mathbf{k}\nu} \hat{b}_{\mathbf{k}\nu} + \sum_{\mathbf{k}} [\omega_{\mathbf{s k}} \hat{a}^{z+}_{\mathbf{k}} \hat{a}^{z}_{\mathbf{k}} + \omega_{\mathbf{c k}3} \hat{b}^+_{\mathbf{k}3} \hat{b}_{\mathbf{k}3}$$

$$- 4q_{\mathbf{k}} (\hat{b}_{\mathbf{k}3} \hat{a}^z_{-\mathbf{k}} - \hat{b}^+_{\mathbf{k}3} \hat{a}^{z+}_{\mathbf{k}} - \hat{b}^+_{-\mathbf{k}3} \hat{a}^z_{-\mathbf{k}} + \hat{b}^+_{-\mathbf{k}3} \hat{a}^{z+}_{\mathbf{k}})] , \quad (8.2.19)$$

where $q_{\mathbf{k}} = (z/2)[\omega_{\mathbf{s k}}\omega_{\mathbf{c k}3}]^{\frac{1}{2}}$, $z = \zeta/(1 + \zeta^2)^{\frac{1}{2}}$, $\omega_{\mathbf{c k}\nu} = ck$, c is the velocity of sound, $\nu = 1,2$. Using the unitary transformations

$$\hat{a}^z_{\mathbf{k}} = u_{zz\mathbf{k}} \hat{c}^z_{\mathbf{k}} + v^*_{zz-\mathbf{k}} \hat{c}^{z+}_{-\mathbf{k}} + u_{z3\mathbf{k}} \hat{d}_{\mathbf{k}3} + v^*_{z3-\mathbf{k}} \hat{d}^+_{-\mathbf{k}3} , \quad (8.2.20)$$

$$\hat{a}^{z+}_{-\mathbf{k}} = u^*_{zz-\mathbf{k}} \hat{c}^{z+}_{-\mathbf{k}} + v_{zz\mathbf{k}} \hat{c}^z_{\mathbf{k}} + u^*_{z3-\mathbf{k}} \hat{d}^+_{-\mathbf{k}3} + v_{z3\mathbf{k}} \hat{d}_{\mathbf{k}3} , \quad (8.2.21)$$

$$\hat{b}_{\mathbf{k}3} = u_{33\mathbf{k}} \hat{d}_{\mathbf{k}3} + v^*_{33-\mathbf{k}} \hat{d}^+_{-\mathbf{k}3} + u_{3z\mathbf{k}} \hat{c}^z_{\mathbf{k}} + v^*_{3z-\mathbf{k}} \hat{c}^{z+}_{-\mathbf{k}} , \quad (8.2.22)$$

$$\hat{b}^+_{-\mathbf{k}3} = u^*_{33-\mathbf{k}} \hat{d}^+_{-\mathbf{k}3} + v_{33\mathbf{k}} \hat{d}_{\mathbf{k}3} + u^*_{3z-\mathbf{k}} \hat{c}^{z+}_{-\mathbf{k}} + v_{3z\mathbf{k}} \hat{c}^z_{\mathbf{k}} . \quad (8.2.23)$$

We can diagonalize (8.2.19):

$$\hat{\mathbb{H}}^*_{s\text{-}ph} = \sum_{\mathbf{k},\nu=1,2} \omega_{\mathbf{c k}\nu} b^+_{\mathbf{k}\nu} b_{\mathbf{k}\nu} + \sum_{\mathbf{k}} (\varepsilon_{\mathbf{s k}} \hat{c}^{z+}_{\mathbf{k}} \hat{c}^z_{\mathbf{k}} + \varepsilon_{3\mathbf{k}} \hat{d}^+_{\mathbf{k}3} \hat{d}_{\mathbf{k}3}) . \quad (8.2.24)$$

In this relation, $\varepsilon_{\mathbf{s k}}$ and $\varepsilon_{3\mathbf{k}}$ are the frequencies of the coupled spin-phonon oscillations:

$$\varepsilon^2_{\mathbf{s k},3\mathbf{k}} = \frac{1}{2}[\omega^2_{\mathbf{s k}} + \omega^2_{\mathbf{c k}3} \pm \{(\omega^2_{\mathbf{s k}} - \omega^2_{\mathbf{c k}3})^2 + 16q^2_{\mathbf{k}}\omega_{\mathbf{s k}}\omega_{\mathbf{c k}3}\}^{\frac{1}{2}}] . (8.2.25)$$

The functions $u_{zz\mathbf{k}}$, $v_{zz\mathbf{k}}$, $u_{z3\mathbf{k}}$, $v_{z3\mathbf{k}}$, $u_{33\mathbf{k}}$, $v_{33\mathbf{k}}$, $u_{3z\mathbf{k}}$, and $v_{3z\mathbf{k}}$ (8.2.20-23) are defined by

$$u_{zz\mathbf{k}} = \frac{q_{\mathbf{k}}[\omega_{ck3}(\varepsilon_{s\mathbf{k}} + \omega_{s\mathbf{k}})]^{\frac{1}{2}}}{[\varepsilon_{s\mathbf{k}}(\varepsilon_{s\mathbf{k}} - \omega_{s\mathbf{k}})(\varepsilon^2_{s\mathbf{k}} - \varepsilon^2_{3\mathbf{k}})]^{\frac{1}{2}}} ;$$

$$v_{zz\mathbf{k}} = \frac{q_{\mathbf{k}}[\omega_{ck3}(\varepsilon_{s\mathbf{k}} - \omega_{s\mathbf{k}})]^{\frac{1}{2}}}{[\varepsilon_{s\mathbf{k}}(\varepsilon_{3\mathbf{k}} + \omega_{ck3})(\varepsilon^2_{s\mathbf{k}} - \varepsilon^2_{3\mathbf{k}})]^{\frac{1}{2}}} ;$$

$$u_{z3\mathbf{k}} = \frac{(\varepsilon_{3\mathbf{k}} + \omega_{s\mathbf{k}})(\varepsilon^2_{3\mathbf{k}} - \omega^2_{3\mathbf{k}})^{\frac{1}{2}}}{2[\varepsilon_{3\mathbf{k}}\omega_{s\mathbf{k}}(\varepsilon^2_{3\mathbf{k}} - \varepsilon^2_{ck3})]^{\frac{1}{2}}} ;$$

$$v_{z3\mathbf{k}} = \frac{(\varepsilon_{3\mathbf{k}} - \omega_{s\mathbf{k}})(\varepsilon^2_{3\mathbf{k}} - \omega^2_{ck3})^{\frac{1}{2}}}{2[\varepsilon_{3\mathbf{k}}\omega_{s\mathbf{k}}(\varepsilon^2_{3\mathbf{k}} - \varepsilon^2_{s\mathbf{k}})]^{\frac{1}{2}}} ;$$

$$u_{33\mathbf{k}} = \frac{q_{\mathbf{k}}[\omega_{s\mathbf{k}}(\varepsilon_{3\mathbf{k}} + \omega_{ck3})]^{\frac{1}{2}}}{[\varepsilon_{3\mathbf{k}}(\varepsilon_{3\mathbf{k}} - \omega_{ck3})(\varepsilon^2_{3\mathbf{k}} - \varepsilon^2_{s\mathbf{k}})]^{\frac{1}{2}}} ;$$

$$v_{33\mathbf{k}} = \frac{q_{\mathbf{k}}[\omega_{s\mathbf{k}}(\varepsilon_{3\mathbf{k}} - \omega_{ck3})]^{\frac{1}{2}}}{[\varepsilon_{3\mathbf{k}}(\varepsilon_{3\mathbf{k}} + \omega_{ck3})(\varepsilon^2_{3\mathbf{k}} - \varepsilon^2_{s\mathbf{k}})]^{\frac{1}{2}}} ;$$

$$u_{3z\mathbf{k}} = \frac{(\varepsilon_{s\mathbf{k}} + \omega_{ck3})(\varepsilon^2_{s\mathbf{k}} - \varepsilon^2_{s\mathbf{k}})^{\frac{1}{2}}}{2[\varepsilon_{s\mathbf{k}}\omega_{ck3}(\varepsilon^2_{s\mathbf{k}} - \varepsilon^2_{3\mathbf{k}})]^{\frac{1}{2}}} ;$$

$$v_{3z\mathbf{k}} = \frac{(\varepsilon_{s\mathbf{k}} - \omega_{ck3})(\varepsilon^2_{s\mathbf{k}} - \omega^2_{s\mathbf{k}})^{\frac{1}{2}}}{2[\varepsilon_{s\mathbf{k}}\omega_{ck3}(\varepsilon^2_{s\mathbf{k}} - \varepsilon^2_{3\mathbf{k}})]^{\frac{1}{2}}} . \qquad (8.2.26)$$

Let us now consider the parameter determining the attraction between electrons that form a superconducting Bose condensate. In the **k**-representation, the effective electron-phonon interaction Hamiltonian is

$$\hat{\mathbb{H}}_{e\text{-}ph} = g_{ph}\frac{1}{\sqrt{V}}\sum_{\substack{\mathbf{k}_1,\mathbf{k}_2\\ \nu=1,2,3}} \frac{ik_{1\nu}e_{\nu}}{(2M_i\omega_{c\mathbf{k}\nu})^{\frac{1}{2}}}(\hat{b}_{\mathbf{k}_1\nu} + \hat{b}^+_{-\mathbf{k}_1\nu})\hat{c}^+_{\alpha\mathbf{k}_2}\,\hat{c}_{\alpha\mathbf{k}_2-\mathbf{k}_1} . \qquad (8.2.27)$$

Since the operators $\hat{b}_{\mathbf{k}\nu}$ and $\hat{b}^+_{\mathbf{k}\nu}$ are connected to the operators $\hat{d}_{\mathbf{k}3}$, $\hat{d}_{\mathbf{k}3}$, $\hat{c}^z_{\mathbf{k}}$, $\hat{c}^{z+}_{\mathbf{k}}$ through unitary transformations, this means that both the quasi-phonon mode $\varepsilon_{3\mathbf{k}}$ and the quasi-spin mode $\varepsilon_{s\mathbf{k}}$ contribute towards the effective electron-phonon coupling parameter. In the simplest case, the effective electron-phonon interaction parameter will be

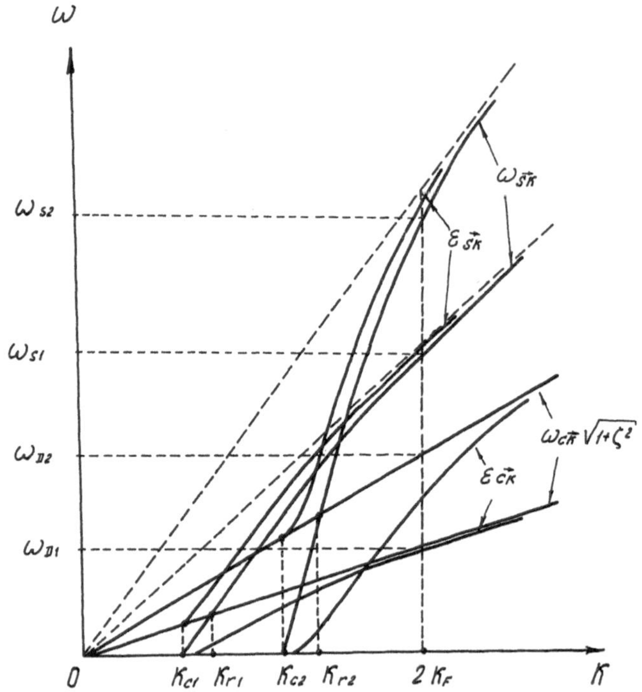

Fig. 8.3. Spectra of coupled spin-phonon oscillations in high-temperature superconductors for different values of the spin-phonon interaction parameter ζ

$$\lambda_{e-ph}(\mathbf{k}) = \lambda_{e-ph}[(2/3) + |u_{33\mathbf{k}} + v_{33\mathbf{k}}|^2(\omega_{c\mathbf{k}3}/3\varepsilon_{3\mathbf{k}}) + |u_{3z\mathbf{k}} + v_{3z\mathbf{k}}|^2 (\omega_{c\mathbf{k}3}/3\varepsilon_{s\mathbf{k}})] \ . \qquad (8.2.28)$$

Figure 8.3 shows the dispersion curves corresponding to coupled spin-phonon oscillations. It can be easily seen that the phonons interact most strongly with the spin fluctuations in the region of resonance intersection of branches $\omega_{s\mathbf{k}}$ and $\omega_{c\mathbf{k}3}$. The wave vector $\mathbf{k}$ corresponding to this intersection is given by

$$\mathbf{k}_r = 2\pi/\langle r_c\rangle\left[1 - \frac{\hbar^2 c^2(1+\zeta^2)}{J_0\langle r_c\rangle^2}\right]^{\frac{1}{2}} \ . \qquad (8.2.29)$$

As the spin-phonon coupling parameter ζ increases, the phonon frequency of the coupled quasi-phonon mode increases (8.2.17). The resonance region is displaced towards larger values of k, and for $\zeta \approx 10$ the resonance may be observed for values of the wave vector on the order of $2k_F$ (k_F is the Fermi momentum), i.e., in the region where the attraction between the pair-forming electrons is the strongest. In this case, the asymptotic values of frequencies corresponding to the quasi-phonon and quasi-spin modes will considerably exceed the Debye frequency for the uncoupled phonon modes. Consequently, the energy region near the Fermi surface, where the superconducting pairing takes place (Fig. 5.2), will be higher than $2\langle\omega_D\rangle$ by an order of magnitude or more. This results in an effective increase in the electron-phonon coupling parameter and to a consequent increase in the critical (superconducting transition) temperature T_c. The critical temperature is determined from the solution of the integral equation

$$\Delta(\omega) = \int_0^{\varepsilon_F} d\omega' \, \frac{\Delta(\omega')}{\omega'} \, Q(\omega,\omega')\tanh(\omega'/2T_c) . \tag{8.2.30}$$

In the simplest case, the kernel of this integral equation can be approximated as a step function which assumes constant values in the intervals $(0\,;\langle\omega_D\rangle)$, $(\langle\omega_D\rangle\,;\langle\omega_D\rangle\sqrt{1+\zeta^2})$, $(\langle\omega_D\rangle\sqrt{1+\zeta^2},\ \omega_s)$, $(\omega_s;\varepsilon_F)$. The condition of solvability of a system of fourth-order linear equations leads to the following expression for the critical temperature T_c:

$$T_c = (2\gamma/\pi)\,\langle\omega_D\rangle \exp\left[-\frac{1}{k_{enh}(\lambda_{e-ph} - \mu^*)}\right] . \tag{8.2.31}$$

The function $k_{enh}(\zeta,\ \lambda_{e-ph},\ \mu^*,\omega_s/\langle\omega_D\rangle)$ is the coefficient of enhancement of the effective electron-phonon interaction $(\lambda_{e-ph} - \mu^*)$ and is a monotonically increasing function of the parameters ζ and $\omega_s/\langle\omega_D\rangle$. For initial values of the electron-phonon interaction parameter typical of complex ternary compounds of rare-earth metals [8.20], $\lambda_{e-ph} \approx 0{,}6$, as well as for $\mu^* \approx 0{,}2$ and $\zeta \simeq 10$, the enhancement factor k_{enh} may attain values on the order of 1.8-1.9, which means that the exchange enhancement effect for $\langle\omega_D\rangle \simeq 300$-$400$ K can ensure, even in the quasi-linear approximation, a high-temperature superconductivity for complex compounds of rare-earth metals and ceramic systems ($T_c \geq 100$ K). In order to consider the possibility of increasing the value of T_c to room temperature, we must take into account the nonlinear interaction of phonons with spin fluctuations.

8.3 Possibility of Increasing the Critical Temperature: Properties of High-Temperature Superconductors

Methods of synthesizing high-temperature superconductors may be suggested by analyzing the expression for the spin-phonon coupling parameter ζ(8.2.18). Indeed, with increasing enhancement factor K_{enh} (8.2.31) which increases monotonically with ζ, the value of the critical temperature T_c also increases. Consequently, higher values of T_c can be attained by synthesizing compounds with a strong spin-phonon coupling. It follows from (8.2.18) that the spin-phonon coupling parameter will be higher, the higher the relative electron-ion potential $g = U/J_0$, the smaller the exchange correlation radius $\langle r_c \rangle$, and the smaller the mass of the ions constituting the crystal lattice and participating in the formation of a covalent bond. A high electron-ion potential can be attained by choosing elements with a low ionization potential. A small exchange correlation radius $\langle r_c \rangle$ will be ensured if the high-temperature superconductor contains elements having a small atomic radius and participating in the formation of a covalent bond. These elements, as well as all the other elements used for synthesizing the HTS material, must have the smallest possible mass. In order to help choose the necessary elements and to construct a homological series of affine elements, we compiled Table 8.1. In this Table the elements are grouped according to the types of bonds they tend to form. For the sake of completeness, Table 8.1 is supplemented by Figs. 8.4,5 showing the dependence of the ionization potential and atomic radius on the atomic number. It can be seen from Fig. 8.5 that oxygen has one of the smallest atomic radii among all the elements. In the new class of HTS compounds, oxygen participates in the formation of a covalent bond (Cu-O bond), i.e., in the formation of exchange interaction between conduction electrons. This means that the exchange correlation radius is determined to a considerable extent by the atomic radii of oxygen and copper. Oxygen is a relatively light element and hence it can be concluded that oxygen is responsible for a strong spin-phonon coupling in HTS ceramics. This explains the reason for the high sensitivity of the new HTS materials to their oxygen concentration. A decrease in the oxygen concentration not only decreases the number density of charge carriers, but also (and this is more important!) leads to an increase in $\langle r_c \rangle$ and hence to a decrease in the critical temperature T_c. The presence of rare-earth elements in new superconductors is important because of the fact that they stabilize the crystal lattice and have low ionization potentials, favoring an increase in the spin-phonon coupling parameter. According to our theory we can also use the elements Tl and Bi for synthesizing high-temperature superconductors since, in spite of the fact that they are twice as heavy as Y, they have a smaller atomic radius than the rare-earth elements. Hence, the spin-

Table 8.1. Properties of the elements. Transition metals are enclosed in *solid rectangles* □, rare-earth elements in *dashed rectangles* ⬚; covalently bonding elements are *underlined* —; *dashed circles* ◌ mark elements that form ionic bonds; the *solid circles* ○ denote nonmetals. Superconducting substances are enclosed in *ovals* ⬭, thin layer superconductors are denoted by ⦸, and high pressure superconductors by ⊘

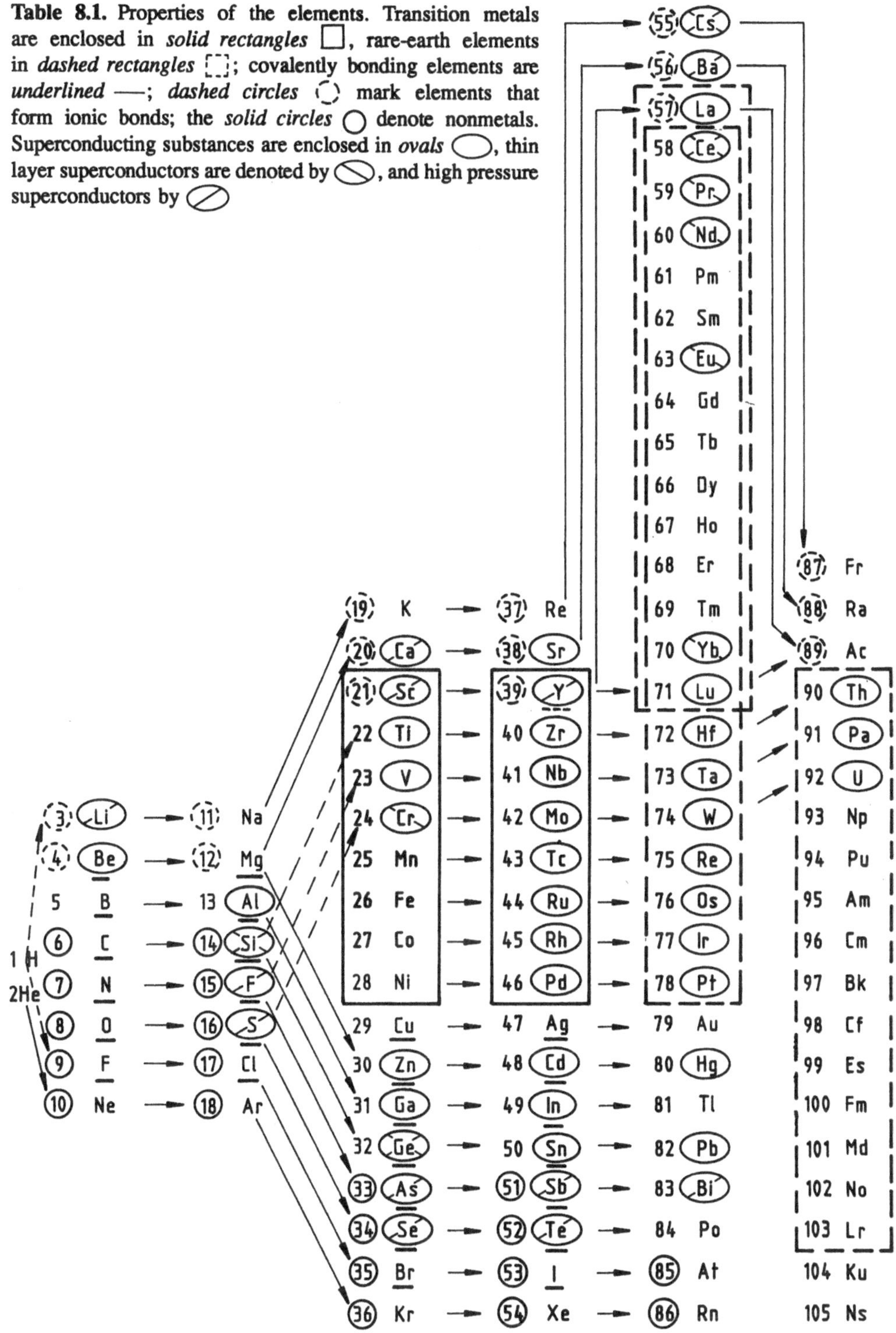

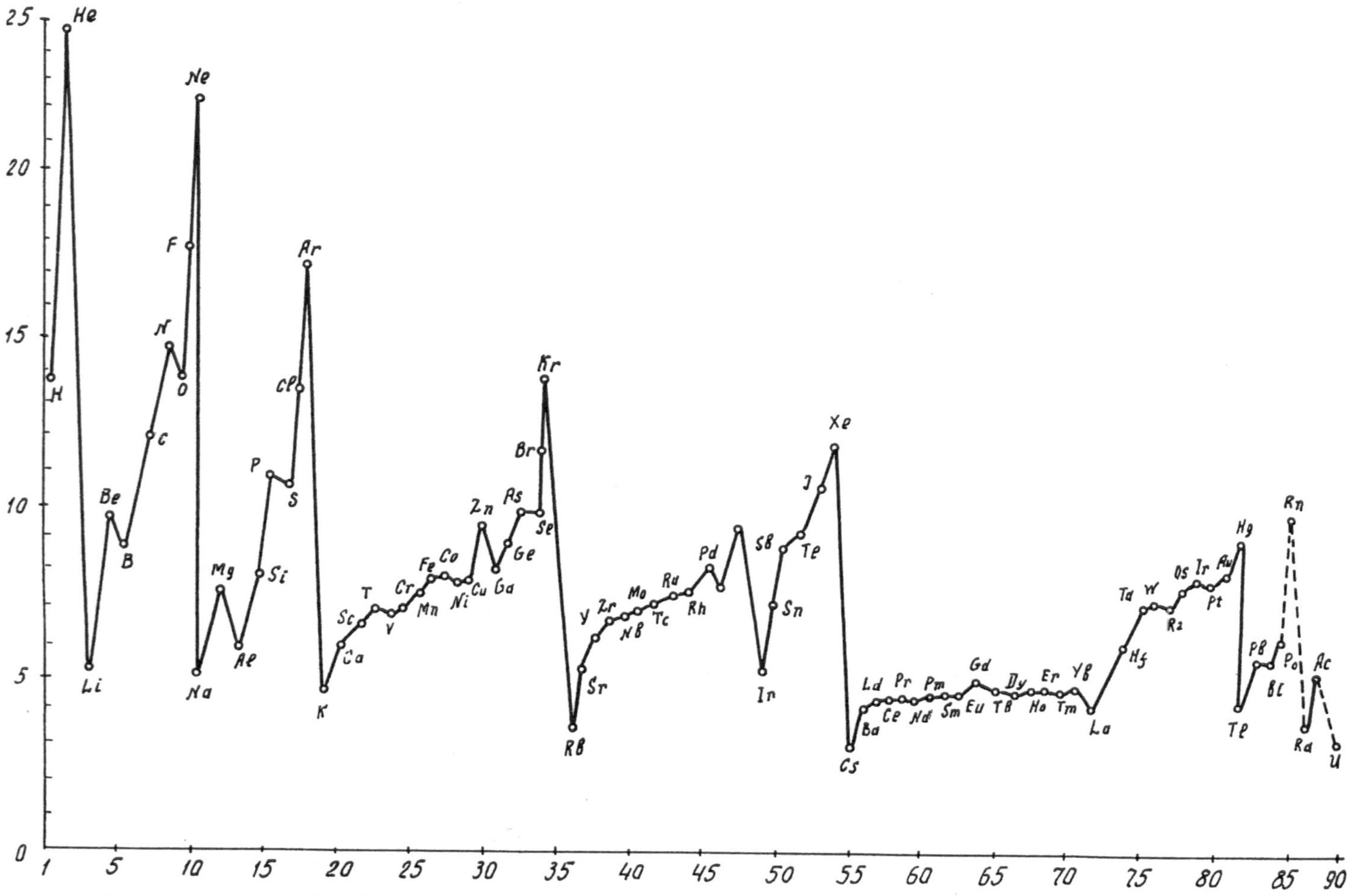

Fig. 8.4. Dependence of the ionization potential of the elements on atomic number

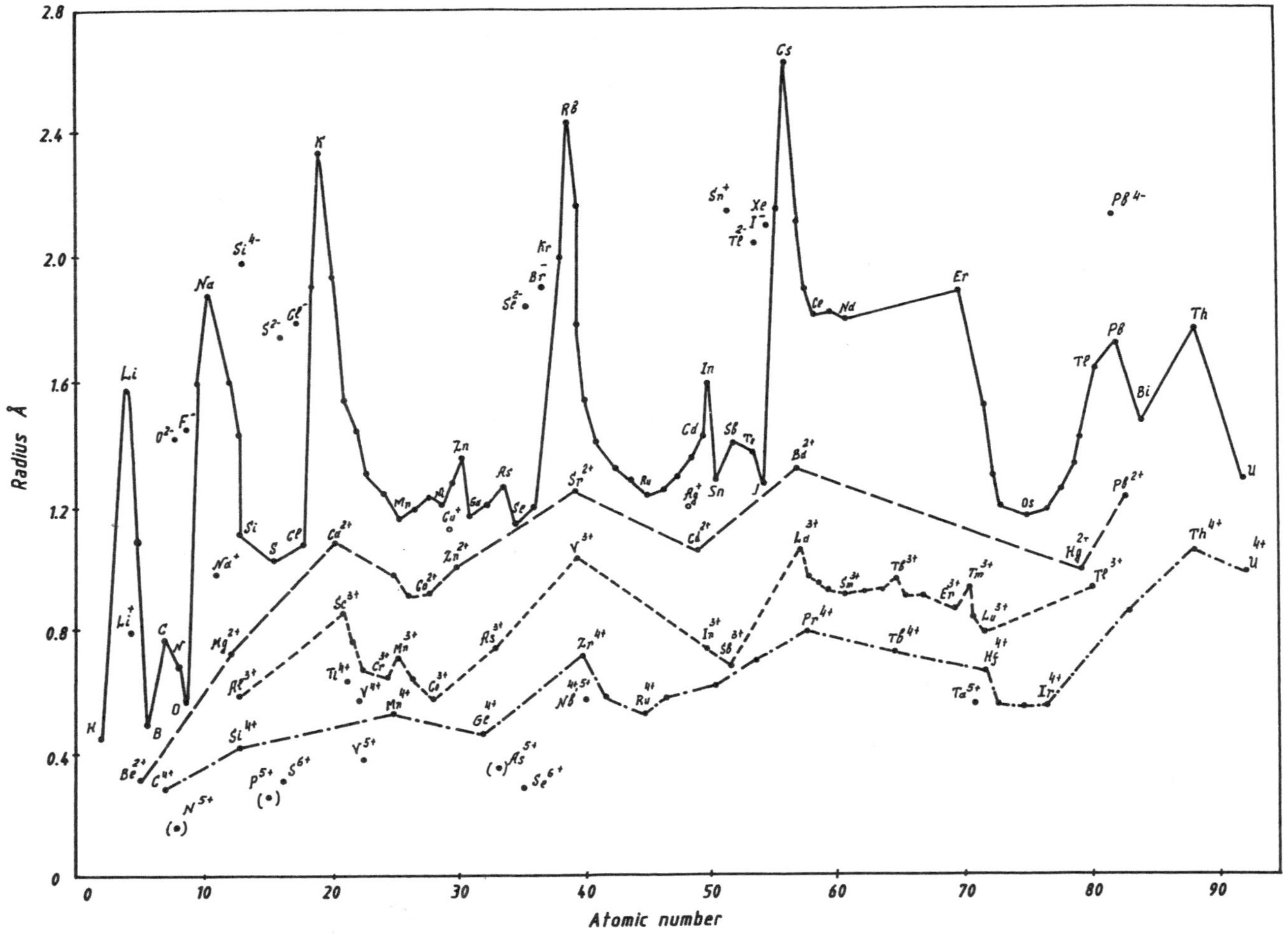

Fig. 8.5. Dependence of the atomic and ionic radii of elements on atomic number

phonon coupling parameter must be quite high for Tl - Ba - Cu - O and Bi - Al - Ca - Sr - Cu - O and may ensure values of $T_c \gtrsim 100$ K. Let us consider Table 8.1. Arrows join the affine elements, which allows us to construct homologic series. The homologic series for oxygen and copper are

O - S - Se - Te - Po
Li - Na - Cu - Ag - Au.

Sulphur and selenium have larger atomic radii than oxygen (Fig. 8.5), and larger masses. Hence, the substitution of an oxygen vacancy in the new superconductor by sulphur or selenium should not cause an increase in the critical temperature. This is confirmed by the experimental results [8.21]. However, in view of the fact that sulphur and selenium are heavier and poorer oxidizers in comparison to oxygen, they may ensure better stability in the HTS [8.21]. Lithium and sodium are much lighter than copper but have a much a larger atomic radius. Thus their substitution in place of copper is also not useful in Y - Ba - Cu - O compounds. It should be interesting to test Li and Na substitution of K in Ba - K - Bi - O ($T_c \simeq 30$ K) salts.

Let us consider the thermodynamic properties of high-temperature superconductors. First, we shall study the magnetic properties of the new compounds. The main problem in this case is to determine the lower and upper critical fields H_{c1} and H_{c2} of new HTS compounds. The field H_{c1} can be calculated by using the results obtained in Chap. 6. If an external magnetic field is applied to the superconductor, the order parameter can be written as

$$\Psi(\mathbf{x}) = u(\mathbf{x})e^{i\Phi(\mathbf{x})}\hat{R}(\mathbf{x})\hat{a} . \tag{8.3.1}$$

Here we have retained the notation of Chap. 6 (6.3.2,23). By using (6.3.19) and (8.3.1), we can write the expression for the superconducting current J_ν^α. The magnetic flux in the superconducting state can be quantized with the help of the condition $\oint_\Gamma dx_\nu J_\nu^\alpha = 0$, where J_ν^α is the superconducting current (see 6.3.19). We shall use the relations:

$$\mathbf{n} = (\cos\psi, \sin\psi, 0) ; \quad \Phi = n\varphi , \quad \psi = m\varphi .$$

In the superconducting state and for a vortex structure of the external magnetic field, the magnetic flux is the sum of the magnetic flux of the external field and the flux $\mathbf{A}$ of the electron spin field, $\mathbf{A} = (0, 0, A_\nu^3)$. The latter is of exchange type, i.e., this flux is electromagnetic in nature and has the same symmetry as the Yang-Mills field. The fluxes are

$$\Phi^0 = \oint_\Gamma dx_\nu A^0_\nu = (hc/4e)(2n - m) ; \tag{8.3.2}$$

$$\Phi^3 = \oint_\Gamma dx_\nu A^3_\nu = -(hc/4e)m ; \tag{8.3.3}$$

$n = 1,2,\ldots;\ m = 1,2,\ldots$ or $1/2,\ 3/2,\ \ldots$ (Sect. 6.3).

It follows from these relations that the total flux in a superconductor will be nonzero if $n = 1$ and $m = \frac{1}{2}$. In order to determine the lower critical field H_{c1} at which the external field starts penetrating a superconductor, we write the free energy of a vortex filament as

$$F_v = \int \frac{d\mathbf{x}}{8\pi} \Big\{\lambda_\ell^2[(\mathrm{curl}\ \mathbf{H}^0)^2 + (\mathrm{curl}\ \mathbf{H}^3)^2] + [(\mathbf{H}^0)^2 + (\mathbf{H}^3)^2 - 2(\mathbf{H}^0\mathbf{H}^3)\cos 2\xi]\Big\} \tag{8.3.4}$$

with $\mathbf{H}^0 = \mathrm{curl}\ \mathbf{A}^0$ being the external magnetic field, $\mathbf{H}^3 = \mathrm{curl}\ \mathbf{A}^3$ the field of the spin system, and λ_ℓ the field penetration depth in the superconductor. In the general case, the fields H^0 and H^3 satisfy a system of complex nonlinear differential equations. The relation (8.3.4) is obtained under the assumption that the functions $u(\rho)$ and $\xi(\rho)$(8.3.1) are slowly varying functions of the distance from the vortex axis ρ. In this case, the free vortex energy can be easily calculated:

$$F_v = [L/(4\pi\lambda_e)^2][(\Phi^0)^2 + (\Phi^3)^2]\ln(\lambda_\ell/\xi_\ell) , \tag{8.3.5}$$

where ξ_ℓ is the coherence length. The lower critical field is calculated from the equation balancing the magnetic energy and the energy of a vortex filament, and is given by

$$H_{c1} = F_v/L\Phi = \Big\{[(\Phi^0)^2 + (\Phi^3)^2]/(4\pi\lambda_\ell)^2\Phi\Big\}\ \ln(\lambda_\ell/\xi_\ell). \tag{8.3.6}$$

It is interesting to study the temperature dependence of H_{c1} in the vicinity of the phase transition to the superconducting state. This dependence is determined by the temperature dependence of the penetration depth λ_ℓ. This mean penetration depth is proportional to $n_s^{-\frac{1}{2}}$, where n_s is the number density of superconducting pairs, determined by the mean amplitude of the order parameter (8.3.1). The superconducting pair number density is defined as

$$n_s = \frac{\pi}{\gamma}\ \frac{1}{v_0}(T_c/\epsilon_F) \quad . \tag{8.3.7}$$

On the other hand, the exchange correlation radius $\langle r_c \rangle$ plays the role of the mean distance between electrons forming the superconducting Bose condensate, since the exchange interaction enhances electron attraction through phonons and hence the number density of superconducting pairs must be inversely proportional to $\langle r_c \rangle^3$. In this case, using formulas (1.1.1) and (8.3.7), we obtain

$$n_s = \frac{\sqrt{A}}{2\sqrt{3}\pi\gamma} \cdot \frac{1}{v_0^{2/3} \xi_\ell} \cong \frac{1}{\langle r_c \rangle^3} , \tag{8.3.8}$$

$$\langle \lambda_\ell \rangle = \left[\frac{\mu_e c^2 \langle r_c \rangle^3}{3e^2} \right]^{\frac{1}{2}} . \tag{8.3.9}$$

Since the fluctuation regime in new superconductors is quite broad and may run into tens or even hundreds of Kelvin [8.14-18], this means that the temperature dependence of the correlation radius below the phase transition point must be the same as in the paramagnetic (normal) phase, i.e., $\langle r_c \rangle = \langle r_{c0} \rangle \tau^{-\frac{1}{2}}$. But this means that $H_{c1} \approx \tau^{-3/2}$. On the other hand, it follows from the self-consistent field theory that $n_s \approx \tau$, $\langle \lambda_\ell \rangle \approx \tau^{-1/2}$ and $H_{c1} \approx \tau$. Hence for weak fluctuations $\langle r_{c0} \rangle \to \infty$, the temperature index of the correlation radius must be approximately 1/3 below the phase transition point. In this case, the fluctuation region is found to be narrow, but much wider than in normal superconductors (see 1.1.2). Using the renormalized group method, we obtain the critical index 2/9 for $\langle r_c \rangle$, and hence $H_{c1} \approx \tau^{2/3}$. However, since the fluctuation region is very wide, the phase transition into the superconducting state turns out to be first-order but close to second-order (Chaps. 4 and 5). Hence, as we approach the phase transition point from below, $H_{c1} \approx \tau^{3/2}$, with the exception of the region in the immediate vicinity of the transition where H_{c1} drops sharply to zero due to a first-order phase transition close to a second-order one. Calculations of the upper critical field H_{c2} at which the superconducting state may be destroyed are carried out on the basis of the rigorous microscopic theory described in Chap. 9. Omitting the cumbersome calculations, we present the final expression for H_{c2} [2)]

[2)] It is well known from the theory of type II superconductors that $H_{c2} = \varphi_0 / 2\pi\xi_\ell^2$, where $\varphi_0 = hc/2e$. But then it follows from (8.3.10) that the exchange correlation radius $\langle r_c \rangle$ is the electron correlation length in the superconducting state. As the field attains the value (8.3.10), exchange interaction between electrons is suppressed. This violates the coherence condition during pairing and hence leads to a destruction of superconductivity in HTS.

Fig. 8.6

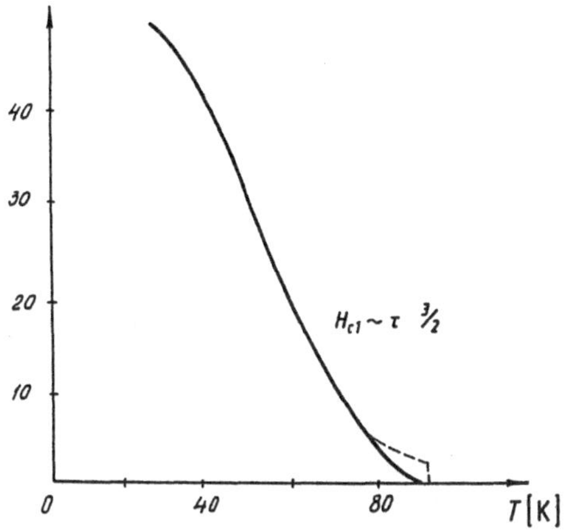

Fig. 8.7

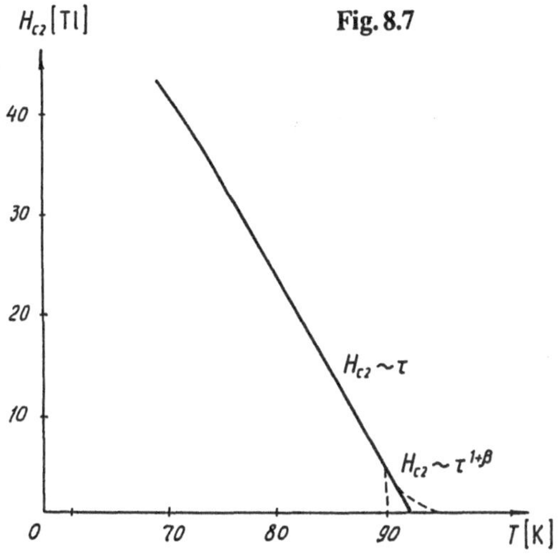

Fig. 8.6. Temperature dependence of the lower critical field H_{c1} of high-temperature superconductors

Fig. 8.7. Theoretical (*solid line*) and experimental (*dashed line*) temperature dependences of the upper critical field H_{c2} of high-temperature superconductors

$$H_{c2} = \Phi/\pi \langle r_c \rangle^2 \qquad (8.3.10)$$

near the phase transition point, $H_{c2} \approx \tau$. However, since the strong magnetic field in the scaling region suppresses the first-order phase transition close to the second and transforms it into a first-order phase transition, we have $H_{c2} \approx \tau^{1+\beta}$, where $\beta = \frac{n+2}{2(n+8)}\varepsilon + \dots$, $\varepsilon \to 1$, $n = 1,2$. Figures 8.6 and 8.7 show the temperature dependences of H_{c1} and H_{c2}. At present, there are no reliable experimental data for H_{c1}, while the dependence $H_{c2}(T/T_c)$ is found to be in good agreement with the experimental results of [8.22]. The penetration depth λ_ℓ and the coherence length ξ_ℓ can be estimated if the exchange correlation radius $\langle r_c \rangle$ is known. This can be done by using (1.1.4) and (5.2.10) and

$$(v_0^{1/3}/\langle r_c \rangle) = \left[\frac{(\chi/\chi_{os})}{(\chi/\chi_{os})^2 (h^2\sqrt{\Delta\tau}/2\mu_e v_0^{2/3} T_c) - 36}\right]^{\frac{1}{2}} . \qquad (8.3.11)$$

Here $\Delta\tau$ is the width of the transition to the superconducting state, determined from the electrical resistivity measurements (Fig. 8.2) and (χ/χ_{os}) is the relative magnetic susceptibility at the transition point $\chi_{os} = \chi(T/T_c \to \infty)$. An estimate for the ceramic $YBa_2Cu_3O_{7-y}$

with the help of (8.3.11) gives the ratio $(v_0^{1/3}/\langle r_c\rangle) \lesssim 10^{-1}$. For $v_0^{1/3} = 5.45$ Å, this leads to the estimates $\lambda_\ell \lesssim 10^3 - 10^4$ Å, $\xi_\ell \lesssim 20\text{-}30$ Å. These values are also found to be in agreement with the experimental results [8.23].

The above analysis shows that superconductors can be classified as follows:

(i) $\lambda_\ell \ll \xi_\ell \ll \langle r_c\rangle$ - type I superconductors,

(ii) $\xi_\ell \ll \langle r_c\rangle \ll \lambda_\ell$ - type II superconductors,

(iii) $\xi_\ell \leq \langle r_c\rangle \ll \lambda_\ell$ - superconductors with strong magnetic fluctuations (HTS, superconductors with heavy fermions, magnetic superconductors),

(iv) $\xi_\ell \ll \lambda_\ell \lesssim \langle r_c\rangle$ - superconducting spin glasses.

It is also interesting to note that if the value of the effective mass is estimated from $\mu_e^* = \mu_e(1 + k_{enh}\lambda_{e\text{-}ph})$, then for $k_{enh} = 1.9$ and $\lambda_{e\text{-}ph} = 0.6\text{-}0.8$ [8.2], we obtain $\mu_e^* = 2\text{-}3\ \mu_e$. This parameter can also be determined from the experimental data on the measurement of the electron specific heat in the normal phase. Using the data presented in [8.24] we obtain $\mu_e^* = 5\mu_e$ which means that the results obtained in the quasilinear approximation do not contradict the experimental results.

Experimental investigations of the magnetoresistance of new superconductors in the normal phase showed that it is negative, which is in agreement with the theoretical results [8.25]. Calculations show that the magnetoresistance is proportional to the quantum correction to the conductivity $\delta\sigma/\sigma_0$ while in the region of weak classical fields $(\tau_e = eH/\mu_e c) \gg 1$, where τ_e is the mean free time of the electron this quantity is proportional to H^2. This is in agreement with the experimental results [8.26]. Since the exchange-type fluctuations play a key role in the enhancement of the electron-phonon interaction in the superconducting state, they must also determine the properties of the system in the normal phase. Indeed, the electrical resistivity in the normal phase is determined by the scattering of electrons by spin fluctuations, and its temperature dependence is

$$\rho(T) = \frac{\mu_e^* v_F}{n_e e^2 \langle r_{co}\rangle}\left(\sqrt{\tau} + \left\langle \cosh\left(\frac{2\pi^2 v_0^{1/3}}{\langle r_{co}\rangle}\right)\right\rangle \frac{h}{\mu_e^* v_F \langle r_{co}\rangle}\,\tau\right) \quad (8.3.12)$$

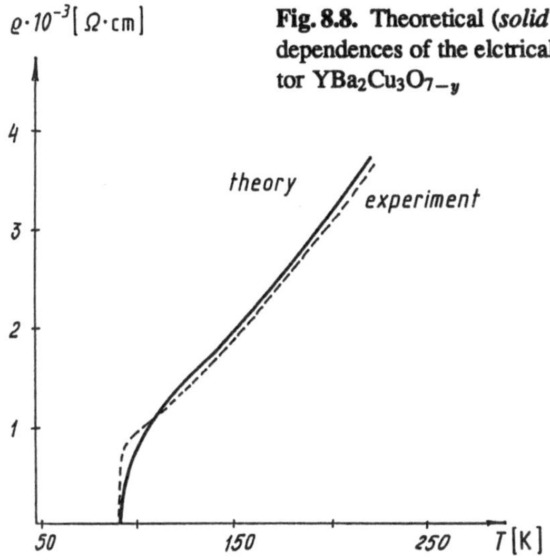

Fig. 8.8. Theoretical (*solid line*) and experimental (*dashed line*) temperature dependences of the elctrical resistivity of the high-temperature superconductor $YBa_2Cu_3O_{7-y}$

In this relation, n_e is the number density of charge carriers, v_F is the electron velocity at the Fermi surface, $\langle\cosh(x)\rangle = \frac{1}{x}\sinh x$. Figure 8.8 shows the experimental and theoretical dependences of the electrical resistivity on temperature. The experimental dependence was obtained for the compound $YBa_2Cu_3O_{7-y}$, while the theoretical dependence was obtained with the help of (8.3.12) for $n_e = 10^{21}\ cm^{-3}$, $v_F = 10^7$ cm/s , $v_0^{1/3} / \langle r_{co}\rangle = 10^{-1}$, and $\mu_e^* = 5\mu_e$. The theoretical dependence fits quite well with the experimental results except for the region in the immediate vicinity of the phase transition point. However, as was shown in Chaps. 4 and 5, the phase transition to the superconducting state is a first-order phase transition close to a second one, and hence the electrical resistivity must change abruptly at the transition point. This allows us to eliminate the discrepancy between the theoretical and experimental results in the immediate vicinity of the phase transition point. Finally, the nonexponential behavior of the specific heat in the superconducting phase [8.27-29]. is attributed to the spin excitations in the electron system. With the help of (8.2.25), it can be shown that the electron specific heat contains a component which is proportional to $T^{(d-1)}$ (d is the space dimensionality, d = 2,3) at low temperatures ($T/T_c \ll 1$). This is found to be in agreement with the experimental results.

The exchange enhancement mechanism of electron-phonon interaction due to spin fluctuations of collective electrons in high-temperature superconductors is also confirmed by the results of Mössbauer effect investigations in $YBa_2(Cu_{1-x}Fe_x)_3O_{7-y}$ (x = 0.010, 0.017, 0.033) [8.30] and by the oxygen isotope effect in Bi-Sr-Ca-Cu-O [8.31].

9. Theory of High-Temperature Superconductivity of Ceramic Systems

9.1 Microscopic Equations for the Superconducting Phase

In Chap. 8, we considered the exchange enhancement of the electron-phonon interaction in a quasilinear approximation. In the general case, however, the interaction of phonons with spin fluctuations is nonlinear, and this fact must be taken into account while constructing the microscopic theory of superconductivity. Moreover, it was shown in Chap. 6 that the phenomenological theory of this class of compounds is a gauge-invariant non-Abelian theory. In the BCS theory, the electrons forming the superconducting Bose-condensate interact with the electromagnetic field, if such a field exists in the system. In this case, the theory of superconductivity is an ordinary Abelian theory. The quasilinear theory of high-temperature superconductivity considered in Chap. 8 is also Abelian because we consider the interaction of phonons with only one spin mode and do not consider the nonlinear interaction of conduction electrons with spin fluctuations. By analyzing the Hamiltonian (8.2.1) in greater detail, we can show that the operation (8.2.3) followed by the introduction of an operator $\hat{A}^{\alpha}_{\nu}(\mathbf{x})$ given by

$$\hat{A}^{\alpha}_{\nu}(\mathbf{x}) = i \frac{\alpha_s \mu_e c}{2e\hbar^2} \frac{\partial \hat{\Omega}^{\alpha}}{\partial x_{\nu}} ,$$

transforms the gradient part of Hamiltonian (8.2.1) as

$$-\frac{\hbar^2}{2\mu_e}\int d\mathbf{x}\ \hat{\psi}^{+}_{\alpha'}(\mathbf{x})\hat{D}^{o2}_{\nu\alpha'\alpha''}\hat{\psi}_{\alpha''}(\mathbf{x}) ,$$

$$\hat{D}^{o}_{\nu} = \nabla_{\nu} - (i/2)\tilde{g}\hat{\tau}^{\alpha} A^{\alpha}_{\nu} , \quad \tilde{g} = 2e/\hbar c . \tag{9.1.1}$$

This expression is analogous to the gauge-invariant gradient term in the expression for the free energy describing a superconductor containing spotaneous vortices (Sect. 6.3). Additionally, the following term appears in the Hamiltonian of (8.2.1):

$$\frac{i\hbar^2}{2\mu_e}\tilde{g}\int d\mathbf{x}\ \hat{\psi}^+_{\alpha'}(\mathbf{x})(\hat{D}^o_\nu\hat{A}_\nu)_{\alpha'\alpha''}\ \hat{\psi}_{\alpha''}(\mathbf{x})\ . \tag{9.1.2}$$

This term can be eliminated by choosing an appropriate gauge. Indeed, the operators $\hat{\psi}(\mathbf{x},\hat{\sigma})$ are column matrices which can be chosen in the form

$$\hat{\psi}(\mathbf{x},\hat{\sigma}) = \Phi(\mathbf{x},\hat{\sigma})\hat{b}\quad ;\quad \hat{b} = \begin{Vmatrix} 1 \\ 1 \end{Vmatrix} \tag{9.1.3}$$

(see also 6.3.2). The operators $\hat{\Phi}(\mathbf{x},\hat{\sigma})$ are linear combinations in the representation of the SU(2) group: $\hat{\Phi}(\mathbf{x},\hat{\sigma}) = \sum_{\sigma=1}^{3} \hat{\tau}^\sigma\ \psi^\sigma(\mathbf{x})$; $\hat{\tau}^\sigma$ are the generators of the SU(2) group (Pauli matrices). We can then construct the Hamiltonian (8.2.1) in the SU(2) group representation. The covariant derivatives of the type in (6.3.16) will be transformed in this case to covariant derivatives of the form

$$\hat{D}^o_\nu \to \hat{D}_\nu = \nabla_\nu - (i/2)\tilde{g}[\hat{A}_\nu\ ]\ . \tag{9.1.4}$$

The term in (9.1.2) can be eliminated by choosing a normal Coulomb gauge of the field $\hat{A}_\nu(\mathbf{x})$: $\nabla_\nu\hat{A}_\nu = 0$. The physical meaning of the non-Abelian field $A^\alpha_\nu(\mathbf{x})$ becomes clear from here. This field is of exchange (i.e., electromagnetic) origin, and is an effective non-uniform field acting on an electron with spin σ. On the basis of these considerations, we can obtain a microscopic model of the superconducting phase and construct the effective Hamiltonian of the system. It was shown in Sect. 8.2 that a system of electron spins has no long-range magnetic order. We shall assume that electron spins are irregularly distributed in space and are randombly oriented. However, such a system has an SO(3)-symmetry (Sect. 6.1). The spin orientation at a point with coordinates $\mathbf{x}$ is defined with the help of Euler angles Θ^α, $\alpha = 1,2,3$ in the local coordinate system [9.1]. The vector can be characterized as the turning vector of this system. We introduce the flexure-torsion tensor in the system of spins $a^\alpha_\nu = \nabla_\nu\Theta^\alpha$. It can be seen that the flexure-torsion tensor will have the same symmetry properties as the "exchange field" potential A^α_ν. Consequently, the non-Abelian exchange field has topological properties and determines the curvature in the spin space. In this case, we can introduce the spin-density tensor $\rho^{\sigma\alpha} = \eta^\sigma_{\mu\nu}\nabla_\mu a^\alpha_\nu$ (6.2.5). The "stress" tensor of the exchange field $F^\alpha_{\mu\nu}$ (6.1.2) is proportional to the spin-density tensor with the exception of nonlinear terms:

$$F^\alpha_{\mu\nu} \approx \eta^\sigma_{\mu\nu}\ \rho^{\sigma\alpha}\ .$$

The energy of the spin system in the simplest case is a bilinear form of the spin-density vectors (tensors), and hence we write it in the effective Hamiltonian as the quadratic form of the tensors $\hat{F}_{\mu\nu}$. It follows from (9.1.1,4) that the field A^{α}_{ν} is the gauge field for the operators $\hat{\psi}^{+}_{\alpha'}$, $\hat{\psi}_{\alpha'}$, $\hat{\Phi}^{+}$ and $\hat{\Phi}$. Indeed, introducing the operator (6.3.6)

$$\hat{R}(\mathbf{x}) = (1/N) \prod_{\nu=1}^{3} \hat{T}_{x'_{\nu}} \quad \exp\left[\int_{x_{\nu 0}}^{x_{\nu}} dx'_{\nu}\, \hat{O}_{\nu}(x'_{\nu})\right] \; ; \; N = 2,3,\ldots$$

we find that its action on the spin part of the operator $\hat{\Phi}(\mathbf{x},\hat{\sigma})$ is a gauge transformation. For the electron operators, we use the representation [9.2]

$$\hat{\psi}_{\alpha'}(\mathbf{x}|\Gamma) = \hat{B}(\Gamma)\, \hat{\psi}_{\alpha'}(\mathbf{x}) \,, \tag{9.1.5}$$

where the operator $\hat{B}(\Gamma)$ has the following form for an arbitrary SU(N) symmetry:

$$\hat{B}(\Gamma) = (1/N)\hat{T}\, \exp\left(\oint_{\Gamma} dx_{\nu}\hat{A}_{\nu}\right) \,, \tag{9.1.6}$$

and Γ is the contour on which the phase of the wave function of an electron with spin $\hat{\sigma}$ changes due to interaction with the exchange field of the surrounding electrons. The operator $\hat{B}(\Gamma)$ is sometimes called a Wilson loop [9.3]. We also make use of the relation

$$\hat{F}_{\mu\nu}\hat{B}(\Gamma) = \frac{\delta}{\delta S_{\mu\nu}}\, \hat{B}(\Gamma) \,, \tag{9.1.7}$$

where $\delta S_{\mu\nu}$ is the dimensionless increase in the area of the contour $\Gamma : \delta S_{\mu\nu} = (i/2)\tilde{g}\delta\sigma_{\mu\nu}$. In this case we can write the effective Hamiltonian of the superconducting phase in the contour representation:

$$\begin{aligned}
\hat{\mathbb{H}}_{\Gamma} = \; & \mathrm{NTr} \int_{\mathbf{x}\in\Gamma} d\mathbf{x} \Big\{ \hat{B}(\Gamma)\, \{\tfrac{1}{2}\, \hat{\psi}^{+}_{\alpha'}(\mathbf{x})\hat{\tilde{\mathbb{D}}}^{2}_{\nu}\, \hat{\psi}_{\alpha'}(\mathbf{x}) + \tfrac{1}{2}[\hat{\psi}^{t}_{\alpha'}(\mathbf{x})\, \hat{\Delta}^{+}_{\alpha'\gamma'}(\mathbf{x})\hat{\psi}_{\gamma'}(\mathbf{x}) \\
& + \hat{\psi}^{+}_{\alpha'}(\mathbf{x})\hat{\Delta}_{\alpha'\gamma'}(\mathbf{x})\, \hat{\psi}^{+t}_{\gamma'}(\mathbf{x})] + \sum_{\alpha,\varkappa} \hat{b}^{+}_{\alpha\varkappa}(\mathbf{x})\hat{\xi}_{\varkappa}(\mathbf{x})\; \hat{b}_{\alpha\varkappa}(\mathbf{x}) \\
& + \frac{i}{\sqrt{2}} \sum_{\alpha,\varkappa} g_{ph}(\mathbf{x})e_{\alpha\varkappa}\; \hat{n}_{\varkappa}(\mathbf{x})[\hat{b}_{\alpha\varkappa}(\mathbf{x}) - \hat{b}^{+}_{\alpha\varkappa}(\mathbf{x})]\hat{\psi}^{+}_{\alpha'}(\mathbf{x})\hat{\psi}_{\alpha'}(\mathbf{x})\Big\}\hat{B}^{+}(\Gamma) \\
& + [\; \tfrac{1}{2}(\frac{\delta}{\delta S_{\mu\nu}})^{2} + \hat{I}(\Gamma)]\hat{B}(\Gamma) \Big\} + \mathrm{NTr} \int_{\mathbf{x},\mathbf{x}'\in\Gamma} d\mathbf{x}d\mathbf{x}'\hat{B}(\Gamma)\psi^{+}_{\alpha'}(\mathbf{x}) \\
& \times \mathrm{Tr}[\hat{\psi}^{+}_{\beta'}(\mathbf{x}')V(\mathbf{x}-\mathbf{x}')\hat{\psi}_{\beta'}(\mathbf{x}')]\hat{\psi}_{\alpha'}(\mathbf{x})\hat{B}^{+}(\Gamma) \,,
\end{aligned}$$

$$\hat{\mathbb{D}}^2 = -\frac{\hbar^2}{\mu_e} \hat{\tilde{\mathbb{D}}}^+_\nu \hat{\tilde{\mathbb{D}}}^-_\nu \quad , \quad \hat{\tilde{\mathbb{D}}}^\pm_\nu = \hat{\mathbb{D}}_\nu \pm i\,(2\mu_e \mu/\hbar^2)^{\frac{1}{2}} \, . \tag{9.1.8}$$

In (9.1.8), $\hat{I}(\Gamma)$ is Lagrange's undetermined multiplier, $\hat{b}_{\alpha\varkappa}(\mathbf{x})$ and $b^+_{\alpha\varkappa}(\mathbf{x})$ are phonon operators in the coordinate representation, and $\varkappa$ is the index of the corresponding isolated unit cell in the crystal. The operators $\hat{\xi}_\varkappa(\mathbf{x})$ and $\hat{\eta}_\varkappa(\mathbf{x})$ satisfy the following equations:

$$\int d\mathbf{x}' \;\delta(\mathbf{x}-\mathbf{x}')\hat{\xi}_\varkappa(\mathbf{x}') \exp[i\mathbf{p}(\mathbf{x}'-\mathbf{x})] = \omega_{\varkappa ph}(\mathbf{p}) \; , \tag{9.1.9}$$

$$\int d\mathbf{x}' \;\delta(\mathbf{x} - \mathbf{x}')\hat{\eta}_\varkappa(\mathbf{x}') \exp[i\mathbf{p}(\mathbf{x}' - \mathbf{x})] = [\omega_{\varkappa ph}(\mathbf{p})]^{\frac{1}{2}} \tag{9.1.10}$$

where $\omega_{\varkappa ph}$ are phonon frequencies, and $\hat{\Delta}_{\alpha'\gamma'}(\mathbf{x})$ is the function determining the order parameter in the superconducting phase. Let us define the generalized δ-function on the contour Γ , since we shall require it in subsequent calculations:

$$\tilde{\delta}(\mathbf{x}-\mathbf{x}') = \oint_{\Gamma_{\mathbf{x}'',\tau''} \in \Gamma_{\mathbf{x},\tau,\; \mathbf{x}',\tau'}} dx''_\nu \; e_\nu \delta(\mathbf{x}-\mathbf{x}'') \;\delta(\mathbf{x}''-\mathbf{x}')$$
$$+ \;\langle\, \delta(\mathbf{x}-\mathbf{x}')\hat{T} \;\ldots\; \frac{\partial}{\partial\tau'} \overleftrightarrow{\hat{T}r}\hat{A}_\nu(\mathbf{x}',\tau') \;\ldots \rangle \; . \tag{9.1.11}$$

The symbol Tr means that the operator Tr acts on all operators within the angle brackets $\langle \ldots \rangle$. Next, we introduce Green functions describing the behavior of the system in the superconducting state:

$$G_{\alpha\beta}(\mathbf{x},\tau|\Gamma|\;\mathbf{x}'.\tau') = \langle\, \hat{T}\hat{\psi}_\alpha(\mathbf{x},\tau)\mathrm{Tr}\hat{B}(\Gamma)\psi^+_\beta(\mathbf{x}',\tau') \,\rangle \; , \tag{9.1.12}$$

$$F_{\alpha\beta}(\mathbf{x},\tau|\Gamma|\;\mathbf{x}',\tau') = \langle\, \hat{T}\hat{\psi}_\alpha(\mathbf{x},\tau)\mathrm{Tr}\hat{B}(\Gamma)\hat{\psi}_\beta(\mathbf{x}',\tau') \rangle \; , \tag{9.1.13}$$

$$F^+_{\alpha\beta}(\mathbf{x},\tau|\Gamma|\;\mathbf{x}',\tau') = \langle\, \hat{T}\psi^+_\alpha(\mathbf{x},\tau)\mathrm{Tr}\hat{B}(\Gamma)\hat{\psi}^+_\beta(\mathbf{x}',\tau') \,\rangle, \tag{9.1.14}$$

$$D_{\alpha\beta\varkappa\varkappa'}(\mathbf{x},\tau|\Gamma|\;\mathbf{x}',\tau') = \langle\, \hat{T}\hat{b}_{\alpha\varkappa}(\mathbf{x},\tau)\mathrm{Tr}\hat{B}(\Gamma)\hat{b}^+_{\beta\varkappa'}(\mathbf{x}',\tau') \,\rangle . \tag{9.1.15}$$

Formulas (9.1.12,15) describe electron and phonon Green functions [9.4], while (9.1.13,14) are used to denote condensate Green functions. Let us write down the equations of motion for the Green functions (9.1.12,14,15), as well as for the field operator $\hat{B}(\Gamma)$:

$$\frac{\partial}{\partial\tau} G_{\alpha\beta}(\mathbf{x},\tau|\Gamma|\ \mathbf{x}',\tau') = \delta_{\alpha\beta}\oint_\Gamma e_\nu d\mathbf{x}'_\nu\ \tilde{\delta}(\mathbf{x}-\mathbf{x}')\langle\ \mathrm{Tr}\hat{B}(\Gamma)\ \rangle$$

$$+\ \langle\hat{T}[\hat{\mathbb{H}}_\Gamma\hat{\psi}_\alpha(\mathbf{x},\tau)]\overleftrightarrow{\mathrm{Tr}}\hat{B}(\Gamma)\ \hat{\psi}^+_\beta(\mathbf{x}',\tau')\ \rangle \quad , \tag{9.1.16}$$

$$\frac{\partial}{\partial\tau}F^+_{\alpha\beta}(\mathbf{x},\tau|\Gamma|\ \mathbf{x}',\tau') = -\ \langle\ \hat{T}[\hat{\mathbb{H}}_\Gamma\ \hat{\psi}^+_\alpha(\mathbf{x},\tau)]\overleftrightarrow{\mathrm{Tr}}\hat{B}(\Gamma)\ \hat{\psi}^+_\beta(\mathbf{x}',\tau')\ \rangle, \tag{9.1.17}$$

$$\frac{\partial}{\partial\tau} D_{\alpha\beta\varkappa\varkappa'}(\mathbf{x},\tau|\Gamma|\ \mathbf{x}',\tau') = \delta_{\alpha\beta}\delta_{\varkappa\varkappa'}\oint_\Gamma dx'_\nu\ e_\nu\ \tilde{\delta}(\mathbf{x}-\mathbf{x}')\langle\ \mathrm{Tr}\hat{B}(\Gamma)\ \rangle$$

$$+\ \langle\hat{T}[\hat{\mathbb{H}}_\Gamma\ \hat{b}_{\alpha\varkappa}(\mathbf{x},\tau)]\overleftrightarrow{\mathrm{Tr}}\hat{B}(\Gamma)\hat{b}^+_{\beta\varkappa}(\mathbf{x}',\tau')\ \rangle \quad , \tag{9.1.18}$$

$$\langle\ \mathrm{Tr}\hat{\mathbb{D}}_\mu\ \hat{F}_{\mu\nu}\hat{B}(\Gamma) = \langle\ \mathrm{Tr}\hat{J}_\nu(\Gamma)\hat{B}(\Gamma)\ \rangle\ . \tag{9.1.19}$$

In this case, it can be easily shown that

$$\langle\ \mathrm{Tr}\hat{\mathbb{D}}_\mu\ \hat{F}_{\mu\nu}\hat{B}(\Gamma)\ \rangle = \nabla_\mu\ \frac{\delta}{\delta S_{\mu\nu}}\ \langle\ \mathrm{Tr}\hat{B}(\Gamma)\ \rangle\ . \tag{9.1.20}$$

Let us now compute the current in the system. We can write (9.1.19) as

$$\nabla_\mu\ \frac{\delta}{\delta S_{\mu\nu}}\langle\ \mathrm{Tr}\hat{B}(\Gamma)\ \rangle = \sum_{\gamma=1}^{N^2-1}\hat{\tau}^\gamma\ \frac{\delta}{\delta A^\gamma_\nu}\Big|\langle\ \hat{\mathbb{H}}_\Gamma\rangle + \langle\ \hat{J}_\nu(\Gamma)\ \rangle_\psi \Big|_{\Gamma\in\hat{I}(\Gamma)\hat{B}(\Gamma),\ \hat{I}(\Gamma)\to\hat{\tau}^0}\ . \tag{9.1.21}$$

The current in the electron subsystem is defined as

$$\langle\hat{J}_\nu(\Gamma)\rangle = -\ \frac{i}{4}\ g_1\ \lim_{\mathbf{x},\tau\to\mathbf{x}',\tau'}\ (\nabla_{x'_\nu} - \nabla_{x_\nu})\ \sum_{\gamma=1}^{N^2-1}\sum_{\delta=1}^{2}\mathrm{Tr}[\hat{G}^\gamma_{\delta\delta}(\mathbf{x},\tau|\Gamma|\mathbf{x}',\tau'),\hat{\tau}^\gamma]$$

$$+\ \sum_{\gamma=1}^{N^2-1}\hat{\tau}^\gamma\ \frac{\delta}{\delta A^\gamma_\nu}\Big|_\Gamma\ \langle\ \overleftrightarrow{\mathrm{Tr}}\hat{B}(\Gamma)\ \rangle\ , \tag{9.1.22}$$

where

$$\hat{G}^\gamma_{\alpha\beta}(\mathbf{x},\tau|\Gamma|\mathbf{x}',\tau') = \langle\ \hat{T}\hat{\tau}^\gamma\ \hat{\psi}_\alpha(\mathbf{x},\tau)\hat{B}(\Gamma)\ \hat{\psi}^+_\beta(\mathbf{x}',\tau')\ \rangle\ .$$

In this case, (9.1.21) can be reduced to

$$\nabla_\mu\ \frac{\delta}{\delta S_{\mu\nu}}\ \langle\ \mathrm{Tr}\hat{B}(\Gamma)\ \rangle + \frac{i}{4}\ g_1\ \lim_{\mathbf{x},\tau\to\mathbf{x}',\tau'}\ (\nabla_{x'_\nu} - \nabla_{x_\nu})$$

$$\times\ \sum_{\gamma=1}^{N^2-1}\sum_{\delta=1}^{2}\mathrm{Tr}[\hat{G}^\gamma_{\delta\delta}(\mathbf{x},\tau|\Gamma|\mathbf{x}',\tau'),\hat{\tau}^\gamma] = \sum_{\gamma=1}^{N^2-1}\hat{\tau}^\gamma\ \frac{\delta}{\delta A^\gamma_\nu}\Big|_\Gamma\langle\overleftrightarrow{\mathrm{Tr}}\hat{B}(\Gamma)\ \rangle\ .$$

The functional derivative is given by

$$\frac{\delta}{\delta A_\nu^\alpha} \langle \mathrm{Tr}\hat{B}(\Gamma) \rangle = N \langle \mathrm{Tr}\hat{B}(C_{\mathbf{x},\mathbf{x}'})\hat{\tau}^\alpha \hat{B}(C_{\mathbf{x}',\mathbf{x}}) \rangle . \qquad (9.1.23)$$

This gives

$$N \sum_{\gamma=1}^{N^2-1} \mathrm{Tr}\hat{\tau}^\gamma \hat{B}(C_{\mathbf{x},\mathbf{x}'})\hat{\tau}^\gamma \hat{B}(C_{\mathbf{x}',\mathbf{x}})$$
$$= N^2 [\langle \mathrm{Tr}\hat{B}(C_{\mathbf{x},\mathbf{x}'})\mathrm{Tr}\hat{B}(C_{\mathbf{x}',\mathbf{x}}) - \frac{1}{N^2}\langle \mathrm{Tr}B(\Gamma) \rangle]. \qquad (9.1.24)$$

Calculating the commutators in (9.1.16-18), we obtain the Green function and the field operator

$$\frac{\partial}{\partial\tau} G_{\alpha\beta}(\mathbf{x},\tau|\Gamma|\mathbf{x}',\tau') - \frac{1}{2} \mathrm{Tr}[\hat{\tilde{\mathbb{D}}}_\nu^2 \hat{G}_{\alpha\beta}(\mathbf{x},\tau|\Gamma| \mathbf{x}',\tau')]$$
$$+ [\Delta_{\alpha\gamma}(\mathbf{x},\tau|\Gamma) - \Delta^t_{\alpha\gamma}(\mathbf{x},\tau|\Gamma)]F^+_{\gamma\beta}(\mathbf{x},\tau|\Gamma|\mathbf{x}',\tau')$$
$$= \delta_{\alpha\beta}\oint_\Gamma e_\nu dx'_\nu \,\tilde{\delta}(\mathbf{x}-\mathbf{x}') \langle \mathrm{Tr}\hat{B}(\Gamma)\rangle$$
$$+ g_{ph}(\mathbf{x}) \langle \hat{T}\hat{\varphi}(\mathbf{x},\tau)\hat{\psi}_\alpha(\mathbf{x},\tau)\overleftrightarrow{\mathrm{Tr}}\hat{B}(\Gamma)\hat{\psi}^+_\beta(\mathbf{x}',\tau') \rangle$$
$$+ N \int d\mathbf{x}'' V(\mathbf{x}-\mathbf{x}'') \langle \hat{T}\, \psi^+_\gamma(\mathbf{x}'',\tau'')\hat{\psi}_\alpha(\mathbf{x}'',\tau'')\hat{\psi}_\alpha(\mathbf{x},\tau)\overleftrightarrow{\mathrm{Tr}}\hat{B}(\Gamma)\psi^+_\beta(\mathbf{x}',\tau')\rangle ,$$
$$(9.1.25)$$

$$\frac{\partial}{\partial\tau} F^+_{\alpha\beta}(\mathbf{x},\tau|\Gamma|\mathbf{x}',\tau') + \frac{1}{2}\mathrm{Tr}[\hat{\tilde{\mathbb{D}}}_\nu^2 \; F^+_{\alpha\beta}(\mathbf{x},\tau|\Gamma|\mathbf{x}',\tau')]$$
$$+ [\Delta^*_{\alpha\gamma}(\mathbf{x},\tau|\Gamma) - \Delta^{*t}_{\alpha\gamma}(\mathbf{x},\tau|\Gamma)]G_{\gamma\beta}(\mathbf{x},\tau|\Gamma|\mathbf{x}',\tau')$$
$$= -g_{ph}(\mathbf{x}) \langle \hat{T}\hat{\varphi}(\mathbf{x},\tau)\hat{\psi}^+_\alpha(\mathbf{x},\tau)\overleftrightarrow{\mathrm{Tr}}\hat{B}(\Gamma)\psi^+_\beta(\mathbf{x}',\tau') \rangle$$
$$- N\int d\mathbf{x}'' V(\mathbf{x}-\mathbf{x}'') \langle \hat{T}\hat{\psi}^+_\alpha(\mathbf{x},\tau)\hat{\psi}^+_\gamma(\mathbf{x}'',\tau'')\hat{\psi}_\gamma(\mathbf{x}'',\tau'')\overleftrightarrow{\mathrm{Tr}}\hat{B}(\Gamma)\; \hat{\psi}^+_\beta(\mathbf{x}',\tau') \rangle ,$$
$$(9.1.26)$$

$$[\frac{\partial}{\partial\tau} + \hat{\xi}_\varkappa(\mathbf{x})]D_{\alpha\beta\varkappa\varkappa'}(\mathbf{x},\tau|\Gamma|\mathbf{x}',\tau') = \delta_{\alpha\beta}\delta_{\varkappa\varkappa'}\oint_\Gamma e_\nu dx'_\nu \,\tilde{\delta}(\mathbf{x}-\mathbf{x}') \langle \mathrm{Tr}\hat{B}(\Gamma)\rangle$$
$$+ \frac{i}{\sqrt{2}} g_{ph}(\mathbf{x}) \langle \hat{T}\hat{\psi}^+_\gamma(\mathbf{x},\tau)\hat{\psi}_\gamma(\mathbf{x},\tau)\hat{\eta}_\varkappa(\mathbf{x})\overleftrightarrow{\mathrm{Tr}}\; \frac{1}{N}\exp[\int_{\varkappa\in\Gamma_{\mathbf{x},\tau}} dx_\nu \hat{\Omega}_{sing}]$$
$$\times \hat{B}(\Gamma)\hat{b}^+_{\beta\varkappa'}(\mathbf{x}',\tau') \rangle , \qquad (9.1.27)$$

$$\nabla_\mu \frac{\delta}{\delta S_{\mu\nu}} \langle \mathrm{Tr}\hat{B}(\Gamma) \rangle + \frac{i}{4} g_1 N^2 \lim_{\mathbf{x},\tau\to\mathbf{x}',\tau'} (\nabla_{x'_\nu} - \nabla_{x_\nu})$$
$$\sum_{\gamma=1}^{N^2-1} \sum_{\delta=1}^{2} \mathrm{Tr}[\hat{\tilde{G}}^\gamma_{\delta\delta}(\mathbf{x},\tau|\Gamma| \mathbf{x}',\tau'),\hat{\tau}^\gamma]$$

$$= \frac{i}{2} g_1 N^2 \oint_\Gamma dx'_\nu \, \delta(\mathbf{x}-\mathbf{x}')[\langle \mathrm{Tr}\hat{B}(C_{\mathbf{x},\mathbf{x}'})\mathrm{Tr}\hat{B}(C_{\mathbf{x}',\mathbf{x}})\rangle$$

$$- \frac{1}{N^2} \langle \mathrm{Tr}\hat{B}(\Gamma)\rangle] . \qquad (9.1.28)$$

The many-particle averages in (9.1.24-27) are defined by the density matrix which contains the Hamiltonian system $\hat{\mathbb{H}}_\Gamma$ (9.1.8). In their existing form, (9.1.25-28) are unsuitable for investigation. Hence we must determine the rules for separation of many-particle averages in order to find the solutions of these equations. We introduce the Fourier contour representations for electron and phonon Green functions:

$$G_{\alpha\beta}(\mathbf{x},\tau|\Gamma|\mathbf{x}',\tau') = \frac{N}{N^2-1} \int_{-\varkappa\in\Gamma} \frac{D\varkappa(\mathbf{x}'',\tau'')}{2\pi C_\varkappa} \frac{d\mathbf{p}}{(2\pi)^3} \frac{d\omega}{2\pi} G_{\alpha\beta\varkappa}(\mathbf{p}, i\omega)$$

$$\times (\langle \mathrm{Tr}\hat{B}^{(1)}(\Gamma)\mathrm{Tr}\hat{B}(\varkappa)\rangle - \frac{1}{N^2} \langle \mathrm{Tr}\hat{B}^{(1)}(\Gamma+\varkappa)\rangle)$$

$$\times \exp\left\{\int_\varkappa dx''_\nu \, p_{o\nu}[\varkappa] + i\mathbf{p}(\mathbf{x}-\mathbf{x}') + i\omega(\tau-\tau')\right\} , \qquad (9.1.29)$$

$$F_{\alpha\beta}(\mathbf{x},\tau|\Gamma|\mathbf{x}',\tau') = \frac{N}{N^2-1} \int_{-\varkappa\in\Gamma} \frac{D\varkappa(\mathbf{x}'',\tau'')}{2\pi C_\varkappa} \frac{d\mathbf{p}}{(2\pi)^3} \frac{d\omega}{2\pi} F_{\alpha\beta\varkappa}(\mathbf{p}, i\omega)$$

$$\times (\langle \mathrm{Tr}\hat{B}^{(2)}(\Gamma)\mathrm{Tr}\hat{B}(\varkappa)\rangle - \frac{1}{N^2} \langle \mathrm{Tr}\hat{B}^{(2)}(\Gamma+\varkappa)\rangle)$$

$$\times \exp\left\{\int_\varkappa dx''_\nu \, p_{o\nu}[\varkappa] + i\mathbf{p}(\mathbf{x}-\mathbf{x}') + i\omega(\tau-\tau')\right\} , \qquad (9.1.30)$$

$$F^*_{\alpha\beta}(\mathbf{x},\tau|\Gamma|\mathbf{x}',\tau') = \frac{N}{N^2-1} \int_{-\varkappa\in\Gamma} \frac{D\varkappa(\mathbf{x}'',\tau'')}{2\pi C_\varkappa} \frac{d\mathbf{p}}{(2\pi)^3} \frac{d\omega}{2\pi} F^*_{\alpha\beta\varkappa}(\mathbf{p}, i\omega)$$

$$\times (\langle \mathrm{Tr}\hat{B}^{(2)}(\Gamma)\mathrm{Tr}\hat{B}(\varkappa)\rangle - \frac{1}{N^2} \langle \mathrm{Tr}\hat{B}^{(2)}(\Gamma+\varkappa)\rangle)$$

$$\times \exp\left\{\int_\varkappa dx''_\nu \, p_{o\nu}[\varkappa] + i\mathbf{p}(\mathbf{x}-\mathbf{x}') + i\omega(\tau-\tau')\right\} , \qquad (9.1.31)$$

$$D_{\alpha\beta\varkappa\varkappa'}(\mathbf{x},\tau|\Gamma|\mathbf{x}',\tau') = \frac{N}{N^2-1} \int_{-\varkappa_1\in\Gamma} \frac{D\varkappa_1(\mathbf{x}'',\tau'')}{2\pi C_\varkappa} \frac{d\mathbf{p}}{(2\pi)^3} \frac{d\omega}{2\pi} D_{\alpha\beta\varkappa\varkappa'\varkappa_1}(\mathbf{p}, i\omega)$$

$$\times (\langle \mathrm{Tr}\hat{B}^{(1)}(\Gamma)\mathrm{Tr}\,\frac{1}{N}\hat{T} \exp(\int_{\varkappa_1} dx''_\nu \hat{\Omega}_\nu \,\mathrm{sing}) \rangle - \frac{1}{N^2}\langle \mathrm{Tr}\hat{B}^{(1)}(\Gamma+\varkappa_1)\rangle)$$

$$\times \exp\left\{\int_{\varkappa_1} dx''_\nu \, p_{o\nu}[\varkappa_1] + i\mathbf{p}(\mathbf{x}-\mathbf{x}') + i\omega(\tau-\tau')\right\} . \qquad (9.1.32)$$

It follows from (9.1.29-32) that the Green functions introduced by us can be conditionally treated as "single-particle" functions. Indeed, the averages contain "dressed" electron and phonon operators due to the interaction of electrons and phonons with the fluctuating field of a disordered system of spins (see also 9.1.5). Hence the mechanism of interaction of "quasi-particles" in the superconducting phase, which leads to the formation of superconducting electrons, will involve the fluctuations of spin- and charge-densities which were shown in Sect. 8.2 to be inseparable. This is the principal difference of our mechanism from that of **Bardeen**, **Cooper**, and **Schrieffer** [9.5].

9.2 Separation of Many-Particle Averages in the Superconducting Phase Equations: Wick's Theorem

Let us consider (9.1.25-28). We can transform these equations to

$$
\begin{aligned}
&\left(\frac{\delta}{\delta C_{\tau\tau'}}+\frac{\partial}{\partial\tau}\right) G_{\alpha\beta}(\mathbf{x},\tau|\Gamma|\mathbf{x}',\tau') - \frac{1}{2}\,\mathrm{Tr}[\hat{\tilde{\mathbb{D}}}{}_\nu^2\; G_{\alpha\beta}(\mathbf{x},\tau|\Gamma|\mathbf{x}',\tau')] \\
&\quad + [\Delta_{\alpha\gamma}(\mathbf{x},\tau|\Gamma) - \Delta^t_{\alpha\gamma}(\mathbf{x},\tau|\Gamma)]F^*_{\gamma\beta}(\mathbf{x},\tau|\Gamma|\mathbf{x}',\tau') \\
&\quad = \delta_{\alpha\beta}\oint_\Gamma e_{\nu'}dx_{\nu'}\tilde{\delta}(\mathbf{x}-\mathbf{x}')\ \langle\, \mathrm{Tr}\hat{B}(\Gamma)\,\rangle \\
&\quad - N^2\int d\mathbf{x}''g_{ph}(\mathbf{x})g_{ph}(\mathbf{x}'')\ \langle\, \hat{T}\ \hat{\varphi}(\mathbf{x},\tau)\hat{\varphi}(\mathbf{x}'',\tau'')\hat{\psi}_\alpha(\mathbf{x},\tau)\overleftrightarrow{\mathrm{Tr}}\hat{B}(\Gamma) \\
&\quad\quad \hat{\psi}^+_\beta(\mathbf{x}',\tau')\hat{\psi}^+_\gamma(\mathbf{x}'',\ '')\hat{\psi}_\gamma(\mathbf{x}'',\tau'')\,\rangle \\
&\quad + N\int d\mathbf{x}''\ V(\mathbf{x}-\mathbf{x}'')\ \langle\, \hat{T}\psi^+_\gamma(\mathbf{x}'',\tau'')\hat{\psi}_\gamma(\mathbf{x}'',\tau'')\hat{\psi}_\alpha(\mathbf{x},\tau)\mathrm{Tr}\hat{B}(\Gamma)\ \hat{\psi}^+_\beta(\mathbf{x}',\tau')\rangle\,,
\end{aligned} \tag{9.2.1}
$$

$$
\begin{aligned}
&\left(\frac{\delta}{\delta C_{\tau\tau'}}+\frac{\partial}{\partial\tau}\right) F^*_{\alpha\beta}(\mathbf{x},\tau|\Gamma|\mathbf{x}',\tau') + \frac{1}{2}\,\mathrm{Tr}[\hat{\tilde{\mathbb{D}}}{}_\nu^2\hat{F}^*_{\alpha\beta}\,(\mathbf{x},\tau|\Gamma|\mathbf{x}',\tau')] \\
&\quad + [\Delta^*_{\alpha\gamma}(\mathbf{x},\tau|\Gamma) - \Delta^{*t}_{\alpha\gamma}(\mathbf{x},\tau|\Gamma)]G_{\gamma\beta}(\mathbf{x},\tau|\Gamma|\mathbf{x}',\tau') \\
&\quad = N^2\int d\mathbf{x}''g_{ph}(\mathbf{x})g_{ph}(\mathbf{x}'')\ \langle\, \hat{T}\hat{\varphi}(\mathbf{x},\tau)\ \hat{\varphi}\ (\mathbf{x}'',\tau'')\hat{\psi}^+_\alpha(\mathbf{x},\tau) \\
&\quad \times \overleftrightarrow{\mathrm{Tr}}\hat{B}(\Gamma)\ \hat{\psi}^+_\beta(\mathbf{x}',\tau')\hat{\psi}^+_\gamma(\mathbf{x}'',\tau'')\hat{\psi}_\gamma(\mathbf{x}'',\tau'')\,\rangle\ - N\int d\mathbf{x}''\ V(\mathbf{x}'-\mathbf{x}'') \\
&\quad \times \langle\, \hat{T}\ \hat{\psi}^+_\alpha(\mathbf{x},\tau)\ \hat{\psi}^+_\gamma(\mathbf{x}'',\tau'')\hat{\psi}_\gamma(\mathbf{x}'',\tau'')\overleftrightarrow{\mathrm{Tr}}\hat{B}(\Gamma)\hat{\psi}^+_\beta(\mathbf{x}',\tau')\,\rangle\,,
\end{aligned} \tag{9.2.2}
$$

$$\left(\frac{\delta}{\delta C_{\tau\tau'}}+\frac{\partial}{\partial\tau}\right) D_{\alpha\beta\varkappa\varkappa'}(\mathbf{x},\tau|\Gamma|\mathbf{x}',\tau') + \hat{\xi}_{\varkappa}(\mathbf{x}) D_{\alpha\beta\varkappa\varkappa'}(\mathbf{x},\tau|\Gamma|\mathbf{x}',\tau')$$

$$= \delta_{\alpha\beta}\delta_{\varkappa\varkappa'}\oint_\Gamma e_{\nu'} dx'_{\nu'}\,\tilde{\delta}(\mathbf{x}-\mathbf{x}')\ \langle \mathrm{Tr}\hat{B}(\Gamma)\rangle - \frac{i}{\sqrt{2}} N^2 \int d\mathbf{x}'' g_{ph}(\mathbf{x}) g_{ph}(\mathbf{x}'')$$

$$\times \langle \hat{T}\hat{\psi}^+_\gamma(\mathbf{x},\tau)\hat{\psi}_\gamma(\mathbf{x},\tau)\hat{\eta}_\varkappa(\mathbf{x})\ \frac{1}{N}\ \overleftrightarrow{\mathrm{Tr}}\hat{T}\ \exp(\oint_\Gamma dx'_\nu\ \hat{\Omega}_{\nu\ \mathrm{sing}})$$

$$\times\ \hat{B}(\Gamma)\hat{b}^+_{\beta\varkappa'}(\mathbf{x}',\tau')\hat{\varphi}(\mathbf{x}'',\tau'')\ \hat{\psi}^+_\delta(\mathbf{x}'',\tau'')\hat{\psi}_\delta(\mathbf{x}'',\tau'')\rangle\ , \qquad (9.2.3)$$

$$\nabla_\mu \frac{\delta}{\delta S_{\mu\nu}} \langle\ \mathrm{Tr}\hat{B}(\Gamma)\ \rangle + \frac{i}{4} g_1\ N^2 \lim_{\mathbf{x},\tau\to\mathbf{x}',\tau'} (\nabla_{x'_\nu} - \nabla_{x_\nu})$$

$$\sum_{\gamma=1}^{N^2-1}\ \sum_{\delta=1}^{2}\ \mathrm{Tr}[\hat{\tilde{G}}^\gamma_{\delta\delta}(\mathbf{x},\tau|\Gamma|\mathbf{x}',\tau'),\hat{\tau}^\gamma]$$

$$= \frac{i}{2} g_1\ N^2 \oint_\Gamma dx'_\nu\ \delta(\mathbf{x}-\mathbf{x}')(\langle \mathrm{Tr}\hat{B}(C_{\mathbf{x},\mathbf{x}'})\mathrm{Tr}\hat{B}(C_{\mathbf{x}',\mathbf{x}}) - \frac{1}{N^2}\langle \mathrm{Tr}\hat{B}(\Gamma)\rangle),$$

$$d\mathbf{x}'' \to d\mathbf{x}'' d\tau''\ , \qquad V(\mathbf{x}-\mathbf{x}'') \to V(\mathbf{x}-\mathbf{x}'')\delta(\tau-\tau'')\ . \qquad (9.2.4)$$

Equations (9.2.1-4) are not closed. For their closure, we must write an infinite chain of equations of motion for many-particle averages on the right-hand side. Such an approach is not effective, however. Hence we formulate the rules according to which the many-particle averages in (9.2.1-3) can be expressed in terms of electron and phonon Green functions.

Let us consider the integral

$$-N^2 \int d\mathbf{x}'' g_{ph}(\mathbf{x}) g_{ph}(\mathbf{x}'')\ \langle \hat{T}\hat{\varphi}(\mathbf{x},\tau)\hat{\psi}_\alpha(\mathbf{x},\tau)\overleftrightarrow{\mathrm{Tr}}\hat{B}(\Gamma)\hat{\psi}^+_\beta(\mathbf{x}',\tau')\hat{\varphi}(\mathbf{x}'',\tau'')$$

$$\times\ \hat{\psi}^+_\gamma(\mathbf{x}'',\tau'')\hat{\psi}_\gamma(\mathbf{x}'',\tau'')\rangle\ = -N^2 \int d\mathbf{x}'' g_{ph}(\mathbf{x}) g_{ph}(\mathbf{x}'')$$

$$\times\ \langle\langle \hat{T}\varphi(\mathbf{x},\tau)\varphi(\mathbf{x}'',\tau'')\hat{\psi}_\alpha(\mathbf{x},\tau)\hat{\psi}^+_\beta(\mathbf{x}',\tau')\hat{\psi}^+_\gamma(\mathbf{x}'',\tau'')\hat{\psi}_\gamma(\mathbf{x}'',\tau'')\rangle$$

$$\times\ \mathrm{Tr}\ \frac{1}{N}\ \hat{T}\ \exp(\oint_{\Gamma_{\mathbf{x},\tau}} dx'_\nu \Omega_{\nu\ \mathrm{sing}})\hat{B}(\Gamma_{\mathbf{x},\tau;\mathbf{x}',\tau'})\ \frac{1}{N}\ \hat{T}\ \exp(\oint_{\Gamma_{\mathbf{x}'',\tau''}} dx_1\hat{\Omega}_{\nu\ \mathrm{sing}})$$

$$\times\ \hat{B}(\Gamma_{\mathbf{x}'',\tau'';\mathbf{x}'',\tau''})\rangle\ . \qquad (9.2.5)$$

Considering that the operators $\hat{\psi}_\alpha(\mathbf{x},\tau)$ and $\hat{\psi}^+_\beta(\mathbf{x}',\tau')$ obey Fermi-Dirac statistics, and that the phonon operators commute with the field operator $\hat{A}_\nu$, (9.2.5) can be transformed as

$$- \frac{1}{2} N^2 \int d\mathbf{x}'' g_{ph}(\mathbf{x}) g_{ph}(\mathbf{x}'') \sum_{\alpha',\alpha'',\varkappa',\varkappa''} [\langle \hat{T}\hat{\eta}_{\varkappa'}(\mathbf{x})\hat{\eta}_{\varkappa''}(\mathbf{x}'')\hat{b}_{\alpha'\varkappa'}(\mathbf{x},\tau)$$

$$\times b^+_{\alpha''\varkappa''}(\mathbf{x}'',\tau'') \rangle + \langle \hat{T}\hat{\eta}_{\varkappa'}(\mathbf{x})\hat{\eta}_{\varkappa''}(\mathbf{x}'')\hat{b}^+_{\alpha'\varkappa'}(\mathbf{x},\tau)\hat{b}_{\alpha''\varkappa''}(\mathbf{x}'',\tau'') \rangle]$$

$$\times [\langle \hat{T}\hat{\psi}_\gamma(\mathbf{x},\tau)\hat{\psi}^+_\gamma(\mathbf{x}'',\tau'') \rangle \langle \hat{T}\hat{\psi}_\gamma(\mathbf{x}'',\tau'')\hat{\psi}^+_\beta(\mathbf{x}',\tau') \rangle$$

$$- \langle \hat{T}\hat{\psi}_\alpha(\mathbf{x},\tau)\hat{\psi}_\gamma(\mathbf{x}'',\tau'')\rangle \langle \hat{T}\, \hat{\psi}^+_\gamma(\mathbf{x}'',\tau'')\hat{\psi}^+_\beta(\mathbf{x}',\tau') \rangle]$$

$$\times \langle \mathrm{Tr}\hat{B}(\Gamma_{\mathbf{x}'',\tau'';\ \mathbf{x},\tau})\hat{B}(\Gamma_{\mathbf{x},\tau;\mathbf{x}'',\tau''})\hat{B}(\Gamma_{\mathbf{x}'',\tau'';\ \mathbf{x}',\tau'}) \rangle \ . \qquad (9.2.6)$$

In order to write this expression in terms of Green functions in their contour representation, we must decouple the averages

$$\langle \mathrm{Tr}\hat{B}(\Gamma_{\mathbf{x}'',\tau'';\mathbf{x},\tau})\hat{B}(\Gamma_{\mathbf{x},\tau;\mathbf{x}'',\tau''})\hat{B}(\Gamma_{\mathbf{x}'',\tau'';\mathbf{x}',\tau'}) \rangle \ .$$

Equation (9.2.6) can be written in the form of a functional integral:

$$- \frac{1}{2} \frac{N^2}{(N^2-1)^3} \int d\mathbf{x}'' \int_{\varkappa''} \frac{D\varkappa''_{\nu'} e_{\nu'}}{2\pi C_\varkappa} \delta(\varkappa' - \varkappa'') g_{ph}(\mathbf{x}) g_{ph}(\mathbf{x}'')$$

$$\times \sum_{\alpha'} \sum_{\alpha_1,\alpha_2,\alpha_3=1}^{N^2-1} \int_{-\varkappa_1\in\Gamma_{\mathbf{x}'',\tau'';\mathbf{x},\tau};\ -\varkappa_2\in\Gamma_{\mathbf{x},\tau;\mathbf{x}'',\tau''};\ -\varkappa_3\in\Gamma_{\mathbf{x}'',\tau'';\mathbf{x}',\tau'}} \frac{D\varkappa_1(\mathbf{x}''_1,\tau''_1)}{2\pi C_\varkappa} \frac{D\varkappa_2(\mathbf{x}''_2,\tau''_2)}{2\pi C_\varkappa} \frac{D\varkappa_3(\mathbf{x}''_3,\tau''_3)}{2\pi C_\varkappa}$$

$$\times [\langle \hat{\eta}_{\varkappa_1}(\mathbf{x})\, \hat{\eta}_{\varkappa_1}(\mathbf{x}'') b_{\alpha'\varkappa_1}(\mathbf{x},\tau) b^*_{\alpha'\varkappa_1}(\mathbf{x}'',\tau'') \rangle$$

$$+ \langle \hat{\eta}_{\varkappa_1}(\mathbf{x})\hat{\eta}_{\varkappa_1}(\mathbf{x}'') b^*_{\alpha'\varkappa_1}(\mathbf{x},\tau) b_{\alpha'\varkappa_1}(\mathbf{x}'',\tau'') \rangle]$$

$$\times [\langle \psi_{\alpha\varkappa_2}(\mathbf{x},\tau)\psi^*_{\gamma\varkappa_2}(\mathbf{x}'',\tau'')\rangle \langle \psi_{\gamma\varkappa_3}(\mathbf{x}'',\tau'')\psi^*_{\beta\varkappa_3}(\mathbf{x}',\tau')$$

$$- \langle \psi_{\alpha\varkappa_2}(\mathbf{x},\tau)\psi_{\gamma\varkappa_2}(\mathbf{x}'',\tau'') \rangle$$

$$\times \langle \psi^*_{\gamma\varkappa_3}(\mathbf{x}'',\tau'')\psi^*_{\beta\varkappa_3}(\mathbf{x}',\tau') \rangle] \langle \mathrm{Tr}[\hat{\tau}^{\alpha_1}\hat{B}(\Gamma_1)\hat{\tau}^{\alpha_1}\hat{B}(\varkappa_1)\hat{\tau}^{\alpha_2}\hat{B}(\Gamma_2)$$

$$\times \hat{\tau}^{\alpha_2} B(\varkappa_2)\, \hat{\tau}^{\alpha_3} \hat{B}(\Gamma_3) \hat{\tau}^{\alpha_3} \hat{B}(\varkappa_3)] \rangle \prod_{i=1}^{3} \exp(\int_{\varkappa_i} dx''_\nu\, p_{io\nu}[\varkappa_i])$$

$$\times \int_{\varkappa_1} \frac{D\varkappa_{1\nu'} e_{\nu'}}{2\pi C_\varkappa} \delta(\varkappa_1 + \varkappa_2 + \varkappa_3) \ . \qquad (9.2.7)$$

We express the field part of (9.2.7) through functional derivatives:

$$\prod_{i=1}^{3} \exp(\int_{\varkappa_i} dx''_\nu\, p_{oi\nu}[\varkappa_i]) \langle \operatorname{Tr}[\hat{\tau}^{\alpha_1} \hat{B}(\Gamma_1) \hat{\tau}^{\alpha_1} \hat{B}(\varkappa_1) \hat{\tau}^{\alpha_2} \hat{B}(\Gamma_2) \hat{\tau}^{\alpha_2} \hat{B}(\varkappa_2)$$

$$\times \hat{\tau}^{\alpha_3} \hat{B}(\Gamma_3) \hat{\tau}^{\alpha_3} \hat{B}(\varkappa_3)] \rangle = \prod_{i=1}^{3} \exp(\int_{\varkappa_i} dx''_\nu\, p_{oi\nu}[\varkappa_i])$$

$$\times \frac{\delta}{\delta A_\nu^{\alpha_3}} \hat{\tau}^{\alpha_3} \Big|_{\Gamma_3, \varkappa_3} \frac{\delta}{\delta A_\nu^{\alpha_2}} \hat{\tau}^{\alpha_2} \Big|_{\Gamma_2, \varkappa_2} \frac{\delta}{\delta A_\nu^{\alpha_1}} \hat{\tau}^{\alpha_1} \Big|_{\Gamma_1, \varkappa_1} \langle \operatorname{Tr} \hat{B}(\Gamma_1) \hat{B}(\Gamma_2) \hat{B}(\Gamma_3) \rangle \ . \qquad (9.2.8)$$

In this expression, the functional derivatives $\frac{\delta}{\delta A_\nu^{\alpha_i}} \hat{\tau}^{\alpha_i}$ act on the operator $\hat{B}(\Gamma_i \pm \Gamma)$ under the symbol Tr. Computing the functional derivatives in (9.2.8), we obtain

$$\prod_{i=1}^{3} \exp(\int_{\varkappa_i} dx''_\nu\, p_{oi\nu}[\varkappa_i]) N [N^2 \langle \operatorname{Tr}\hat{B}(\Gamma_1) \operatorname{Tr}\hat{B}(\Gamma_2) \operatorname{Tr}\hat{B}(\Gamma_3) \operatorname{Tr}\hat{B}(\varkappa_1 + \varkappa_2 + \varkappa_3) \rangle$$

$$- N \langle \operatorname{Tr}\hat{B}(\Gamma_1) \operatorname{Tr}\hat{B}(\Gamma_2) \operatorname{Tr}\hat{B}(\Gamma_3 + \varkappa_1 + \varkappa_2 + \varkappa_3) \rangle - N \langle \operatorname{Tr}\hat{B}(\Gamma_1) \operatorname{Tr}\hat{B}(\Gamma_2 + \varkappa_1 + \varkappa_2 + \varkappa_3)$$

$$\times \operatorname{Tr}\hat{B}(\Gamma_3) \rangle - N \langle \operatorname{Tr}\hat{B}(\Gamma_1 + \varkappa_1 + \varkappa_2 + \varkappa_3) \operatorname{Tr}\hat{B}(\Gamma_2) \operatorname{Tr}\hat{B}(\Gamma_3) \rangle$$

$$+ \langle \operatorname{Tr}\hat{B}(\Gamma_1) \operatorname{Tr}\hat{B}(\Gamma_2 + \Gamma_3 + \varkappa_1 + \varkappa_2 + \varkappa_3) \rangle + \langle \operatorname{Tr}\hat{B}(\Gamma_1 + \Gamma_3 + \varkappa_1 + \varkappa_2 + \varkappa_3) \operatorname{Tr}\hat{B}(\Gamma_2) \rangle$$

$$+ \langle \operatorname{Tr}\hat{B}(\Gamma_1 + \Gamma_2 + \varkappa_1 + \varkappa_2 + \varkappa_3) \operatorname{Tr}\hat{B}(\Gamma_3) \rangle - \frac{1}{N} \langle \operatorname{Tr}\hat{B}(\Gamma_1 + \Gamma_2 + \Gamma_3 + \varkappa_1 + \varkappa_2 + \varkappa_3) \rangle]$$

$$\int_{\varkappa_1} \frac{D\varkappa_{1\nu'} e_{\nu'}}{2\pi C_\varkappa} \delta(\varkappa_1 + \varkappa_2 + \varkappa_3) \ . \qquad (9.2.9)$$

Using (9.2.9), we can write the final expression for a "three-particle" mean:

$$- \frac{1}{2} N^2 \left(\frac{N}{N^2 - 1}\right)^3 \int d\mathbf{x}'' g_{ph}(\mathbf{x}) g_{ph}(\mathbf{x}'')$$

$$\times \sum_{\alpha'} \int_{-\varkappa_1 \in \Gamma_{\mathbf{x}'',\tau'';\mathbf{x},\tau};\ -\varkappa_2 \in \Gamma_{\mathbf{x},\tau;\mathbf{x}'',\tau''};\ -\varkappa_3 \in \Gamma_{\mathbf{x}'',\tau'';\mathbf{x}',\tau'}} \frac{D\varkappa_1(\mathbf{x}_1'',\tau_1'')}{2\pi C_\varkappa} \frac{D\varkappa_2(\mathbf{x}_2'',\tau_2'')}{2\pi C_\varkappa} \frac{D\varkappa_3(\mathbf{x}_3'',\tau_3'')}{2\pi C_\varkappa}$$

$$\times [\langle \hat{\eta}_{\varkappa_1}(\mathbf{x}) \hat{\eta}_{\varkappa_1}(\mathbf{x}'') b_{\alpha'\varkappa_1}(\mathbf{x},\tau) b^*_{\alpha'\varkappa_1}(\mathbf{x}'',\tau'') \rangle$$

$$+ \langle \hat{\eta}_{\varkappa_1}(\mathbf{x}) \hat{\eta}_{\varkappa_1}(\mathbf{x}'') b^*_{\alpha'\varkappa_1}(\mathbf{x},\tau) b_{\alpha'\varkappa_1}(\mathbf{x}'',\tau'') \rangle]$$

$$\times [\langle \psi_{\alpha\varkappa_2}(\mathbf{x},\tau) \psi^*_{\gamma\varkappa_2}(\mathbf{x}'',\tau'') \rangle \langle \psi_{\gamma\varkappa_3}(\mathbf{x}'',\tau'') \psi^*_{\beta\varkappa_3}(\mathbf{x}',\tau') \rangle$$

$$- \langle \psi_{\alpha\varkappa_2}(\mathbf{x},\tau)\ \psi_{\gamma\varkappa_2}(\mathbf{x}'',\tau'') \rangle \langle \psi^*_{\gamma\varkappa_3}(\mathbf{x}'',\tau'') \psi^*_{\beta\varkappa_3}(\mathbf{x}',\tau') \rangle]$$

$$\times \prod_{i=1}^{3} \exp\left(\int_{\varkappa_i} dx''_\nu\, p_{oi\nu}[\varkappa_i]\right) \Big\{ \left(N - \frac{3}{N}\right) \langle \mathrm{Tr}\hat{B}(\Gamma_{1\mathbf{x},\tau;\mathbf{x}'',\tau''})$$

$$\times \mathrm{Tr}\hat{B}(\Gamma_{2\mathbf{x},\tau;\mathbf{x}'',\tau''}) \mathrm{Tr}\hat{B}(\Gamma_{3\mathbf{x}'',\tau'';\mathbf{x}',\tau'}) \rangle$$

$$+ \frac{1}{N^2} [\langle \mathrm{Tr}\hat{B}(\Gamma_{1\mathbf{x},\tau;\mathbf{x}'',\tau''}) \mathrm{Tr}\hat{B}(\Gamma_{2\mathbf{x},\tau;\mathbf{x}'';\tau''} + \Gamma_{3\mathbf{x}'',\tau'';\mathbf{x}',\tau'}) \rangle$$

$$+ \langle \mathrm{Tr}\hat{B}(\Gamma_{1\mathbf{x},\tau;\mathbf{x}'',\tau''} + \Gamma_{3\mathbf{x}'',\tau'';\ \mathbf{x}',\tau'}) \mathrm{Tr}\hat{B}(\Gamma_{2\mathbf{x},\tau;\ \mathbf{x}'',\tau''}) \rangle$$

$$+ \langle \mathrm{Tr}\hat{B}(\Gamma_{1\mathbf{x},\tau;\ \mathbf{x}'',\tau''} + \Gamma_{2\mathbf{x},\tau;\ \mathbf{x}'',\tau''}) \mathrm{Tr}\hat{B}(\Gamma_{3\mathbf{x}'',\tau'';\mathbf{x}',\tau'}) \rangle]$$

$$- \frac{1}{N^3} \langle \mathrm{Tr}\hat{B}(\Gamma_{1\mathbf{x},\tau;\mathbf{x}'',\tau''} + \Gamma_{2\mathbf{x},\tau;\mathbf{x}'',\tau''} + \Gamma_{3\mathbf{x}'',\tau'';\mathbf{x}',\tau'}) \rangle \Big\} .$$

(9.2.10)

Recalling the definition of the electron and phonon Green functions (9.1.2-5) and considering the topological equivalence of contours with different unit cells $\varkappa_i$, we can write (9.2.10) in the required form:

$$- \frac{1}{2} N^2 f_3(N) \int_{\mathbf{x}'',\tau''\in\Gamma} d\mathbf{x}'' \; g_{ph}(\mathbf{x}) g_{ph}(\mathbf{x}'') \sum_{\alpha'} [\hat{\eta}(\mathbf{x})\hat{\eta}(\mathbf{x}'') D_{\alpha'\alpha'}(\mathbf{x},\tau|\Gamma|\mathbf{x}'',\tau'')$$

$$+ \hat{\eta}(\mathbf{x})\hat{\eta}(\mathbf{x}'') D^*_{\alpha'\alpha'}(\mathbf{x},\tau|\Gamma|\mathbf{x}'',\tau'')][G_{\alpha\gamma}(\mathbf{x},\tau|\Gamma|\mathbf{x}'',\tau'') G_{\gamma\beta}(\mathbf{x}'',\tau''|\Gamma|\mathbf{x}',\tau')$$

$$- F_{\alpha\gamma}(\mathbf{x},\tau|\Gamma|\mathbf{x}'',\tau'') F^*_{\gamma\beta}(\mathbf{x}'',\tau''|\Gamma|\mathbf{x}',\tau')] \; . \qquad (9.2.11)$$

The function $f_3(N)$ can be presented as a symbolic sequence of numerical series:

$$f_3(N) = \left(\frac{N}{N^2-1}\right)^3 \left\{(N - \frac{3}{N}) + \frac{3^*}{N^2} - \frac{1^{**}}{N^3}\right\}$$

$$= \left(\frac{N}{N^2-1}\right)^3 \left\{(N - \frac{3}{N}) + \frac{3}{(N^2-1)^2} (N - \frac{2}{N} + \frac{1^*}{N^2}) \right.$$

$$\left. - \frac{1}{(N^2-1)^3} \left[(N - \frac{3}{N}) + \frac{3^*}{N^2} - \frac{1^{**}}{N^3}\right]\right\} = \ldots \qquad (9.2.12)$$

In these expressions, the coefficients marked by asterisks should be treated as coefficients in front of the averages of the field operator derivatives which depend on the sums of the contours (see 9.2.9,10). Such averages require further separation, and hence we consider the sequence in (9.2.12). The series appearing in each sequence can be truncated in the parameter $1/(N^2 - 1)^2$ which is raised to a higher power in each subsequent iteration. The choice of the approximation is determined by the quantity N. Truncating a chain, we obtain a numerical factor on the order of unity in front of the product of the Green functions. The discrete function $f_3(N)$ has the asymptotic properties

$$\lim_{N\to 1} f_3(N) = 1 \; , \qquad \lim_{N\to\infty} f_3(N) = 1/N^2 \; .$$

The separation of the average

$$N \int d\mathbf{x}'' V(\mathbf{x}-\mathbf{x}'') \langle \; \hat{T}\hat{\psi}^+_\gamma(\mathbf{x}'',\tau'')\hat{\psi}_\gamma(\mathbf{x}'',\tau'')\hat{\psi}_\alpha(\mathbf{x},\tau) \mathrm{Tr}\hat{B}(\Gamma)\hat{\psi}^+_\beta(\mathbf{x}',\tau') \; \rangle$$

in (9.2.1,2) is carried out in the same way as for the average corresponding to the electron-phonon interaction. As a result we obtain

$$N \int d\mathbf{x}'' \; V(\mathbf{x}-\mathbf{x}'') \; \langle \; \hat{T}\hat{\psi}^+_\gamma(\mathbf{x}'',\tau'')\hat{\psi}_\gamma(\mathbf{x}'',\tau'')\hat{\psi}_\alpha(\mathbf{x},\tau) \overleftrightarrow{\mathrm{Tr}}\hat{B}(\Gamma)\hat{\psi}^+_\beta(\mathbf{x}',\tau')$$

$$= N\, f_2(N) \int_{\mathbf{x}'',\tau''\in\Gamma} d\mathbf{x}'' V(\mathbf{x}-\mathbf{x}'')[G_{\alpha\gamma}(\mathbf{x},\tau|\Gamma|\mathbf{x}'',\tau'')G_{\gamma\beta}(\mathbf{x}'',\tau''|\Gamma|\mathbf{x}',\tau')$$

$$- F_{\alpha\gamma}(\mathbf{x},\tau|\Gamma|\mathbf{x}'',\tau'')F^*_{\gamma\beta}(\mathbf{x}'',\tau''|\Gamma|\,\mathbf{x}',\tau')] \ , \qquad (9.2.13)$$

$$f_2(N) = \left(\frac{N}{N^2-1}\right)^2 \left(N - \frac{2}{N} + \frac{1^*}{N^2}\right) = \left(\frac{N}{N^2-1}\right)^2$$

$$\times \left[N - \frac{2}{N} + \frac{1}{(N^2-1)^2}\left(N - \frac{2}{N} + \frac{1^*}{N^2}\right)\right] = \ldots \qquad (9.2.14)$$

The function $f_2(N)$ has the following asymptotics:

$$\lim_{N\to 1} f_2(N) = 1 \ ; \qquad \lim_{N\to\infty} f_2(N) = 1/N \ .$$

If we separate the averages from a large number of operators, they will be distributed in a similar manner, and each derivative of a Green function will be preceded by a discrete function of the form

$$N^{k-1} f_k(N)$$

where

$$f_k(N) = \left(\frac{N}{N^2-1}\right)^k \left\{(N-1) + \sum_{\ell=0}^{k} (-1)^\ell C_k^\ell \ N^{\overbrace{\ell-1+\delta_{\ell 0}}^{-\ell\,***\ldots*}}\right\}, \qquad (9.2.15)$$

C_k^ℓ being the binomial coefficient. The asymptotic forms of the functions are apparent:

$$\lim_{N\to 1} f_k(N) = 1 \ , \qquad \lim_{N\to\infty} f_k(N) = 1/N^{k-1} \ .$$

Formulas (9.2.11,13) represent Wick's theorem for the electron and phonon operators in the controu representation.

Let us now write the closed equations for the Green functions. For this purpose, we make use of two more relations (9.1.11)

$$\frac{\partial}{\partial\tau} G_{\alpha\beta}(\mathbf{x},\tau|\Gamma|\mathbf{x}',\tau') = \frac{\partial^{(\omega)}}{\partial\tau} G_{\alpha\beta}(\mathbf{x},\tau|\Gamma|\mathbf{x}',\tau')$$

$$+ \mathrm{Tr}[\hat{A}_\tau \ , \ \hat{G}_{\alpha\beta}(\mathbf{x},\tau|\Gamma|\mathbf{x}',\tau')] \ , \qquad (9.2.16)$$

$$\oint_{\Gamma} dx'_{\nu'} e_{\nu'} \, \tilde{\delta}(\mathbf{x}-\mathbf{x}') \, \langle \, \mathrm{Tr}\hat{B}(\Gamma) \, \rangle_{\alpha\beta} = \delta_{\alpha\beta} \oint_{\Gamma_{\mathbf{x}'',\tau''}\in\Gamma_{\mathbf{x},\tau;\mathbf{x}',\tau'}} e_{\nu'} dx''_{\nu'} \, \delta(\mathbf{x}-\mathbf{x}'')\delta(\mathbf{x}''-\mathbf{x}')$$
$$\times \, \langle \mathrm{Tr}\hat{B}(\Gamma) \rangle + \mathrm{Tr}[\hat{A}_{\tau}, \hat{G}_{\alpha\beta}(\mathbf{x},\tau|\Gamma|\mathbf{x}',\tau')] \, . \qquad (9.2.17)$$

Here, $\frac{\partial^{(\omega)}}{\partial\tau}$ is the derivative with respect to time τ on which the electron operators $\hat{\psi}_{\alpha}(\mathbf{x},\tau)$ depend. A relation similar to (9.2.16) can be written for the phonon Green function $D_{\alpha\beta\varkappa\varkappa'}(\mathbf{x},\tau|\Gamma|\mathbf{x}',\tau')$ as well. Substituting (9.2.11,13,16,17) into (9.2.1-3), we obtain

$$\left(\frac{\delta}{\delta C_{\tau\tau'}} + \frac{\partial^{(\omega)}}{\partial\tau}\right) G_{\alpha\beta}(\mathbf{x},\tau|\Gamma|\mathbf{x}',\tau') - \frac{1}{2}\,\mathrm{Tr}[\hat{\tilde{\mathbb{D}}}{}^2_{\nu} \; G_{\alpha\beta}(\mathbf{x},\tau|\Gamma|\mathbf{x}',\tau')]$$
$$+ \, [\Delta_{\alpha\gamma}(\mathbf{x},\tau|\Gamma) - \Delta^{t}_{\alpha\gamma}(\mathbf{x},\tau|\Gamma)] F^{*}_{\alpha\beta}(\mathbf{x},\tau|\Gamma|\mathbf{x}',\tau')$$
$$= \delta_{\alpha\beta} \oint_{\Gamma} e_{\nu'} dx'_{\nu'} \delta(\mathbf{x}-\mathbf{x}') \langle \, \mathrm{Tr}\hat{B}(\Gamma) \, \rangle - \frac{1}{2} N^2 f_3(N) \int_{\mathbf{x}'',\tau''\in\Gamma} d\mathbf{x}'' g_{ph}(\mathbf{x}) g_{ph}(\mathbf{x}'')$$
$$\times \sum_{\alpha'} \hat{\eta}(\mathbf{x})\hat{\eta}(\mathbf{x}'')[D_{\alpha'\alpha'}(\mathbf{x},\tau|\Gamma|\mathbf{x}'',\tau'') + D^{*}_{\alpha'\alpha'}(\mathbf{x},\tau|\Gamma|\mathbf{x}'',\tau'')]$$
$$\times \, [G_{\alpha\gamma}(\mathbf{x},\tau|\Gamma|\mathbf{x}'',\tau'') G_{\gamma\beta}(\mathbf{x}'',\tau''|\Gamma| \, \mathbf{x}',\tau') - F_{\alpha\gamma}(\mathbf{x},\tau|\Gamma|\mathbf{x}'',\tau'')$$
$$\times \, F^{*}_{\gamma\beta}(\mathbf{x}'',\tau''|\Gamma| \, \mathbf{x}',\tau')] + N \, f_2(N) \int_{\mathbf{x}'',\tau''\in\Gamma} d\mathbf{x}'' V(\mathbf{x}-\mathbf{x}'')[G_{\alpha\gamma}(\mathbf{x},\tau|\Gamma| \, \mathbf{x}'',\tau'')$$
$$\times \, G_{\gamma\beta}(\mathbf{x}'',\tau''|\Gamma| \, \mathbf{x}',\tau') - F_{\alpha\gamma}(\mathbf{x},\tau|\Gamma|\mathbf{x}'',\tau'') F^{*}_{\gamma\beta}(\mathbf{x}'',\tau''|\Gamma|\mathbf{x}',\tau')] , \qquad (9.2.18)$$

$$\left(\frac{\delta}{\delta C_{\tau\tau'}} + \frac{\partial^{(\omega)}}{\partial\tau}\right) F^{*}_{\alpha\beta}(\mathbf{x},\tau|\Gamma|\mathbf{x}',\tau') + \frac{1}{2}\,\mathrm{Tr}[\hat{\tilde{\mathbb{D}}}{}^2 \, F^{*}_{\alpha\beta}(\mathbf{x},\tau|\Gamma|\mathbf{x}',\tau')]$$
$$+ \, [\Delta^{*}_{\alpha\gamma}(\mathbf{x},\tau|\Gamma) - \Delta^{*t}_{\alpha\gamma}(\mathbf{x},\tau|\Gamma)] G_{\gamma\beta}(\mathbf{x},\tau|\Gamma|\mathbf{x}',\tau')$$
$$= \frac{1}{2} N^2 f_3(N) \int_{\mathbf{x}'',\tau''\in\Gamma} d\mathbf{x}'' g_{ph}(\mathbf{x}) g_{ph}(\mathbf{x}'') \sum_{\alpha'} \hat{\eta}(\mathbf{x})\hat{\eta}(\mathbf{x}'')$$
$$\times \, [D_{\alpha'\alpha'}(\mathbf{x},\tau|\Gamma|\mathbf{x}'',\tau'') + D^{*}_{\alpha'\alpha'}(\mathbf{x},\tau|\Gamma|\mathbf{x}'',\tau'')]$$
$$\times \, [F^{*}_{\alpha\gamma}(\mathbf{x},\tau|\Gamma|\mathbf{x}'',\tau'') G_{\gamma\beta}(\mathbf{x}'',\tau''|\Gamma|\mathbf{x}',\tau')$$
$$- \, G^{*}_{\alpha\gamma}(\mathbf{x},\tau|\Gamma|\mathbf{x}'',\tau'') F^{*}_{\gamma\beta}(\mathbf{x}'',\tau''|\Gamma| \, \mathbf{x}',\tau')] \, , \qquad (9.2.19)$$

$$\left(\frac{\delta}{\delta C_{\tau\tau'}} + \frac{\partial^{(\omega)}}{\partial\tau}\right) D_{\alpha\beta}(\mathbf{x},\tau|\Gamma|\mathbf{x}',\tau') + \hat{\xi}(\mathbf{x})D_{\alpha\beta}(\mathbf{x},\tau|\Gamma|\mathbf{x}',\tau')$$

$$=\delta_{\alpha\beta}\oint_\Gamma e_{\nu'}dx'_{\nu'}\delta(\mathbf{x}-\mathbf{x}')\langle \mathrm{Tr}\hat{B}(\Gamma)\rangle - \tfrac{1}{2}N^2 f_3(N) \int_{\mathbf{x}'',\tau''\in\Gamma} d\mathbf{x}'' g_{ph}(\mathbf{x}) g_{ph}(\mathbf{x}'')$$

$$\times \hat{\eta}(\mathbf{x})[G_{\gamma\delta}(\mathbf{x},\tau|\Gamma|\mathbf{x}'',\tau'')G_{\delta\gamma}(\mathbf{x}'',\tau''|\Gamma|\mathbf{x},\tau)$$

$$- F_{\gamma\delta}(\mathbf{x},\tau|\Gamma|\mathbf{x}'',\tau'')F^*_{\delta\gamma}(\mathbf{x}'',\tau''|\Gamma|\mathbf{x},\tau)]\hat{\eta}(\mathbf{x}'')D_{\alpha\beta}(\mathbf{x}'',\tau''|\Gamma|\mathbf{x}',\tau') ,\tag{9.2.20}$$

$\delta(\mathbf{x}-\mathbf{x}') \to \delta(\mathbf{x}-\mathbf{x}')\ \delta(\tau-\tau')$; $d\mathbf{x}'' \to d\mathbf{x}''d\tau''$;

$V(\mathbf{x}-\mathbf{x}'') \to V(\mathbf{x}-\mathbf{x}'')\ \delta(\tau-\tau'')$; $N = 2,\quad N^2 f_3(N) \to 1;\ Nf_2(N) \to 1.$

$$\nabla_\mu \frac{\delta}{\delta S_{\mu\nu}} \langle \mathrm{Tr}\hat{B}(\Gamma)\rangle + \frac{i}{4} g_1 N^2 \lim_{\substack{\mathbf{x},\tau\to\mathbf{x}',\tau'\\ \varepsilon\to 0}} \int_{-\varkappa\in\Gamma_{\mathbf{x},\tau;\mathbf{x}',\tau'}} \frac{D\varkappa(\mathbf{x}'',\tau'')}{2\pi C_\varkappa}(\nabla_{x'_\nu} - \nabla_{x_\nu})$$

$$\times \sum_{\gamma=1}^{2} G_{\gamma\gamma\varkappa}(\mathbf{x},\tau|\varepsilon|\mathbf{x}',\tau')\exp\left(\int_\varkappa d\mathbf{x}''_\nu\, p_{o\nu}[\varkappa]\right)$$

$$\times\ [\langle \mathrm{Tr}\hat{B}(\Gamma+\varkappa)\rangle - \langle \mathrm{Tr}\hat{B}(\Gamma)\mathrm{Tr}\hat{B}(\varkappa)\rangle]$$

$$= \frac{i}{2} g_1 N^2 \oint_\Gamma dx'_\nu\ \delta(\mathbf{x}-\mathbf{x}') \times [\langle \mathrm{Tr}\hat{B}(C_{\mathbf{x},\mathbf{x}'})\mathrm{Tr}\hat{B}(C_{\mathbf{x}',\mathbf{x}})\rangle$$

$$- \frac{1}{N^2}\langle \mathrm{Tr}\hat{B}(\Gamma)\rangle] .\tag{9.2.21}$$

From (9.2.21), we can write down an expression for current including both superconducting and normal components. Indeed, if we make $\hat{A}_\nu$ tend to zero, we obtain an expression for current density that is well known in quantum mechanics [9.6]. In the absence of superconductivity and an external electromagnetic field, the current is equal to zero. Let us consider in greater detail the right-hand side of (9.2.21). Its appearance is formally connected with a change in the area of the contour Γ . By changing this area by $\delta S_{\mu\nu}$, we change the flux of the fluctuating field across the contour. A change in the flux leads to the induction of an effective electromotive force in the contour, which creates an electromagnetic field able to act on the quasi-particles. Consequently, the right-hand side in (9.2.21) can be interpreted as a "displacement current" resulting from the fluctuations of spin and charge densities. Thus,

$$\langle \hat{J}_\nu(\mathbf{x}) \rangle = \frac{i}{4} g_1 N^2 \lim_{\mathbf{x},\tau\to\mathbf{x}',\tau'} (\nabla_{x'_\nu} - \nabla_{x_\nu}) \sum_{\alpha=1}^{N^2-1} \sum_{\gamma=1}^{2} \mathrm{Tr}[\hat{\tau}^\alpha, \hat{\tilde{G}}^\alpha_{\gamma\gamma}(\mathbf{x},\tau|\Gamma|\mathbf{x}',\tau')]$$

$$+ \frac{i}{2} g_1 N^2 \oint_\Gamma dx'_\nu\, \delta(\mathbf{x}-\mathbf{x}') \sum_{\alpha=1}^{N^2-1} \mathrm{Tr}\,\hat{\tau}^\alpha \hat{B}(C_{\mathbf{x},\mathbf{x}'}) \hat{\tau}^\alpha \hat{B}(C_{\mathbf{x}',\mathbf{x}}) \rangle . \qquad (9.2.22)$$

In the tensor form, the expression for the current can be written as

$$\langle J^\alpha_\gamma(\mathbf{x}) \rangle = \frac{i}{4} g_1 N^2 \lim_{\mathbf{x},\tau\to\mathbf{x}',\tau'} (\nabla_{x'_\nu} - \nabla_{x_\nu})$$

$$\sum_{\gamma=1}^{2} \mathrm{Tr}[\hat{\tau}^\alpha, \hat{\tilde{G}}^\alpha_{\gamma\gamma}(\mathbf{x},\tau|\Gamma|\mathbf{x}',\tau')]$$

$$+ \frac{i}{2} g_1 N^2 \oint_\Gamma dx'_\nu\, \delta(\mathbf{x}-\mathbf{x}') \langle \mathrm{Tr}\,\hat{\tau}^\alpha \hat{B}(C_{\mathbf{x},\mathbf{x}'}) \hat{\tau}^\alpha \hat{B}(C_{\mathbf{x}',\mathbf{x}}) \rangle . \qquad (9.2.23)$$

Equations (9.2.18-21) have been written for the group SU(N/2), which is non-Abelian. If we consider the case in which the potential $\hat{A}_\nu$ is transformed only in the coordinate space (and not in the spin space), the corresponding gauge-field Hamiltonian is found to be invariant relative to the transformations of the Abelian symmetry group U(1). In this case, (9.2.18-21) are simplified and assume the form

$$\left(\frac{\delta}{\delta C_{\tau\tau'}} + \frac{\partial^{(\omega)}}{\partial\tau} - \frac{1}{2}\mathbb{D}^{02}_\nu\right) G_{\alpha\beta}(\mathbf{x},\tau|\Gamma|\mathbf{x}',\tau') + [\Delta_{\alpha\gamma}(\mathbf{x},\tau|\Gamma) - \Delta^t_{\alpha\gamma}(\mathbf{x},\tau|\Gamma)]$$

$$\times F^*_{\gamma\beta}(\mathbf{x},\tau|\Gamma|\mathbf{x}',\tau') = \delta_{\alpha\beta} \oint_\Gamma e_{\nu'} dx'_{\nu'}\, \delta(\mathbf{x}-\mathbf{x}') \langle \mathrm{Tr}\hat{B}(\Gamma) \rangle$$

$$- \frac{1}{2} \int_{\mathbf{x}'',\tau''\in\Gamma} d\mathbf{x}'' \Big\{ g_{ph}(\mathbf{x}) g_{ph}(\mathbf{x}'') \sum_{\alpha'} \hat{\eta}(\mathbf{x})\hat{\eta}(\mathbf{x}'')[D_{\alpha'\alpha'}(\mathbf{x},\tau|\Gamma|\mathbf{x}'',\tau'')$$

$$+ D^*_{\alpha'\alpha'}(\mathbf{x}',\tau|\Gamma|\mathbf{x}'',\tau'')] - 2V(\mathbf{x}-\mathbf{x}'') \Big\} [G_{\alpha\gamma}(\mathbf{x},\tau|\Gamma|\mathbf{x}'',\tau'') G_{\gamma\beta}(\mathbf{x}'',\tau''|\Gamma|\mathbf{x}',\tau')$$

$$- F_{\alpha\gamma}(\mathbf{x},\tau|\Gamma|\mathbf{x}'',\tau'') F^*_{\gamma\beta}(\mathbf{x}'',\tau''|\Gamma|\mathbf{x}',\tau')] , \qquad (9.2.24)$$

$$\left(\frac{\delta}{\delta C_{\tau\tau'}} + \frac{\partial^{(\omega)}}{\partial\tau} + \frac{1}{2}\mathbb{D}^{02}_\nu\right) F^*_{\gamma\beta}(\mathbf{x},\tau|\Gamma|\mathbf{x}',\tau') + [\Delta^*_{\alpha\gamma}(\mathbf{x},\tau|\Gamma)$$

$$- \Delta^{*t}_{\alpha\gamma}(\mathbf{x},\tau|\Gamma)] G_{\gamma\beta}(\mathbf{x},\tau|\Gamma|\mathbf{x}',\tau') = \frac{1}{2} \int_{\mathbf{x}'',\tau''\in\Gamma} d\mathbf{x}'' \Big\{ g_{ph}(\mathbf{x}) g_{ph}(\mathbf{x}'')$$

$$\times \sum_{\alpha'} \hat{\eta}(\mathbf{x})\hat{\eta}(\mathbf{x}'')[D_{\alpha'\alpha'}(\mathbf{x},\tau|\Gamma|\mathbf{x}'',\tau'') + D^{*}_{\alpha'\alpha'}(\mathbf{x},\tau|\Gamma|\mathbf{x}'',\tau'')]$$

$$- 2V(\mathbf{x}-\mathbf{x}'') \Big\} [F^{*}_{\alpha\gamma}(\mathbf{x},\tau|\Gamma|\mathbf{x}'',\tau'')G_{\gamma\beta}(\mathbf{x},\tau|\Gamma|\mathbf{x}'',\tau'')$$

$$- G^{*}_{\alpha\gamma}(\mathbf{x},\tau|\Gamma|\mathbf{x}'',\tau'')F^{*}_{\gamma\beta}(\mathbf{x}'',\tau''|\Gamma|\mathbf{x}',\tau')] , \qquad (9.2.25)$$

$$\left(\frac{\delta}{\delta C_{\tau\tau'}} + \frac{\partial^{(\omega)}}{\partial\tau} + \hat{\xi}(\mathbf{x})\right) D_{\alpha\beta}(\mathbf{x},\tau|\Gamma|\mathbf{x}',\tau') = \delta_{\alpha\beta}\oint_{\Gamma} e_{\nu'}dx'_{\nu'}\,\delta(\mathbf{x}-\mathbf{x}')$$

$$\times \langle \mathrm{Tr}\hat{B}(\Gamma)\rangle - \frac{1}{2}\int_{\mathbf{x}'',\tau''\in\Gamma} dx'' g_{ph}(\mathbf{x})g_{ph}(\mathbf{x}'')\, \hat{\eta}(\mathbf{x})[G_{\gamma\delta}(\mathbf{x},\tau|\Gamma|\mathbf{x}'',\tau'')$$

$$\times G_{\delta\gamma}(\mathbf{x}'',\tau''|\Gamma|\mathbf{x},\tau) - F_{\gamma\delta}(\mathbf{x},\tau|\Gamma|\mathbf{x}'',\tau'')F^{*}_{\delta\gamma}(\mathbf{x}'',\tau''|\Gamma|\mathbf{x},\tau)]\hat{\eta}(\mathbf{x}'')$$

$$\times D_{\alpha\beta}(\mathbf{x}'',\tau''|\Gamma|\mathbf{x}',\tau') , \qquad (9.2.26)$$

$$\langle \mathrm{Tr}\hat{B}(\Gamma)\rangle \to \langle B(\Gamma)\rangle \; ; \quad \delta(\mathbf{x}-\mathbf{x}') \to \delta(\mathbf{x}-\mathbf{x}')\,\delta(\tau-\tau') \; ;$$

$$dx'' \to d\mathbf{x}''d\tau'' \; ; \quad V(\mathbf{x}-\mathbf{x}'') \to V(\mathbf{x}-\mathbf{x}'')\,\delta(\tau-\tau'') .$$

$$\nabla_{\mu}\frac{\delta}{\delta S_{\mu\nu}}\langle B(\Gamma)\rangle + \frac{i}{4}g_1 \lim_{\substack{\mathbf{x},\tau\to\mathbf{x}',\tau' \\ \varepsilon\to 0}} \int_{-\varkappa\in\Gamma} \left.\frac{D\varkappa(\mathbf{x}'',\tau'')}{2\pi C_{\varkappa}}\right|_{\mathbf{x},\tau;\mathbf{x}',\tau'}$$

$$(\nabla_{x'_{\nu}} - \nabla_{x_{\nu}}) \sum_{\gamma=1}^{2} G_{\gamma\gamma\varkappa}(\mathbf{x},\tau|\varepsilon|\mathbf{x}',\tau')\exp\left(\int_{\varkappa} d\mathbf{x}''_{\nu}\, p_{o\nu}[\varkappa]\right)\langle B(\Gamma+\varkappa)\rangle$$

$$= \frac{i}{2}g_1 \oint_{\Gamma} dx'_{\nu}\;\delta(\mathbf{x}-\mathbf{x}')\langle B(\Gamma)\rangle .$$

Here,

$$\mathbb{D}^{02}_{\nu} \quad \mathbb{D}^{0+}_{\nu}\;\mathbb{D}^{0-}_{\nu} \; ; \; \mathbb{D}^{0\pm}_{\nu} = \frac{1}{\sqrt{\mu_e}}(\nabla_{\nu} - \frac{i}{2}g_1[A_{\nu}\,,\;])\pm i\sqrt{2\mu} \; ; \quad \mu\to\varepsilon_F .$$

These equations describe the state of a superconductor in the absence of a magnetic subsystem. The potential A_{ν} is just the vector potential of the electromagnetic field. These equations do not coincide with the equations for a superconductor obtained in [9.7]. This is because the s , d(f)-electrons forming coupled states may be paired, the momenta of pairing electrons being **p**, **p'**, $|\mathbf{p}| \neq |\mathbf{p}'|$. This is due to their interaction with the fluctuating electromagnetic field A_{ν} defined by the electron-ion system.

Consequently, the formation of electron pairs will involve, as before, the participation of fluctuations. Hence electron correlation functions have a structure analogous to the one defined by (9.1.19-22), while the equations for these functions will be more complex than those obtained in [9.7]. If the electromagnetic field fluctuations are not taken into account, (9.2.24-26) must be transformed into Gor'kov equations for the phonon model. In order to perform the limiting transition, we must "cut" the contour Γ along the line joining points $\mathbf{x},\tau$ and $\mathbf{x}',\tau$. The operator $\hat{B}(\Gamma)$ can be written as

$$\hat{B}(\Gamma) = \hat{B}(C_{\mathbf{x},\mathbf{x}'})\hat{B}(C_{\mathbf{x}',\mathbf{x}}) ,$$

$$G_{\alpha\beta}(\mathbf{x},\tau|\Gamma|\mathbf{x}',\tau') = \hat{B}(C_{\mathbf{x},\mathbf{x}'})\hat{G}_{\alpha\beta}(\mathbf{x},\tau|C_{\mathbf{x}',\mathbf{x}}|\ \mathbf{x}',\tau') .$$

Hence it can be saily shown that the function $\hat{G}_{\alpha\beta}(\mathbf{x},\tau \mid C_{\mathbf{x}',\mathbf{x}}|\mathbf{x}',\tau')$ coincides with the Green function in the presence of an external field introduced by Gor'kov [9.7]. Substituting $G_{\alpha\beta}(\mathbf{x},\tau|\Gamma|\mathbf{x}',\tau')$ into (9.2.24) and acting on it from left and right by the operator $\hat{B}^{-1}(C_{\mathbf{x}',\mathbf{x}})$, we obtain an equation for the usual Green function in an electromagnetic field. Equations for the condensate Green function and phonon function are obtained in an analogous manner. In the absence of fluctuations, (9.2.27) is transformed into Maxwell's equation in covariant form.

9.3 Electron-Phonon Interaction in Superconducting Phases: Superconducting Transition Temperature

Let us now analyze the superconducting phase equations (9.2.18-21). Adopting Fourier's contour representation (9.1.29-32) and averaging over the configuration of the cells $\varkappa_i$ belonging to the contours Γ_j, we obtain the system of equations for the Green functions

$$[i\omega - \varepsilon(\langle p^2_{o\nu}[\varkappa] \rangle|\Gamma|\mathbf{p})]G_{\alpha\beta\varkappa}(\mathbf{p},i\omega)\langle \mathrm{Tr}\hat{\tau}^{\alpha_1}\hat{B}^{(1)}(\Gamma)\hat{\tau}^{\alpha_1}\hat{B}(\varkappa)$$

$$+ \varphi_1(N) \sum_{\alpha_2=1}^{N^2-1} \int_{-\varkappa_2\in\Gamma} \frac{D\varkappa_2(\mathbf{x}'',\tau'')}{2\pi C_\varkappa} \exp\left(\int_{\varkappa_2} dx''_\nu p_{o\nu}[\varkappa_2]\right)$$

$$\times \langle \mathrm{Tr}\hat{\tau}^{\alpha_2}\hat{B}(\Gamma)\hat{\tau}^{\alpha_2}\hat{B}(\varkappa_2)\rangle[\Delta_{\alpha\gamma}(\varkappa_2) - \Delta^t_{\alpha\gamma}(\varkappa_2)]F^*_{\gamma\beta\varkappa}(\mathbf{p},\ i\omega)$$

$$\times \langle \mathrm{Tr}\hat{\tau}^{\alpha_1}\hat{B}^{(2)}(\Gamma)\hat{\tau}^{\alpha_1}\hat{B}(\varkappa)\rangle = \frac{\varphi_2(N)}{\tilde{\varphi}_1(N)}\langle \mathrm{Tr}\hat{B}(\Gamma)\rangle$$

$$- \frac{1}{2} g_{ph}^2 N^2 f_3(N) \langle Tr\hat{B}(\Gamma)\rangle \varphi_1(N) \sum_{\alpha_2=1}^{N^2-1} \int_{-\varkappa_2} \frac{D\varkappa_2(\mathbf{x}'',\tau'')}{2\pi C_\varkappa}$$

$$\times\Big\{ \sum_{\alpha'} [\omega_{ph}(\mathbf{p}_o)\omega_{ph}(-\mathbf{p}_o)]^{\frac{1}{2}} \Big[\langle D_{\alpha'\alpha'\varkappa_o}(\mathbf{p}_o, i\omega)\rangle_{\rho(\varkappa_o)}$$

$$+ \langle D^*_{\alpha'\alpha'}(\mathbf{p}_o, i\omega)\rangle_{\rho(\varkappa_o)}\Big] \langle \exp(\int_{\varkappa_o} d\mathbf{x}''_\nu p_{o\nu ph}[\varkappa_o])\rangle \langle Tr\hat{B}(\Gamma)$$

$$- \frac{2Nf_2(N)}{N^2 f_3(N)} \frac{\langle V_{\varkappa_o}(2\mathbf{p}_o)\rangle_{\rho(\varkappa_o)}}{g_{ph}^2}\Big\} \exp(\int_{\varkappa_2} dx''_\nu\, p_{o\nu}[\varkappa_2])$$

$$\times [G_{\alpha\gamma}(\varkappa_2)G_{\gamma\beta\varkappa}(\mathbf{p}, i\omega) \langle Tr\hat{\tau}^{\alpha_1}\hat{B}^{(1)}(\Gamma)\hat{\tau}^{\alpha_1}\hat{B}(\varkappa)\rangle$$

$$\langle Tr\hat{\tau}^{\alpha_2}\hat{B}^{(1)}(\Gamma)\hat{\tau}^{\alpha_2}\hat{B}(\varkappa_2)\rangle - F_{\alpha\gamma}(\varkappa_2)F^*_{\gamma\beta\varkappa}(\mathbf{p}, i\omega)$$

$$\langle Tr\hat{\tau}^{\alpha_1}\hat{B}^{(2)}(\Gamma)\hat{\tau}^{\alpha_1}\hat{B}(\varkappa)\rangle \langle Tr\hat{\tau}^{\alpha_2}\hat{B}^{(2)}(\Gamma)\hat{\tau}^{\alpha_2}\hat{B}(\varkappa_2)\rangle \quad , \qquad (9.3.1)$$

$$[i\omega + \varepsilon(\langle p_{o\nu}^2[\varkappa]\rangle\ |\Gamma|\mathbf{p})]F^*_{\alpha\beta}(\mathbf{p}, i\omega)\langle Tr\hat{\tau}^{\alpha_1}\hat{B}^{(2)}(\Gamma)\hat{\tau}^{\alpha_1}\hat{B}(\varkappa)$$

$$+ \varphi_1(N) \sum_{\alpha_2=1}^{N^2-1} \int_{-\varkappa_2\in\Gamma} \frac{D\varkappa_2(\mathbf{x}'',\tau'')}{2\pi C_\varkappa} \exp(\int_{\varkappa_2} dx''_\nu p_{o\nu}[\varkappa_2])$$

$$\times \langle Tr\hat{\tau}^{\alpha_2}\hat{B}(\Gamma)\hat{\tau}^{\alpha_2}\hat{B}(\varkappa_2)\rangle \quad [\Delta^*_{\alpha\gamma}(\varkappa_2) - \Delta^{*t}_{\alpha\gamma}(\varkappa_2)]$$

$$\times G_{\gamma\beta\varkappa}(\mathbf{p}, i\omega) \langle Tr\hat{\tau}^{\alpha_1}\hat{B}^{(1)}(\Gamma)\hat{\tau}^{\alpha_1}\hat{B}(\varkappa)\rangle$$

$$= \frac{1}{2} g_{ph}^2 N^2 f_3(N) \langle Tr\hat{B}(\Gamma)\rangle \varphi_1(N) \sum_{\alpha_2=1}^{N^2-1} \int_{-\varkappa_2\in\Gamma} \frac{D\varkappa_2(\mathbf{x}'',\tau'')}{2\pi C_\varkappa}$$

$$\times \Big\{ \sum_{\alpha'} [\omega_{ph}(\mathbf{p}_o)\omega_{ph}(-\mathbf{p}_o)]^{\frac{1}{2}} [\langle D_{\alpha'\alpha'\varkappa_o}(\mathbf{p}_o, i\omega)\rangle_{\rho(\varkappa_o)}$$

$$+ \langle D^*_{\alpha'\alpha'\varkappa_o}(\mathbf{p}_o, i\omega)\rangle_{\rho(\varkappa_o)}] \langle \exp(\int_{\varkappa_o} dx''_\nu\, p_{o\nu ph}[\varkappa_o])\rangle \langle Tr\hat{B}(\Gamma)\rangle$$

$$- \frac{2Nf_2(N)}{N^2 f_3(N)} \frac{\langle V_{\varkappa_o}(2\mathbf{p}_o) \rangle_{\rho(\varkappa_o)}}{g^2_{ph}} \Big\} \exp(\int_{\varkappa_2} dx''_\nu \, p_{o\nu}[\varkappa_2])$$

$$[F^*_{\alpha\gamma}(\varkappa_2) G_{\gamma\beta\varkappa}(\mathbf{p}, i\omega) \; \langle Tr \hat{\tau}^{\alpha_1} \hat{B}^{(1)}(\Gamma) \hat{\tau}^{\alpha_1} \hat{B}(\varkappa)$$

$$\times \langle Tr \hat{\tau}^{\alpha_2} \hat{B}^{(2)}(\Gamma) \hat{\tau}^{\alpha_2} \hat{B}(\varkappa) \rangle - G^*_{\alpha\gamma}(\varkappa_2) F^*_{\gamma\beta\varkappa}(\mathbf{p}, i\omega)$$

$$\times \langle Tr \hat{\tau}^{\alpha_1} \hat{B}^{(2)}(\Gamma) \hat{\tau}^{\alpha_1} B(\varkappa) \rangle \; \langle Tr \hat{\tau}^{\alpha_2} \hat{B}^{(1)}(\Gamma) \hat{\tau}^{\alpha_2} \hat{B}(\varkappa_2) \rangle] , \qquad (9.3.2)$$

$$\varepsilon(\langle p^2_{o\nu}[\varkappa] \rangle | \Gamma | \mathbf{p}) = - \frac{p^2}{2\mu_e} + \mu + \frac{1}{2\mu_e} \langle p^2_{o\nu}[\varkappa] \rangle_{\rho(\varkappa)}$$

$$[i\omega + \omega_{ph}(\mathbf{p})] \langle D_{\alpha\beta\varkappa}(\mathbf{p}, i\omega) \rangle_{\rho(\varkappa)} \langle Tr \hat{\tau}^{\alpha_1} \hat{B}(\Gamma) \hat{\tau}^{\alpha_1} \frac{1}{N} \hat{T} \exp(\oint_\varkappa dx'_\nu \hat{\Omega}_{\nu sing}) \rangle$$

$$= \frac{\varphi_2(N)}{\tilde{\varphi}_1(N)} \delta_{\alpha\beta} \langle Tr \hat{B}(\Gamma) \rangle - \frac{1}{2} N^2 f_3(N) g^2_{ph} \varphi^2_1(N)$$

$$\times \sum_{\alpha'_1, \alpha_2 = 1}^{N^2-1} \int_{-\varkappa_2 \in \Gamma} \frac{D\varkappa_2(\mathbf{x}'', \tau'')}{2\pi C_\varkappa} \frac{d\mathbf{p}_2}{(2\pi)^3} \frac{d\omega_2}{2\pi} \exp(\int_{\varkappa_2} dx''_\nu p_{o\nu}[\varkappa_2])$$

$$\times [\omega_{ph}(-\mathbf{p}) \omega_{ph}(\mathbf{p}_2)]^{\frac{1}{2}} \langle D_{\alpha\beta\varkappa}(\mathbf{p}, i\omega) \rangle_{\rho(\varkappa)} \langle Tr \hat{\tau}^{\alpha_1} \hat{B}(\Gamma) \hat{\tau}^{\alpha_1} \frac{1}{N} \hat{T}$$

$$\times \exp(\oint_\varkappa dx'_\nu \Omega_{\nu \, sing}) \rangle \; [G_{\gamma\delta\,\varkappa_2}(\mathbf{p}_2, i\omega_2) G_{\delta\gamma\varkappa\,-\varkappa_2}(\mathbf{p}-\mathbf{p}_2, i(\omega-\omega_2))$$

$$\times \langle Tr \hat{\tau}^{\alpha'_1} \hat{B}^{(1)}(\Gamma) \hat{\tau}^{\alpha'_1} B(\varkappa_2) \rangle \langle Tr \hat{\tau}^{\alpha_2} \hat{B}^{(1)}(\Gamma) \hat{\tau}^{\alpha_2} B(\varkappa - \varkappa_2) \rangle$$

$$- F_{\gamma\delta\varkappa_2}(\mathbf{p}_2, i\omega_2) F^*_{\delta\gamma\varkappa-\varkappa_2}(\mathbf{p}-\mathbf{p}_2, i(\omega - \omega_2))$$

$$\times \langle Tr \hat{\tau}^{\alpha'_1} \hat{B}^{(2)}(\Gamma) \hat{\tau}^{\alpha'_1} B(\varkappa_2) \rangle \; \langle Tr \hat{\tau}^{\alpha_2} \hat{B}^{(2)}(\Gamma) \hat{\tau}^{\alpha_2} \hat{B}(\varkappa - \varkappa_2) \rangle] , \qquad (9.3.3)$$

$$N = 2 , \quad N^2 f_3(N) \to 1 \; ; \quad N f_2(N) \to 1 ,$$

where we have used the notation

$$G_{\alpha\beta}(\varkappa) = \int \frac{d\mathbf{p}}{(2\pi)^3} \frac{d\omega}{2\pi} G_{\alpha\beta\varkappa}(\mathbf{p}, i\omega) , \quad F_{\alpha\beta}(\varkappa) = \int \frac{d\mathbf{p}}{(2\pi)^3} \frac{d\omega}{2\pi} F_{\alpha\beta\varkappa}(\mathbf{p}, i\omega).$$

If the system has a nonzero density of plastic disclinations, the contour integral of the field Ω^{α}_{ν} will also be nonzero:

$$\oint_{\varkappa} dx_{\nu}\, \Omega_{\nu\ \mathrm{sing}} = 2\pi C_{\varkappa}\, n_d |\mathbf{\Omega}_d| (\hat{\tau}\mathbf{n}) . \qquad (9.3.4)$$

Here, n_d is the density of disclinations, $\mathbf{\Omega}_d$ is Frank's vector of an isolated disclination, $\mathbf{n}$ is a unit vector oriented along Frank's vector,

$$\varphi_1(N) = N/(N^2 - 1) ; \quad \varphi_2(N) = 1/N ; \ \tilde{\varphi}_1(N) = 1/(N^2 - 1).$$

In deriving (9.3.1-3), we assumed that

$$g_{ph}(\mathbf{x}) = \lim_{\boldsymbol{\xi}\to -\infty} g_{ph}\, \Theta(\mathbf{x}-\boldsymbol{\xi}) ; \quad \Theta(\mathbf{x}) = \begin{cases} 1 , & x_i > 0 \\ 0, & x_i < 0 \end{cases}$$

and $G_{\alpha\beta\varkappa}(\mathbf{p},i\omega)$, $F_{\alpha\beta\varkappa}(\mathbf{p},i\omega)$, $D_{\alpha\beta\varkappa}(\mathbf{p},i\omega)$ is a slowly varying function of the momentum exchanged by quasi-particles in the cells $\varkappa$ and $\varkappa^*$. In addition, we assumed a slow evolution of contour Γ in time τ. In this case, $\hat{A}_{\nu}$ will be a slowly varying function of τ and the D'Alembertian $\Box$ $(-p^2_{\nu}$, $\nu = 1\text{-}4)$ on the left-hand side of (9.3.1-3) is transformed into the Laplacian Δ $(\nu = 1\text{-}3)$, since the characteristic frequencies $\omega \lesssim \langle \omega_D \rangle$. Indeed, the time derivative $\frac{\partial}{\partial x_o} = \frac{1}{v^*}\frac{\partial}{\partial\tau}$, where $v^* \to v_F$ (the electron velocity on the Fermi surface), and hence $\omega/\varepsilon_F \ll 1$ and $\omega[\varkappa]/\varepsilon_F \ll 1$.

$$\varphi_1(N) \sum_{\alpha_1=1}^{N^2-1} G_{\alpha\beta\varkappa}(\mathbf{p},i\omega|\Gamma_{\mathbf{x},\tau}) \langle \mathrm{Tr}\hat{\tau}^{\alpha_1}\hat{B}^{(1)}(\Gamma)\hat{\tau}^{\alpha_1}\hat{B}(\varkappa) \rangle$$

$$= - \delta_{\alpha\beta} \langle \mathrm{Tr}\hat{B}(\Gamma) \rangle [i\omega - \varepsilon(\langle p^2_{o\nu}[\varkappa]\rangle|\Gamma|\mathbf{p})] \Big\{\omega^2 + \varepsilon^2(\langle p^2_{o\nu}[\varkappa]|\Gamma|\,\mathbf{p})$$

$$+ \frac{1}{8}\mathrm{Tr}\Big\{[\hat{\Delta}(\Gamma_{\mathbf{x},\tau}) - \hat{\Delta}^t(\Gamma_{\mathbf{x},\tau})][\hat{\Delta}(\Gamma_{\mathbf{x},\tau}) - \hat{\Delta}^t(\Gamma_{\mathbf{x},\tau})]^+\Big\}\Big\}^{-1}, \qquad (9.3.5)$$

$$\varphi_1(N) \sum_{\alpha_1=1}^{N^2-1} F^*_{\alpha\beta\varkappa}(\mathbf{p},i\omega|\Gamma_{\mathbf{x},\tau}) \langle \mathrm{Tr}\hat{\tau}^{\alpha_1}\hat{B}^{(2)}(\Gamma)\hat{\tau}^{\alpha_1}\hat{B}(\varkappa) \rangle$$

$$= \frac{1}{2} \langle \mathrm{Tr}\hat{B}(\Gamma) \rangle [\Delta^*_{\alpha\beta}(\Gamma_{\mathbf{x},\tau}) - \Delta^{*t}_{\alpha\beta}(\Gamma_{\mathbf{x},\tau})]$$

$$\times\left\{\omega^2 + \varepsilon^2(\langle p^2_{o\nu}[\varkappa]\rangle|\Gamma|\mathbf{p}) + \frac{1}{8}\,\mathrm{Tr}\left\{[\hat{\Delta}(\Gamma_{\mathbf{x},\tau}) - \hat{\Delta}^t(\Gamma_{\mathbf{x},\tau})]\right.\right.$$

$$\left.\left.\times[\hat{\Delta}(\Gamma_{\mathbf{x},\tau}) - \hat{\Delta}^t(\Gamma_{\mathbf{x},\tau})]^+\right\}\right\}^{-1}, \qquad (9.3.6)$$

$$\varphi_1(N)\sum_{\alpha_1=1}^{N^2-1} F_{\alpha\beta\varkappa}(\mathbf{p},i\omega|\Gamma_{\mathbf{x},\tau})\ \langle \mathrm{Tr}\,\hat{\tau}^{\alpha_1}\hat{B}^{(2)}(\Gamma)\hat{\tau}^{\alpha_1}\hat{B}(\varkappa)\rangle$$

$$= [\varphi_1(N)\sum_{\alpha_1=1}^{N^2-1} F^*_{\alpha\beta\varkappa}(\mathbf{p},i\omega|\,\Gamma_{\mathbf{x},\tau})\ \langle \mathrm{Tr}\,\hat{\tau}^{\alpha_1}\hat{B}^{(2)}(\Gamma)\hat{\tau}^{\alpha_1}\hat{B}(\varkappa)\rangle]^+,$$

(9.3.7)

$$\varphi_1(N)\sum_{\alpha_1=1}^{N^2-1}\langle D_{\alpha\beta\varkappa}(\mathbf{p},i\omega|\Gamma_{\mathbf{x},\tau})\rangle_{\rho(\varkappa)}\ \langle \mathrm{Tr}\,\hat{\tau}^{\alpha_1}\hat{B}(\Gamma)\hat{\tau}^{\alpha_1}\frac{1}{N}\hat{T}$$

$$\times\exp(\oint_{\varkappa} dx'_\nu\ \hat{\Omega}_{\nu\mathrm{sing}})\rangle = \delta_{\alpha\beta}\langle \mathrm{Tr}\hat{B}(\Gamma)\rangle[i\omega + \omega_{ph}(p)]^{-1}, \qquad (9.3.8)$$

where $\hat{\Delta}(\Gamma_{\mathbf{x},\tau}) - \hat{\Delta}^t(\Gamma_{\mathbf{x},\tau}) = [i\sigma^y - (i\sigma^y)^t]\ \Delta(\Gamma_{\mathbf{x},\tau})$, $\sigma^y = \begin{pmatrix}0, & -i\\ i, & 0\end{pmatrix}$ and the matrix $\hat{\Delta}(\Gamma_{\mathbf{x},\tau}) - \hat{\Delta}^t(\Gamma_{\mathbf{x},\tau})$, which determines the gap in the spectrum of quasi-particles, may be determined from the solution of the integral equation

$$\hat{\Delta}(\Gamma_{\mathbf{x},\tau}) - \Delta^t(\Gamma_{\mathbf{x},\tau}) = \frac{1}{2}\langle \mathrm{Tr}\hat{B}(\Gamma_{\mathbf{x},\tau})\rangle\ g^2_{ph}\int\frac{d\mathbf{p}}{(2\pi)^3}\frac{d\omega}{2\pi}$$

$$\times\left\{\sum_{\alpha'}[\omega_{ph}(\mathbf{p}_o)\ \omega_{ph}(-\mathbf{p}_o)]^{\frac{1}{2}}[\langle D_{\alpha'\alpha'\varkappa_o}(\mathbf{p}_o,\ i\omega)\rangle_{\rho(\varkappa_o)}\right.$$

$$+\langle D^*_{\alpha'\alpha'\varkappa_o}(\mathbf{p}_o,\ i\omega)\rangle_{\rho(\varkappa_o)}]\ \langle\mathrm{Tr}\hat{B}(\Gamma_{\mathbf{x},\tau})\rangle\langle\exp(\int_{\varkappa_o} dx_\nu p_{o\nu ph}[\varkappa_o])\rangle$$

$$\left.-\frac{2\langle V_{\varkappa_o}(2\mathbf{p}_o)\rangle}{g^2_{ph}}\right\}[\hat{\Delta}(\Gamma_{\mathbf{x},\tau}) - \hat{\Delta}^t(\Gamma_{\mathbf{x},\tau})]\ \langle\mathrm{Tr}\hat{B}(\delta\Gamma_{\mathbf{x},\tau})\rangle$$

$$\times\left\{\omega^2 + \varepsilon^2(\langle p^2_{o\nu}[\varkappa]\rangle|\Gamma_{\mathbf{x},\tau}|\mathbf{p}) + \frac{1}{8}\,\mathrm{Tr}\left\{[\hat{\Delta}(\Gamma_{\mathbf{x},\tau}) - \hat{\Delta}^t(\Gamma_{\mathbf{x},\tau})]\right.\right.$$

$$\left.\left.\times[\hat{\Delta}(\Gamma_{\mathbf{x},\tau}) - \hat{\Delta}^t(\Gamma_{\mathbf{x},\tau})]^+\right\}\right\}^{-1}, \qquad (9.3.9)$$

where $\delta\Gamma_{\mathbf{x},\tau}$ is a small contour enveloping the neighborhood of the point $(\mathbf{x},\tau)$. If the field $A_\nu^\alpha(\mathbf{x},\tau)$ does not have a singularity in the vicinity of $(\mathbf{x},\tau)$, the contour $\delta\Gamma_{\mathbf{x},\tau}$ can be contracted to a point, and in this case, $\langle \mathrm{Tr}\hat{B}(\delta\Gamma_{\mathbf{x},\tau}) \to 1$. Computing the integral with respect to the four-dimensional momentum $\mathbf{p}$, ω, we obtain the following expression for the superconducting "gap":

$$\Delta(\Gamma_{\mathbf{x},\tau}) = 2\,\langle\omega_D\rangle_{\rho(\varkappa)} \exp\left[-\frac{1}{\nu(\varepsilon_F)\langle\lambda_{eph}(\mathbf{p}_o|\Gamma_{\mathbf{x},\tau}|\bar{\mathbf{p}}_{\perp o})\rangle_{\rho(\varkappa_o,\bar{p}_{\perp o})}}\right], \tag{9.3.10}$$

$$\begin{aligned}\langle\lambda_{eph}(\mathbf{p}_o|\Gamma_{\mathbf{x},\tau}|\bar{p}_{\perp o})\rangle_{\rho(\varkappa_o,\bar{p}_{\perp o})} &= \frac{1}{2}\,\langle\,\mathrm{Tr}\hat{B}(\Gamma_{\mathbf{x},\tau})\,\rangle\, g_{ph}^2\\ &\times\Big\{\sum_{\alpha'}\,[\omega_{ph}(\mathbf{p}_o)\omega_{ph}(-\mathbf{p}_o)]^{\frac{1}{2}}\,[\langle D_{\alpha'\alpha'\varkappa_o}(\mathbf{p}_o,0)\rangle_{\rho(\varkappa_o)}\\ &+\langle D^*_{\alpha'\alpha'}(\mathbf{p}_o,0)\rangle_{\rho(\varkappa_o)}]\,\langle\,\mathrm{Tr}\hat{B}(\Gamma_{\mathbf{x},\tau})\rangle\,\langle\exp(\int_{\varkappa_o}dx''_\nu\,p_{o\nu ph}[\varkappa_o])\rangle_{\rho(\varkappa_o,\bar{p}_{\perp o})}\\ &-\frac{2\langle V_{\varkappa_o}(2\mathbf{p}_o)\rangle_{\rho(\varkappa_o)}}{g_{ph}^2}\Big\},\end{aligned} \tag{9.3.11}$$

where $\nu(\varepsilon_F)$ is the density of electron states on the Fermi surface, and $\langle\,\omega_D\,\rangle_{\rho(\varkappa)}$ is the Debye energy averaged over all possible configurations of the cell $\varkappa$.

Let us consider the average $\langle\exp(\int_{\varkappa_o}dx''_\nu\,p_{o\nu ph}[\varkappa_o]\rangle_{\rho(\varkappa_o,\bar{p}_{\perp o})}$. This average is taken over the phonon momentum exchanged by the electrons in $\varkappa_o$ and $\varkappa_o^*$ cells, and over all configurations of the cell $\varkappa_o$. This quantity can be expressed in the form

$$\begin{aligned}&\langle\exp(\int_{\varkappa_o}dx''_\nu\,p_{o\nu ph}[\varkappa_o])\rangle_{\rho(\varkappa_o,\bar{p}_{\perp o})}\\ &=\langle\int_{-\varkappa_o,-\varkappa_o^*\in\Gamma^\varkappa}\frac{D\varkappa_o}{2\pi C_\varkappa}\frac{D\varkappa_o^*}{2\pi C_\varkappa}\exp\int_{\varkappa_o}dx'_\nu\int_{\varkappa_o^*}dx''_\nu\,\mathbb{P}_\nu(\mathbf{x}-\mathbf{x}''))\rangle_{\rho(\varkappa_o,\varkappa_o^*,\bar{p}_{\perp o})}\,.\end{aligned} \tag{9.3.12}$$

Here, $\bar{p}_{\perp o}$ is the momentum exchanged by electrons in the $\varkappa_o$ and $\varkappa_o^*$ cells, and $\mathbb{P}_\nu(\mathbf{x}'-\mathbf{x}'') = \langle\,p_{o\nu ph}(\mathbf{x}')p_{o\nu ph}(\mathbf{x}'')\,\rangle$ is the regularized Green function which is proportional to the correlation function of the fields $\delta\hat{A}_\nu(\mathbf{x})$, $\langle\,\mathrm{Tr}\delta\hat{A}_\nu(\mathbf{x}')\delta\,\hat{A}_\nu(\mathbf{x}'')\,\rangle$. This function can be evaluated and is found to be proportional to the fluctuational correlation radius $\langle r_c \rangle$. In this case, we can write

$$\int_{\varkappa_o} dx'_\nu \int_{\varkappa_o^*} dx''_\nu \, \mathbb{P}_\nu (\mathbf{x}'-\mathbf{x}'') = k_F^2 \langle r_c \rangle^2 \, C \, \{ \langle \mathrm{Tr}\hat{\mathbb{B}}(\Gamma) \rangle, \chi/\chi_{os}, L\varkappa_o \}.$$

Here, C is a functional of the static magnetic susceptibility of the system and effective radius of interaction of quasi-particles in the cells $\varkappa_o$ and $\varkappa_o^*$. The effective radius of interaction of quasi-particles $L_{\varkappa_o}$ is found to be identical to the average size of the contour Γ, pierced by a nonzero exchange field flux A_ν^α. But this means that the quantity $L_{\varkappa_o}$ must be less than or on the order of, the exchange correlation radius $\langle r_c \rangle$ of the conduction electrons (see 1.1.5). The quantity C is found to be a quasiperiodic function of $L_{\varkappa_o}$, and $\lim_{L_{\varkappa_o} \to \infty} C = 0$. We introduce the notation

$$k_F \langle r_c \rangle^2 \, C = 2\pi^2 \ell_{\varkappa_o} / L_{\varkappa_o} .$$

Since the size $L_{\varkappa_o}$ of a contour is much larger than the reciprocal of the Fermi momentum k_F^{-1}, we can treat the interaction between electrons as a four-particle, and not pair, interaction. Indeed, the electrons forming pairs exchange phonons with momentum $k \simeq 2k_F$, and at the same time, interact with each other at distances of the order of $\langle r_c \rangle$ on account of spin fluctuations. But in this case we can interpret $\ell_{\varkappa_o}$ as the effective radius of interaction of electrons forming pairs with different magnitudes of momenta $\mathbf{p}$ and $\mathbf{p}'$ ($|\mathbf{p}| \neq |\mathbf{p}'|$). The momentum deficit in this case is compensated by the absorption of an additional virtual phonon emitted by another pair of electrons (located in the cell $\varkappa_o^*$). The possibility of multiparticle interaction in superconductors was discussed by **Heine** and **Pippard** [9.8]. The process of electron interaction in superconductors considered above does not contradict the results obtained in Chap. 8 (Sect. 8.2). In the quasilinear theory, the pair-forming electrons exchange virtual quasi-phonons whose energy considerably exceeds the energy of a normal phonon due to resonant interaction with spin fluctuations. This results in an enhancement of the electron-phonon interaction. In the case under consideration, however, the enhancement occurs as a result of the absorption of an additional phonon due to the fact that exchange interaction takes place not only between electrons forming a pair, but also between electrons from different pairs. Thus, the spin fluctuations of exchange type cause an additional interaction of electrons through phonons with a momentum $p_{\perp o} = 2\pi/L_{\varkappa_o}$ ($L_{\varkappa_o} \lesssim \langle r_c \rangle$). This is equivalent to an increase in the Debye energy of the quasi-particle exchanged by the electrons in a pair. We shall assume that the value of the momentum $p_{\perp o}$ varies in the interval $(-\bar{p}_{\perp o}, \bar{p}_{\perp o})$, and that all the values of $p_{\perp o}$ in this interval are equally probable. In this case, the mean value of the exponent in (9.3.12) can be easily calculated. This quantity is found to be equal to

$$\langle \exp(\int_{\varkappa_o} dx_\nu p_{o\nu ph}[\varkappa_o]) \rangle_{\rho(\varkappa_o, \bar{p}_{\perp o})} = \langle \cosh(\pi \ell_\varkappa \bar{p}_{\perp o}) \rangle_{\rho(\varkappa_o, \bar{p}_{\perp o})}$$

$$= (1/2\pi \ell_{\varkappa_o} \bar{p}_{\perp o}) \sinh(2\pi \ell_{\varkappa_o} \bar{p}_{\perp o}) . \qquad (9.3.13)$$

The above approximation is applicable if $\langle \cosh(x) \rangle \lesssim 10$, i.e., $(\ell_{\varkappa_o}/L_{\varkappa_o}) \lesssim 1/6$. The applicability of this approximation is determined by the condition for screening of the Coulomb repulsion between electrons, according to which $(\langle \omega_D \rangle \langle \cosh x \rangle / \varepsilon_F) \ll 1$. In view of the fact that $k_F^2 \langle r_c \rangle^2 C = \pi \ell_{\varkappa_o} \bar{p}_{\perp o}$ and $C \to 0$ for $\langle r_c \rangle \to \infty$, and considering that $\ell_{\varkappa_o} \approx k_F^{-1}$, we can estimate the quantity C, which is found to be much smaller than unity. Its asymptotic form can be chosen as $C \approx (k_F \langle r_c \rangle)^{-3}$, i.e., $\pi \ell_{\varkappa_o} \bar{p}_{\perp o} \approx 2\pi^2 A (v_o^{1/3} / \langle r_c \rangle$,
$A = \dfrac{\sqrt{3}\, g\, \hbar}{\pi v_o^{1/3} \sqrt{J_o s M}} \approx I$ (Chap. 8). Recalling (8.2.18), we find that $v_o^{1/3} / \langle r_c \rangle$ is the same parameter that determines the linear spin-phonon interaction. If the system contains no plastic defects of the crystal lattice, i.e., disclinations, the expression (9.3.3) for the phonon Green function simplifies to

$$\langle D_{\alpha\beta\varkappa}(\mathbf{p}) \rangle_{\rho(\varkappa)} = \frac{\delta_{\alpha\beta}}{i\omega + \tilde{\omega}_{ph}(\mathbf{p})} , \qquad (9.3.14)$$

where $\tilde{\omega}_{ph}$ is the phonon frequency taking into account the nonlinear renormalization due to interaction with the electron system on account of spin fluctuations (Chap. 5). However, calculations show that a consideration of the electron-phonon interaction in higher orders of the perturbation theory does not significantly distort the phonon spectrum [9.9]. Hence we shall confine ourselves, as before, to a step approximation of the kernel of the integral equation for the superconducting order parameter. Equation (9.3.11) remains valid, and by proceeding to the limit $\omega \to 0$, we obtain an expression for the effective electron-phonon interaction parameter:[1)]

$$\langle \lambda_{e\text{-}ph}(k_F|\Gamma_{\mathbf{x},\tau}|\bar{p}_{\perp o})\rangle_{\rho(\varkappa_o,\bar{p}_{\perp o})} = 3\langle \mathrm{Tr}\hat{B}(\Gamma_{\mathbf{x},\tau})\rangle g_{ph}^2$$

$$\times\left\{\langle \mathrm{Tr}\hat{B}(\Gamma_{\mathbf{x},\tau})\rangle \langle \cosh(\pi \ell \varkappa_o \bar{p}_{\perp o})\rangle_{\rho(\varkappa_o,\bar{p}_{\perp o})} - \frac{\langle V_{\varkappa_o}(2\mathbf{k}_F)\rangle}{g_{ph}^2}\right\} . \tag{9.3.15}$$

Since we are considering a case in which the function $\langle \mathrm{Tr}\hat{B}(\Gamma_{\mathbf{x},\tau})$ can be assumed to be independent of τ , (9.3.15) describes the spatial distribution of the effective electron-phonon interaction parameter. For $\langle \mathrm{Tr}\hat{B}(\Gamma)\rangle \to 1$, (9.3.15) determines the gap in the conduction electron spectrum and the superconducting transition temperature T_c. We introduce the function

$$T_c(\Gamma_{\mathbf{x},\tau}) = \frac{\gamma}{\pi}\Delta(\Gamma_{\mathbf{x},\tau}) . \tag{9.3.16}$$

By averaging this expression over the spatial variable Γ , we obtain

$$\langle T_c(\Gamma_{\mathbf{x},\tau})\rangle_{(\Gamma_V)} = \left\langle \frac{2\gamma}{\pi}\langle\omega_D\rangle_{\rho(\varkappa)}\right.$$

$$\left.\times \exp\left[-\frac{1}{\nu(\varepsilon_F)\langle \lambda_{e\text{-}ph}(k_F|\Gamma_{\mathbf{x},\tau}|\bar{p}_{\perp o})\rangle_{\rho(\varkappa_o,\bar{p}_{\perp o})}}\right]\right\rangle_{(\Gamma_V)} . \tag{9.3.17}$$

This relation defines the temperature of the phase transition into the superconducting state. The results obtained are found to be in accord with the conclusions drawn in Chap. 8, since the expression for the effective electron-phonon interaction parameter contains the quantity $\langle\cosh(x)\rangle$ which depends on the ratio $(v_o^{1/3}/\langle r_c\rangle)$, i.e., it is the enhancement factor for the electron-phonon interaction (8.2.31). The quantity $\langle\cosh(x)\rangle$ may turn out to be sufficiently large to ensure a high value of the superconducting transition tempeature T_c. For example, by putting

$$v_o^{1/3}/\langle r_c\rangle \approx 10^{-1} , \qquad \langle\cosh(x)\rangle \approx 2 ,$$

1) Equation (9.3.15) is obtained by a series summation which describes the nonlinear spin-phonon interaction.

and using the data of [9.9, see also 9.10], we can estimate the critical temperature, which is found to be 100 K or more. This is in accord with the experimental results. It follows from (9.3.15) that an increase in the critical temperature is possible mainly through a decrease in the exchange correlation radius of the conduction electron system. It can be seen from Table 8.1 and Fig. 8.5 that this can be attained by substituting for the oxygen vacancies elements of a smaller atomic radius and a comparable mass, for example, nitrogen and fluorine. It is interesting to note that the anions O^{2-} and F^{1-} have similar radii (Fig. 8.5).

9.4 Phase Lamination in Superconducting States

In Sects. 9.1,3, we constructed the microscopic theory of superconducting phases in rare-earth metal compounds without taking into consideration the specific structure of the superconducting state. The superconducting phase structure was considered in Chap. 6 on the basis of a phenomenological theory. It was shown that this phase splits into a number of intermediate phases with a vortex structure, i.e., the system undergoes a phase lamination in the same way as in the magnetically ordered crystals considered in Sect. 7.1. We calculated the spatial distribution of the order parameter $\Psi(\rho,z)$ and of the current $J^{\alpha}_{\nu}(\rho,z)$ by using (6.3.32) (Fig. 6.3).

In this section, we shall show that the phase lamination effect can be derived from a microscopic theory. We assume that the function $\langle \mathrm{Tr}\hat{B}(\Gamma_{\mathbf{x},\tau})\rangle$ can describe the spatial distribution of the superconducting order parameter. For this purpose, we must know the form of this function. This can be done by solving the field equation. Taking into account the solutions of the Green function equations for electrons and phonons, we can present this equation as

$$\nabla_{\mu}\frac{\delta}{\delta S_{\mu\nu}}\langle \mathrm{Tr}\hat{B}(\Gamma)\rangle - \frac{i}{4}g_1 N^2 \lim_{\mathbf{x}',\tau\to\mathbf{x},\tau}(\nabla_{x'_{\nu}} - \nabla_{x_{\nu}})\int_{-\varkappa\in\Gamma}\frac{D\varkappa(\mathbf{x}'',\tau'')}{2\pi C_{\varkappa}}$$

$$\frac{d\mathbf{p}}{(2\pi)^3}\frac{d\omega}{2\pi}\exp\Big[\int_{\varkappa}dx''_{\nu}\,p_{o\nu}[\varkappa] + i\mathbf{p}(\mathbf{x}-\mathbf{x}') + i\omega(\tau-\tau')\Big]$$

$$\times\,[\langle \mathrm{Tr}\hat{B}(\Gamma+\varkappa)\rangle - \langle \mathrm{Tr}\hat{B}(\Gamma)\mathrm{Tr}\hat{B}(\varkappa)\rangle]\langle \mathrm{Tr}\hat{B}(\Gamma)\rangle$$

$$\times\,\frac{i\omega - \varepsilon(\langle p^2_{o\nu}[\varkappa]\rangle|\Gamma|\mathbf{p})}{\varphi_1(N)\sum_{\alpha_1=1}^{N^2-1}\langle \mathrm{Tr}\hat{\tau}^{\alpha_1}\hat{B}^{(1)}(\Gamma)\hat{\tau}^{\alpha_1}\hat{B}(\varkappa)\rangle[\omega^2+\varepsilon^2(\langle p_{o\nu}[\varkappa]\rangle|\Gamma|\mathbf{p})+|\Delta(\Gamma)|^2]}$$

$$= \frac{i}{2} g_1 N^2 \oint_\Gamma dx'_\nu \, \delta(\mathbf{x}-\mathbf{x}')[\langle \mathrm{Tr}\hat{B}(C_{\mathbf{x},\mathbf{x}'})\mathrm{Tr}\hat{B}(C_{\mathbf{x}',\mathbf{x}}) \rangle - \frac{1}{N^2} \langle \mathrm{Tr}\hat{B}(\Gamma) \rangle]. \tag{9.4.1}$$

Let us consider a system of superconducting vortices of cylindrical symmetry, whose rotation axis of infinite order is oriented along the isolated z-axis of the crystal. In this case, the operator $\hat{A}_\nu$ has only two components $\hat{A}_x$ and $\hat{A}_y$. The integral over a contour in the (x,y) plane can be written as

$$\oint_\Gamma dx_\nu \hat{A}_\nu = - i2\pi n(\hat{\tau}\mathbf{n})\sin^2\xi , \tag{9.4.2}$$

where $\mathbf{n}$ is a unit vector, $n = (m, m/2)$, m being the quanta of current circulation around the vortex axis. The function $\sin^2\xi$ depends on the distance ρ from the vortex axis and will be in the simplest case:

$$\sin^2\xi = \mathrm{sech}^2[n \ln(\rho/\rho_o)] .$$

Consequently,

$$\langle \mathrm{Tr}\hat{B}(\Gamma_\rho) = \langle \cos[2\pi n \sin^2\xi_n(\rho)] \rangle , \tag{9.4.3}$$

or, if we assume that the contour Γ is a circle,

$$\langle \mathrm{Tr}\hat{B}(\Gamma_\rho) \rangle = \cos \{2\pi n \, \mathrm{sech}^2[n \ln(\rho/\rho_o)]\} . \tag{9.4.4}$$

For the constant ρ_o, we can choose the total fluctuation correlation radius $\langle r_c(T) \rangle$ which is a function of temperature. The form of the solutions for the functions $\langle \mathrm{Tr}\hat{B}(\Gamma_{\mathbf{x},\tau}) \rangle$ is chosen directly from the symmetry considerations for the vortex structure of the superconducting phase. However, it can be easily verified that it satisfies (9.4.1). As a matter of fact, this is because the function (9.4.4) does not contain fluctuations in an explicit form, i.e., it depends on averaged parameters determined by the fluctuations (for example, ρ_o). Hence, in the vicinity of the point to which the loop $\varkappa$ belongs, the function (9.4.4) varies slowly. Hence the loops $\varkappa$ can be contracted to a point and the second term in (9.4.1) vanishes. As a result, we obtain an ordinary field equation

$$\nabla_\mu \frac{\delta}{\delta S_{\mu\nu}} \langle \mathrm{Tr}\hat{B}(\Gamma) \rangle = \frac{i}{2} g_1 N^2 \oint dx'_\nu \, \delta(\mathbf{x}-\mathbf{x}')[\langle \mathrm{Tr}\hat{B}(C_{\mathbf{x},\mathbf{x}'})\mathrm{Tr}\hat{B}(C_{\mathbf{x}',\mathbf{x}})$$
$$- \frac{1}{N^2} \langle \mathrm{Tr}\hat{B}(\Gamma) \rangle] .$$

The functional derivative on the left-hand side of the field equation vanishes since the function $\langle \mathrm{Tr}\hat{B}(\Gamma) \rangle$ depends only on ρ. For the same reason, the right-hand side also vanishes. In this case, we can write the following expression for the effective electron-phonon interaction parameter:

$$\langle \lambda_{eph}(\mathbf{p}_o|\Gamma_\rho,n\ |\bar{p}_{\perp o}) \rangle = 3g^2_{ph}\ \cos(2\pi\ n\ \mathrm{sech}^2[\ln(\rho/\rho_o)])$$
$$\times \left\{\cos(2\pi n\ \ \mathrm{sech}^2[\ln(\rho/\rho_o)])\langle\ \cosh(2\pi^2 v_o^{1/3}/\langle r_c\rangle) - \frac{\langle V_{\varkappa_o}(2\mathbf{p}_o)\rangle\,\rho(\varkappa)}{g^2_{ph}}\right\}. \tag{9.4.5}$$

The function (9.4.4) oscillates in space, and this leads to a nonuniform spatial distribution of the superconducting order parameter. Indeed, for $\langle \lambda_{e-ph}(\mathbf{p}_o|\Gamma_\rho,n|\bar{\mathbf{p}}_{\perp o}) \rangle > 0$, we obtain the superconducting state, while for $\langle \lambda_{e-ph}(\mathbf{p}_o|\Gamma_\rho,n|\bar{p}_{\perp o}) \rangle < 0$ the superconductivity disappears and a normal metal region with a metastable complex magnetic structure emerges, since an exchange interaction exists between the spins of the s- and d(f)-electrons (Sect. 6.4). However, in view of the fact that

$$\frac{\langle V_{\varkappa_o}(2\mathbf{p}_o)\rangle\,\rho(\varkappa_o)}{g^2_{ph}} < 1 ,$$

the range of existence of the normal phase is found to be quite narrow, i.e., $\delta R << T_\varkappa$, where $T_\varkappa$ is the characteristic size of the oscillations. Moreover, with decreasing ρ, the strength of the Coulomb repulsion between conduction electrons increases, since it is mathematically associated with a decrease in the size of the contour Γ. A decrease in ρ means that if $\rho < \rho_o$, the fluctuation correlation radius will exceed the effective radii $\ell_{\varkappa_o}$ and $L_{\varkappa_o}$ for interactions between quasi-particles. Hence the quasi-particles will be situated in a field of longwave fluctuations of spin and charge densities. It was shown in Chap. 4 that such fluctuations may affect the process of electron pairing only near the phase transition point ($\langle r_c(\tau)\rangle \to \infty$). In this case, ρ_o is finite, since such fluctuations do not affect the pairing process in any way at a point away from the phase transition. Hence there will be no exchange enhancement effect near the vortex core. There will be no superconducting pairs either, and we shall be left with the longwave magnetic fluctuations. Consequently, we can expect a longwave metastable magnetic structure to appear at the core of a vortex. Hence, while the phenomenological approach to the vortex structure (Sect. 6.3) led to a monotonic increase in the amplitude of the order parameter $u(\rho)$ with increasing distance from the vortex core, we now obtain an increase with oscillations. This noncoincidence can be explained by the fact that in the phenomenological approach we were not able to determine the influence of the spatial distribution of the effective electron-phonon interaction parameter on the spatial distribution of the order parameter of the superconducting phase. For $\langle V_{\varkappa_o}(2\mathbf{p}_o)\rangle_{\rho(\varkappa_o)}/g^2_{ph} \ll 1$, the function (9.3.10) vanishes if $\cos(2\pi n\, \mathrm{sech}^2[n \ln(\rho/\rho_o)]) \to 0$. It can be easily shown that in this case,

$$(\rho/\rho_o)^n = \frac{\sqrt{4n}}{\sqrt{4k \pm 1}} \pm \sqrt{\frac{4(n-k) \mp 1}{4k \pm 1}}\,, \qquad (9.4.6)$$

where $k = 0,1,\ldots,n\,;\quad n = m\,,$

$k = 0,1,\ldots,2n-1\,;\quad n = m/2\,.$

In this case, if $n = 1$, the roots of the equation $\cos(2\pi n\, \mathrm{sech}^2[n \ln(\rho/\rho_o)]) = 0$ are

$$\rho_1 = (2+\sqrt{3})\rho_o\,,\qquad \rho_2 = (2-\sqrt{3})\rho_o\,,$$
$$\rho_3 = \sqrt{3}\,\rho_o\,,\qquad \rho_4 = \frac{1}{\sqrt{3}}\rho_o\,. \qquad (9.4.7)$$

The quantity ρ_1 is close to the vortex size R_0 determined from the phenomenological theory ($R_0 \simeq 3\rho_0$) (see 6.3.33). However, for $\rho = \rho_1$, we obtain a normal region, and hence the saturation region occurs for $\rho > \rho_1$. Consequently, the vortex size is found to be larger than $3\rho_o$, but not to such an extent as to change the order of magnitude of the vortex flux quantum $K_o = \frac{\pi\hbar c}{4e}\frac{\rho_o}{R_o}$. For $n = \frac{1}{2}$, we obtain

$$\rho_1 = (1+\sqrt{2})^2\rho_o\,,\quad \rho_2 = (-1+\sqrt{2})^2\rho_o\,.$$

Figures 9.1 and 9.2 show the dependences of the functions $\langle \mathrm{Tr}\hat{B}(\Gamma_\rho, n)\rangle$ on ρ for $n = 1$ and $\frac{1}{2}$. The spatial regions in which the superconductivity can be suppressed and a metastable magnetic structure can emerge are indicated in these figures by wavy lines. The effective value of the phase transition temperature in this structure can be determined by spatial averaging of the function determining the superconducting order parameter (9.3.17) and is found to be a function of n, i.e., $T_c[n] = T_{cn}$. Thus, each type of vortex has a transition temperature corresponding to it, and the following relation holds between the transition temperatures:

$$f(\varrho) = \cos Z(\varrho) \left[\cos Z(\varrho) - \tfrac{\mu^*}{\lambda}\right]$$

$$Z(\varrho) = 2\pi n \sin^2 \xi(\varrho), \quad n = 1$$

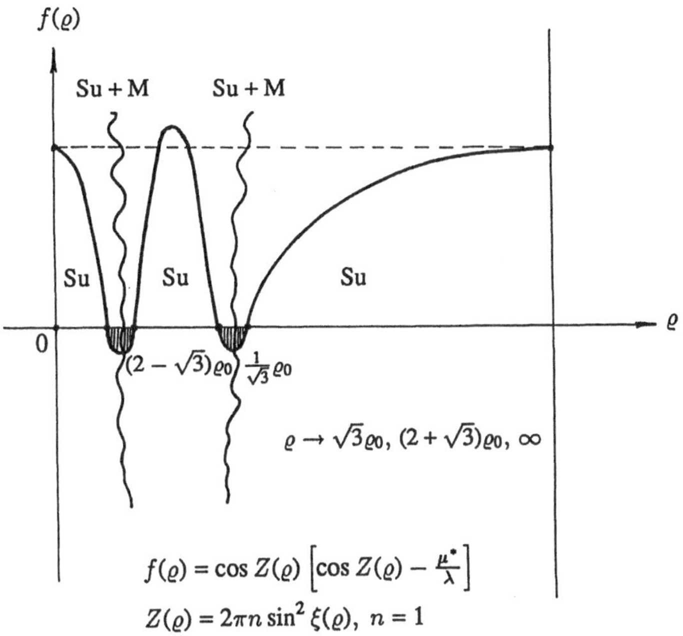

Fig. 9.1. Spatial distribution of the field function which determines the structure of inhomogeneous superconducting states with a vortex lattice for $n = 1$; $f(\varrho) = \cos Z(\varrho)[\cos Z(\varrho)\mu^*/\lambda]$; $Z(\varrho) = 2\pi n \sin^2 \xi(\varrho)$. Su – superconducting state, M – normal phase region with strong magnetic fluctuations, Su + M – mixed region

$$f(\varrho) = \cos Z(\varrho) \left[\cos Z(\varrho) - \tfrac{\mu^*}{\lambda}\right]$$

$$Z(\varrho) = 2\pi n \sin^2 \xi(\varrho), \quad n = \tfrac{1}{2}$$

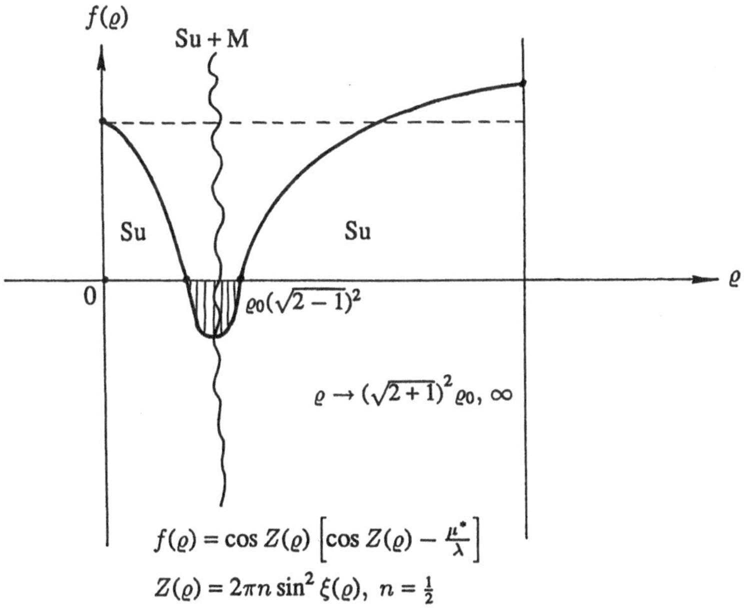

Fig. 9.2. Spatial distribution of the field function for a system containing vortices with $n = 1/2$, $f(\varrho)$ and labels on phases as in Fig. 9.1; $Z(\varrho) = 2\pi n \sin^2 \xi(\varrho)$

$$\ldots < T_{cn+1} < T_{cn} < T_{cn-1} < \ldots .$$

Thus, we have obtained phase lamination in the superconducting state itself. It follows from (9.4.8) that the homogeneous superconducting state has the highest phase transition temperature. In this context, it should be interesting to estimate the critical temperature for a ceramic of the type $YBa_2Cu_3O_{7-y}$. By using (9.3.17) for the homogeneous superconducting state, we can estimate the critical temperature by using the values $\lambda_{e-ph} \approx 0{,}6\text{-}0{,}8$; $\mu^* \approx 0{,}2\text{-}0{,}3$; $\cosh(2\pi^2 v_o^{2,1/3}/\langle r_c\rangle) \approx 2$, $\langle \omega_D \rangle \approx 400$ K. This value may be as high as 130 K. Such a result was obtained in [9.11], where the electrical resistivity of the oxygen-enriched compound $YBa_2Cu_3O_{7-y}$ was investigated. Such a low critical temperature for this compound may be due to the existence of oxygen vacancies. In other words, the possibilities of increasing the critical temperature for these compounds have not been exhausted completely.

9.5 Microscopic Equations near the Phase Transition Point: Relation to the Phenomenological Theory

The final section of this chapter will be devoted to a comparison of the microscopic and phenomenological theories of the superconducting phase in magnetic superconductors. For this purpose, we must write down the equation for the superconducting order parameter for $\tau << 1$ as well as the expression for the current from bound electrons. The equations for the electron Green functions, averaged over the cell configuration, can be written as

$$\Big\{[i\omega + \varepsilon(\mathbf{p}|\Gamma|\langle p_{o\nu}^2[\varkappa]\rangle]\delta_{\alpha\beta'} + \langle \lambda_{eph}(\mathbf{p}_o|\Gamma|\bar{p}_{\perp o})\rangle_{\rho(\varkappa_o,\bar{p}_{\perp o})}$$

$$\varphi_1(N) \sum_{\alpha_2=1}^{N^2-1} \int_{-\varkappa_2\in\Gamma} \frac{D\varkappa_2(\mathbf{x}'',\tau'')}{2\pi C_\varkappa} \exp\Big(\int_{\varkappa_2} dx_\nu'' p_{o\nu}[\varkappa_2]\Big)$$

$$\times \langle \mathrm{Tr}\hat{\tau}^{\alpha_2}\hat{B}^{(1)}(\Gamma)\hat{\tau}^{\alpha_2}\hat{B}(\varkappa_2)\rangle \quad G_{\alpha\beta'}(\varkappa_2)\Big\} \, G_{\beta'\beta\varkappa}(\mathbf{p},i\omega)$$

$$\times \langle \mathrm{Tr}\hat{\tau}^{\alpha_1}\hat{B}^{(1)}(\Gamma)\hat{\tau}^{\alpha_1}\hat{B}(\varkappa)\rangle + \frac{1}{2}\varphi_1(N) \sum_{\alpha_2=1}^{N^2-1} \int_{-\varkappa_2\in\Gamma} \frac{D\varkappa_2(\mathbf{x}'',\tau'')}{2\pi C_\varkappa}$$

$$\times \exp\Big(\int_{\varkappa_2} dx_\nu'' \, p_{o\nu}[\varkappa_2]\Big) \quad \langle \mathrm{Tr}\hat{\tau}^{\alpha_2}\hat{B}(\Gamma)\hat{\tau}^{\alpha_2}\hat{B}(\varkappa_2)\rangle$$

$$\times[\Delta_{\alpha\beta'}(\varkappa_2) - \Delta^t_{\alpha\beta'}(\varkappa_2)]F^*_{\beta'\beta}(\mathbf{p},i\omega)\langle \mathrm{Tr}\hat{\tau}^{\alpha_1}\hat{B}^{(2)}(\Gamma)\hat{\tau}^{\alpha_1}B(\varkappa)\rangle$$

$$= \delta_{\alpha\beta}\langle \mathrm{Tr}\hat{B}(\Gamma)\rangle \quad , \tag{9.5.1}$$

$$\varepsilon(\mathbf{p}|\Gamma|\langle p^2_{o\nu}[\varkappa]\rangle) = -\varepsilon(\langle p^2_{o\nu}[\varkappa]\rangle|\Gamma|\mathbf{p}) \ ,$$

$$\Big\{[i\omega - \varepsilon(\mathbf{p}|\Gamma|\langle p^2_{o\nu}[\varkappa]\rangle)]\delta_{\alpha\beta'} + \langle \lambda_{eph}(\mathbf{p}_o|\Gamma|\bar{p}_{\perp o})\rangle_{\rho(\varkappa_o,\bar{p}_{\perp o})}$$

$$\varphi_1(N)\sum_{\alpha_2=1}^{N^2-1}\int_{-\varkappa_2\in\Gamma}\frac{D\varkappa_2(\mathbf{x}'',\tau'')}{2\pi C_\varkappa}\quad \exp(\int_{\varkappa_2} dx''_\nu\, p_{o\nu}[\varkappa_2])$$

$$\times\langle \mathrm{Tr}\hat{\tau}^{\alpha_2}\hat{B}^{(1)}(\Gamma)\hat{\tau}^{\alpha_2}\hat{B}(\varkappa_2)\rangle \quad G^*_{\alpha\beta'}(\varkappa_2)\Big\}\ \mathrm{Tr}\hat{\tau}^{\alpha_1}\hat{B}^{(2)}(\Gamma)\hat{\tau}^{\alpha_1}\hat{B}(\varkappa)\rangle$$

$$\times\ F^*_{\beta'\beta}(\mathbf{p},i\omega) + \frac{1}{2}\varphi_1(N)\sum_{\alpha_2=1}^{N^2-1}\int_{-\varkappa_2\in\Gamma}\frac{D\varkappa_2(\mathbf{x}'',\tau'')}{2\pi C_\varkappa}\ \exp(\int_{\varkappa_2} dx''_\nu p_{o\nu}[\varkappa_2])$$

$$\times\langle \mathrm{Tr}\hat{\tau}^{\alpha_2}\hat{B}(\Gamma)\hat{\tau}^{\alpha_2}\hat{B}(\varkappa_2)\rangle[\Delta^*_{\alpha\beta'}(\varkappa_2) - \Delta^{*t}_{\alpha\beta'}(\varkappa_2)]$$

$$\times\ G_{\beta'\beta\varkappa}(\mathbf{p},i\omega)\ \ \langle \mathrm{Tr}\hat{\tau}^{\alpha_1}\hat{B}^{(1)}(\Gamma)\hat{\tau}^{\alpha_1}\hat{B}(\varkappa)\rangle \quad . \tag{9.5.2}$$

In a more compact form, these equations can be presented as

$$[i\omega\delta_{\alpha\beta'} + \mu_{\alpha\beta'}(\mathbf{p}_o|\Gamma|\langle p^2_{o\nu}[\varkappa]\rangle)]G_{\omega\beta'\beta}(\mathbf{x}|\Gamma|\mathbf{x}')$$

$$-\ \frac{1}{2}\ \mathrm{Tr}[\hat{\mathbb{D}}^2_\nu\ \hat{G}_{\omega\alpha\beta}(\mathbf{x}|\Gamma|\mathbf{x}')] + \frac{1}{2}[\Delta_{\alpha\beta'}(\Gamma) - \Delta^t_{\alpha\beta'}(\Gamma)]F^*_{\omega\beta'\beta}(\mathbf{x}|\Gamma|\mathbf{x}')$$

$$= \delta_{\alpha\beta}\oint_\Gamma\ e_{\nu'}dx'_{\nu'}\ \delta(\mathbf{x}-\mathbf{x}')\ \langle \mathrm{Tr}\hat{B}(\Gamma)\rangle \quad , \tag{9.5.3}$$

$$[i\omega\delta_{\alpha\beta'} + \mu^*_{\alpha\beta'}(\mathbf{p}_o|\Gamma|\langle p^2_{o\nu}[\varkappa]\rangle)]F^*_{\omega\beta'\beta}(\mathbf{x}|\Gamma|\mathbf{x}')$$

$$+\ \frac{1}{2}\ \mathrm{Tr}[\hat{\mathbb{D}}^2_\nu\ \hat{F}^*_{\omega\alpha\beta}(\mathbf{x}|\Gamma|\mathbf{x}')] + \frac{1}{2}[\Delta^*_{\alpha\beta'}(\Gamma) - \Delta^{*t}_{\alpha\beta'}(\Gamma)]G_{\omega\beta'\beta}(\mathbf{x}|\Gamma|\mathbf{x}')=0. \tag{9.5.4}$$

In the following, we shall use the notation

$$\frac{1}{2}[\Delta_{\alpha\beta}(\Gamma_{\mathbf{x},\tau}) - \Delta^t_{\alpha\beta}(\Gamma_{\mathbf{x},\tau})] = \Phi_{\alpha\beta}(\Gamma_{\mathbf{x},\tau}).$$

It was shown in Sect. 9.3 that the functions $\Phi_{\alpha\beta}(\Gamma_{\mathbf{x},\tau})$ satisfy the integral equation

$$\Phi_{\alpha\beta}(\Gamma_{\mathbf{x},\tau}) = \langle \lambda_{eph}(\mathbf{p}_o|\Gamma_{\mathbf{x},\tau}|\bar{p}_{\perp o})\rangle_{\rho(\varkappa_o,\bar{p}_{\perp o})}$$

$$\times\ \varphi_1(N) \sum_{\alpha_2=1}^{N^2-1} \int_{-\varkappa_2\in\Gamma} \frac{D\varkappa_2(\mathbf{x}'',\tau'')}{2\pi C_\varkappa} \frac{d\mathbf{p}}{(2\pi)^3}\frac{d\omega}{2\pi} \exp\left(\int_{\varkappa_2} dx''_\nu p_{o\nu}[\varkappa_2]\right)$$

$$\times \Phi_{\alpha\beta}(\varkappa_2|\Gamma_{\mathbf{x},\tau}) \frac{\langle \mathrm{Tr}\hat{\tau}^{\alpha_2}\hat{B}(\Gamma)\hat{\tau}^{\alpha_2}\hat{B}(\varkappa_2)\rangle}{\beta^2[\omega^2+\varepsilon^2(\mathbf{p}|\Gamma|\langle p^2_{o\nu}[\varkappa_2]\rangle) + |\Phi(\Gamma_{\mathbf{x},\tau})|^2]}$$

$$\beta = 1/T.$$

At finite temperatures $T > 0$, the integration with respect to the frequency in (9.5.5) should be replaced by summation. As a result, we obtain

$$\Phi_{\alpha\beta}(\Gamma_{\mathbf{x},\tau}) = \langle \lambda_{eph}(\mathbf{p}_o|\Gamma_{\mathbf{x},\tau}|\bar{p}_{\perp o})\rangle'_{\rho(\varkappa_o,\bar{p}_{\perp o})} \sum_\omega F_{\omega\alpha\beta}(\mathbf{x}|\Gamma_{\mathbf{x},\tau}|\mathbf{x}). \tag{9.5.6}$$

We expand $F_{\omega\alpha\beta}(\mathbf{x}|\Gamma|\mathbf{x}')$ into a functional series in $\Phi_{\alpha\beta}(\Gamma)$. For this purpose, we define the integral relations for electron Green functions:

$$G_{\omega\alpha\beta}(\mathbf{x}|\Gamma|\mathbf{x}') = \tilde{G}^{(0)}_{\omega\alpha\beta}(\mathbf{x}|\Gamma|\mathbf{x}') - \beta \int_{-\varkappa\in\Gamma} \frac{D\tilde{\varkappa}(\mathbf{x}'',\tau'')}{2\pi C_\varkappa}$$

$$\times\ \tilde{G}^{(0)}_{\omega\alpha\beta'}(\mathbf{x}|\Gamma+\varkappa|\mathbf{x}'')\Phi_{\beta'\beta''}(\tilde{\varkappa})F_{\omega\beta''\beta}(\mathbf{x}''|\Gamma+\varkappa|\mathbf{x}'), \tag{9.5.7}$$

$$F_{\omega\alpha\beta}(\mathbf{x}|\Gamma|\mathbf{x}') = \beta \int_{-\varkappa\in\Gamma} \frac{D\tilde{\varkappa}(\mathbf{x}'',\tau'')}{2\pi C_\varkappa} \tilde{G}^{(0)}_{\omega\alpha\beta''}(\mathbf{x}|\Gamma+\varkappa|\mathbf{x}'')\Phi_{\beta'\beta''}(\tilde{\varkappa})$$

$$\times\ G_{\omega\beta''\beta}(\mathbf{x}''|\Gamma+\varkappa|\mathbf{x}'), \tag{9.5.8}$$

$$D\tilde{\varkappa} = D\varkappa d\mathbf{x}''.$$

In (9.5.7,8), the functions $\tilde{G}^{(0)}_{\omega\alpha\beta}(\mathbf{x}|\Gamma|\mathbf{x}')$ satisfy the condition

$$[i\omega\delta_{\alpha\beta'} + \mu_{\alpha\beta'}(\mathbf{p}_o|\Gamma|\langle p^2_{o\nu}[\varkappa]\rangle)]\tilde{G}^{(0)}_{\omega\beta'\beta}(\mathbf{x}|\Gamma|\mathbf{x}')$$

$$-\frac{1}{2}\mathrm{Tr}[\hat{\mathbb{D}}^2_\nu \tilde{G}^{(0)}_{\omega\alpha\beta}(\mathbf{x}|\Gamma|\mathbf{x}')] = \delta_{\alpha\beta}\oint_\Gamma e_{\nu'}dx'_{\nu'}\delta(\mathbf{x}-\mathbf{x}')\langle \mathrm{Tr}\hat{B}(\Gamma)\rangle. \tag{9.5.9}$$

It can be easily shown that $\tilde{G}^{(0)*}_{\omega\alpha\beta}(\mathbf{x}|\Gamma|\mathbf{x}') = \tilde{G}^{(0)}_{-\omega\alpha\beta}(\mathbf{x}|\Gamma|\mathbf{x}')$. This leads to the relation for the function $F^*_{\omega\alpha\beta}(\mathbf{x}|\Gamma|\mathbf{x}')$:

$$F^{*}_{\omega\alpha\beta}(\mathbf{x}|\Gamma|\mathbf{x}') = \beta \int\limits_{-\varkappa} \frac{D\tilde{\varkappa}(\mathbf{x}'',\tau'')}{2\pi C_{\varkappa}} \tilde{G}^{(0)}_{-\omega\alpha\beta'}(\mathbf{x}|\Gamma+\varkappa|\mathbf{x}'')\Phi^{*}_{\beta'\beta''}(\tilde{\varkappa})$$

$$\times\ G_{\omega\beta''\beta}(\mathbf{x}''|\Gamma+\varkappa|\mathbf{x}') \quad . \tag{9.5.10}$$

Taking into consideration (9.5.6,10), we can write the equation for the function defining the superconducting order parameter as

$$\Phi_{\alpha\beta}(\mathbf{x},\tau\ |\Gamma) = \langle\ \lambda_{eph}(\mathbf{p}_{o}|\ \Gamma_{\mathbf{x},\tau}\ |\bar{p}_{\perp o})\ \rangle_{\rho(\varkappa_{o},\bar{p}_{\perp o})}$$

$$\times \sum_{\omega}\Big\{\ \beta\int\limits_{-\varkappa\in\Gamma} \frac{D\tilde{\varkappa}(\mathbf{x}'',\tau'')}{2\pi C_{\varkappa}}\ \tilde{G}^{(0)}_{\omega\alpha\beta}\ (\mathbf{x}|\Gamma+\varkappa|\mathbf{x}'')\Phi_{\beta'\beta''}(\mathbf{x}'',\tau''|\varkappa)$$

$$\times\ \tilde{G}^{(0)}_{-\omega\beta''\beta}(\mathbf{x}''|\Gamma+\varkappa|\mathbf{x})\ -\ \beta^{3}\int\limits_{-\varkappa_1,-\varkappa_2,-\varkappa_3\in\Gamma} \frac{D\tilde{\varkappa}_1(\mathbf{x}''_1,\tau''_1)}{2\pi C_{\varkappa}}\ \frac{D\tilde{\varkappa}_2(\mathbf{x}''_2,\ \tau''_2)}{2\pi C_{\varkappa}}\ \frac{D\tilde{\varkappa}_3(\mathbf{x}''_3,\tau''_3)}{2\pi C_{\varkappa}}$$

$$\times\ \tilde{G}^{(0)}_{\omega\alpha\beta'_1}\ (\mathbf{x}|\Gamma+\varkappa_1|\mathbf{x}'')\Phi_{\beta'_1\beta''_1}\ (\mathbf{x}''_1,\ \tau''_1|\varkappa_1)\tilde{G}^{(0)}_{-\omega\beta''_1\beta'_2}\ (\mathbf{x}''_1|\Gamma+\varkappa_1+\varkappa_2|\mathbf{x}''_2)$$

$$\Phi^{*}_{\beta'_2\beta''_2}\ (\mathbf{x}''_2,\tau''_2|\varkappa_2)\ \tilde{G}^{(0)}_{\omega\beta''_2\beta'_3}\ (\mathbf{x}''_2|\Gamma+\varkappa+\varkappa_2+\varkappa_3\ \ |\mathbf{x}''_3)$$

$$\times\ \Phi_{\beta'_3\beta''_3}(\mathbf{x}''_3,\tau''_3|\varkappa_3)G_{\omega\beta''_3\beta}(\mathbf{x}''_3|\Gamma+\varkappa_1+\varkappa_2+\varkappa_3\ \ |\mathbf{x})\Big\} \quad . \tag{9.5.11}$$

Substituting the variables into the contour integrals over the loops belonging to the contours $\Gamma_{\mathbf{x},\mathbf{x}''_i}$; $\Gamma_{\mathbf{x}''_j,\mathbf{x}}$, we obtain

$$\Phi_{\alpha\beta}(\mathbf{x},\tau|\Gamma) = \langle\ \lambda_{e\text{-}ph}(\mathbf{p}_{o}|\Gamma_{\mathbf{x},\tau}|\bar{p}_{\perp o})\rangle_{\rho(\varkappa_{o},\bar{p}_{\perp o})}$$

$$\times \sum_{\omega}\Big\{\ \beta\int\limits_{-\varkappa\in\Gamma} \frac{D\tilde{\varkappa}(\mathbf{x}'',\tau'')}{2\pi C_{\varkappa}}\ \tilde{G}^{(0)}_{\omega\alpha\beta'}(\mathbf{x}|\varkappa|\mathbf{x}'')\ \Phi_{\beta'\beta''}(\mathbf{x}'',\tau''|\Gamma+\Gamma\ +\varkappa)$$

$$\times\ \tilde{G}^{(0)}_{-\omega\beta''\beta}(\mathbf{x}''\ |\varkappa|\mathbf{x})\ -\ \beta^{3}\int\limits_{-\varkappa_1,-\varkappa_2,-\varkappa_3} \frac{D\tilde{\varkappa}_1(\mathbf{x}''_1,\ \tau''_1)}{2\pi C_{\varkappa}}\ \frac{D\tilde{\varkappa}_2(\mathbf{x}''_2,\tau''_2)}{2\pi C_{\varkappa}}\ \frac{D\tilde{\varkappa}_3(\mathbf{x}''_3,\tau''_2)}{2\pi C_{\varkappa}}$$

$$\times\ \tilde{G}^{(0)}_{\alpha\beta'_1\omega}\ (\mathbf{x}|\varkappa_1|\mathbf{x}''_1)\Phi_{\beta'_1\beta''_1}(\mathbf{x}''_1,\tau''_1|\Gamma+\Gamma+\varkappa_1)\tilde{G}^{(0)}_{-\omega\beta''_1\beta'_2}(\mathbf{x}''_1|\varkappa_1+\varkappa_2|\mathbf{x}''_2)$$

$$\times\Phi^{*}_{\beta'_2\beta''_2}(\mathbf{x}'',\tau''_2|\Gamma\ +\varkappa_2)\ \tilde{G}^{(0)}_{\omega\beta''_2\beta'_3}(\mathbf{x}''_2|\varkappa_1+\varkappa_2+\varkappa_3|\mathbf{x}''_3)$$

$$\times\ \Phi_{\beta'_3\beta''_3}\ (\mathbf{x}''_3,\tau''_3|\varkappa_3)G^{*}_{\omega\beta''_3\beta}\ (\mathbf{x}''_3|\varkappa_1+\varkappa_2+\varkappa_3|\mathbf{x})\Big\} \ . \tag{9.5.12}$$

In this equation, the Green function $\tilde{G}^{(0)}_{\pm\omega}(\mathbf{x}|\varkappa|\mathbf{x}')$ has the form

$$\tilde{G}^{(0)}_{\pm\omega\alpha\beta}(\mathbf{x}|\varkappa|\ \mathbf{x}') = \delta_{\alpha\beta}\int\frac{d\mathbf{p}}{(2\pi)^3}\ \frac{\langle\ \mathrm{Tr}\hat{B}(\varkappa)\ \rangle\exp[i\mathbf{p}(\mathbf{x}-\mathbf{x}')]}{\beta[\varepsilon(\mathbf{p}|\Gamma\to\varkappa|\langle p^2_{o\nu}[\varkappa]\ \rangle\pm\ i\omega]} \tag{9.5.13}$$

where

$$\varepsilon(\mathbf{p}|\Gamma\to\varkappa|\ \langle\ p^2_{o\nu}[\varkappa]\ \rangle) \ = \ v_o(p - p_o) - \mu_\varkappa,$$

$$\mu_\varkappa = \frac{1}{2\mu_e}\ \langle\ p^2_{o\nu}[\varkappa]\ \rangle;\quad v_o \approx v_F\ ;\quad p_o \to p_F\ .$$

The dependence of the function $G^{(0)}_{\omega\alpha\beta}\ (\mathbf{x}|\varkappa|\ \mathbf{x}')$ on frequency is given by

$$\tilde{G}^{(0)}_{\omega\alpha\beta}(R|\varkappa) = \delta_{\alpha\beta}\langle\ \mathrm{Tr}\hat{B}(\varkappa)\ \frac{\mu_e T v_o}{2\pi R h^2}\left\{\begin{array}{ll}\exp(ik_F + \frac{|\omega|}{v_F})R & ,\ \omega > 0\\ \exp(ik_F - \frac{|\omega|}{v_F})R & ,\ \omega < 0\end{array}\right.$$

$$\omega\in(\varepsilon_o - \mu_\varkappa,\ \ -\varepsilon_o - \mu_\varkappa)\ ,$$

$$\varepsilon_o \approx \langle\ \omega_D\ \rangle_{\rho(\varkappa)}\ , \tag{9.5.14}$$

where $k_F = p_F/\hbar$.

With the help of (9.5.14), we can reduce the part of (9.5.12), which is linear in $\Phi_{\alpha\beta}(\mathbf{x},\tau|\varkappa)$ to a form which assumes a slow variation of $\Phi_{\alpha\beta}(\mathbf{x},\tau\ |\Gamma)$ as a function of the contour variable $\Gamma_{\mathbf{x},\tau}$:

$$\Phi^{(0)}_{\alpha\beta}(\mathbf{x},\tau|\Gamma) = \langle\ \lambda_{eph}(\mathbf{p}_o|\Gamma_{\mathbf{x},\tau}\ |\bar{p}_{\perp o}\rangle\rangle_{\rho(\varkappa_o,\bar{p}_{\perp o})}\ \langle\ \mathrm{Tr}\hat{B}(\varkappa_o)\ \rangle$$

$$\times\left\{\int d\mathbf{R}\Lambda(R)\Phi_{\alpha\beta}\ (\mathbf{x},\tau|\Gamma) + \frac{1}{2}\ \mathrm{Tr}[\hat{\tilde{\mathbb{D}}}^2_\nu\ \hat{\Phi}_{\alpha\beta}(\mathbf{x},\tau|\ \Gamma)]\ \int\ d\mathbf{R}\ R^2\ \Lambda(R)\right\}$$

$$\hat{\tilde{\mathbb{D}}}_\nu = \hat{\tau}^{\,o}\nabla_\nu - 2[\hat{A}_{\nu,}]\ . \tag{9.5.15}$$

The function $\Lambda(R)$ is defined by the relation

$$\Lambda(R) = \left(\frac{\mu_e v_o T^{\frac{1}{2}}}{2\ \hbar^2 R}\right)^2\ \cosh(\frac{2\pi RT}{hv_F})[\mathrm{cotanh}(\frac{2\pi RT}{hv_F} - 1)]\ . \tag{9.5.16}$$

Equation (9.5.15) describes the distribution of the order parameter in the region of values of the contour variable $\Gamma_{\mathbf{x},\tau}(\rho)$ (9.4.5), for which the order parameter attains saturation, i.e., for $\rho/\rho_o >> 1$ (Sect. 9.4). Evaluating the integrals in (9.5.15), we obtain

$$\tilde{\Phi}^{(0)}_{\alpha\beta}(\mathbf{x},\tau|\Gamma) = \frac{\hbar^2}{4\mu_e}\mathrm{Tr}[\hat{\mathbb{D}}_\nu\hat{\Phi}_{\alpha\beta}(\mathbf{x},\tau|\Gamma)] + \frac{3\pi^2}{\beta^2\varepsilon_F A}\Phi_{\alpha\beta}(\mathbf{x},\tau|\Gamma)$$

$$\ln\left\{\left[\frac{\langle\omega_D\rangle^2\rho(\varkappa) - \mu_\varkappa^2}{T^2 - \mu_\varkappa^2}\right]^{\frac{1}{2}} \exp\left[-\nu_{\varkappa_o}(\varepsilon_F)\langle\lambda_{eph}(\mathbf{p}_o|\Gamma_{\mathbf{x},\tau}|\bar{p}_{\perp o})\rangle_{\rho(\varkappa_o,\bar{p}_{\perp o})}\right]^{-1}\right\},$$

(9.5.17)

where

$$A = \sum_{n=0}^{\infty}\frac{\Gamma(n+3)\zeta(n+3)}{2^n(2n!)}, \quad \nu_{\varkappa_o}(\varepsilon_F) = \nu(\varepsilon_F)\langle\mathrm{Tr}\hat{B}(\varkappa_o)\rangle .$$

At $T\to T_c$, $T_c\to T_N$, it follows that $T_N > \mu_\varkappa$, $\mu_\varkappa \to \varepsilon_F(v_o^{1/3}/\langle r_c\rangle)^2$, i.e., we have the restriction (1.8). The term with a cubic nonlinearity has the form

$$-\langle\mathrm{Tr}\hat{B}(\varkappa_o)\rangle^2 \frac{7\mu_e v_o P_F}{(2\pi)^4 h^3}\zeta(3)\Phi_{\alpha\beta}(\mathbf{x},\tau|\Gamma)|\Phi(\mathbf{x},\tau|\Gamma)|^2 . \qquad (9.5.18)$$

In order to write the equation for the order parameter in the superconducting phase, let us determine the current in the system (Sect. 9.3). Recalling the equation for the field operator $\hat{B}(\Gamma)$(9.2.22), we observe that the current has two components which will be denoted by $\langle\hat{J}_\nu(\mathbf{x})\rangle_S$ and $\langle\hat{J}_\nu(\mathbf{x})\rangle_N$. The first of these components corresponds to the current of electrons forming the bound states, and is nonzero only in the superconducting phase. The second term corresponds to the normal component of the current induced by the fluctuations of the s-, d(f)-electron spins in the system (Sect. 9.3). Indeed,

$$\langle\hat{J}_\nu(\mathbf{x})\rangle_S = -\frac{i}{4}g_1N^2 \lim_{\substack{\mathbf{x},\tau\to\mathbf{x}',\tau'\\ \varepsilon\to 0}} \sum_{\alpha=1}^{N^2-1}\sum_{\gamma=1}^{2}\int_{-\varkappa\in\Gamma}\frac{D\varkappa(\mathbf{x}'',\tau'')}{2\pi C_\varkappa}$$

$$\times(\nabla_{x'_\nu} - \nabla_{x_\nu})G_{\gamma\gamma\varkappa}(\mathbf{x}|\varepsilon|\mathbf{x}')\langle\mathrm{Tr}[\hat{\tau}^\alpha,\hat{\tau}^\alpha\hat{B}(\varkappa)]\hat{B}(\Gamma)\rangle$$

$$\times\exp\left(\int_\varkappa dx''_\nu\, p_{o\nu}[\varkappa]\right), \qquad (9.5.19)$$

$$\langle\hat{J}_\nu(\mathbf{x})\rangle_N = \frac{i}{2}g_1N^2\oint_\Gamma dx'_\nu\,\delta(\mathbf{x}-\mathbf{x}')\sum_{\alpha=1}^{N^2-1}\langle\mathrm{Tr}\hat{\tau}^\alpha\hat{B}(C_{\mathbf{x},\mathbf{x}'})\hat{\tau}^\alpha\hat{B}(C_{\mathbf{x}',\mathbf{x}})\rangle.$$

(9.5.20)

Using (9.5.7,8) for the Green functions, we can transform (9.5.19) as

$$\langle \hat{J}_\nu(\mathbf{x}) \rangle_S = -\frac{i}{4} g_1 N^2 \lim_{\substack{\mathbf{x},\tau\to\mathbf{x}',\tau' \\ \varepsilon\to 0}} \beta^2 \int_{-\varkappa,-\varkappa_1,-\varkappa_2^{\varkappa}\in\Gamma} \frac{D\varkappa(\mathbf{x}'',\tau'')}{2\pi C_\varkappa} \frac{D\varkappa_1(\mathbf{x}_1'',\tau_1'')}{2\pi C_\varkappa}$$

$$\times \frac{D\varkappa_2(\mathbf{x}_2'', \tau_2'')}{2\pi C_\varkappa} (\nabla_{x_\nu'} - \nabla_{x_\nu}) \sum_{\alpha=1}^{N^2-1} \sum_\omega \tilde{G}^{(0)}_{\omega\gamma\beta'} (\mathbf{x}|\varepsilon+\varkappa+\varkappa_1|\mathbf{x}_1'')\Phi_{\beta'\beta''}(\tilde{\varkappa}_1)$$

$$\times \tilde{G}^{(0)}_{-\omega\beta''\beta_1'} (\mathbf{x}_1''|\varepsilon+\varkappa+\varkappa_1+\varkappa_2|\mathbf{x}_2'')\Phi^*_{\beta_1'\beta_1''}(\varkappa_2)$$

$$\times \tilde{G}^{(0)}_{\omega\beta_1''\gamma} (\mathbf{x}_1''|\varepsilon+\varkappa+\varkappa_1+\varkappa_2|\mathbf{x}')\exp\Big(\int_\varkappa dx_\nu'' p_{o\nu}[\varkappa]\Big) \langle \mathrm{Tr}[\hat{\tau}^\alpha,\hat{\tau}^\alpha \hat{B}(\varkappa)]\hat{B}(\Gamma) \rangle .$$

(9.5.21)

In the following analysis, it will be more convenient to write (9.5.21) in differential form. For this purpose, we shall require the explicit form of the operator

$$\lim_{\substack{\mathbf{x},\tau\to\mathbf{x}',\tau' \\ \varepsilon\to 0}} (\nabla_{x_\nu'} - \nabla_{x_\nu}) \sum_\omega \beta^2 \int d\mathbf{x}_1'' d\mathbf{x}_2'' \, \tilde{G}^{(0)}_{\omega\gamma\beta'} (\mathbf{x}|\varepsilon+\varkappa+\varkappa_1|\mathbf{x}_1'')$$

$$\times \tilde{G}^{(0)}_{-\omega\beta''\beta_1'} (\mathbf{x}_1''|\varepsilon+\varkappa+\varkappa_1+\varkappa_2|\mathbf{x}_2'')\tilde{G}^{(0)}_{\omega\beta_1''\gamma} (\mathbf{x}_2''|\varepsilon+\varkappa+\varkappa_1+\varkappa_2|\mathbf{x}_1')$$

$$= \lim_{\substack{\mathbf{x},\tau\to\mathbf{x}',\tau' \\ \varepsilon\to 0}} \beta^2 \delta_{\gamma\beta'}\delta_{\beta''\beta_1'}\delta_{\beta_1''\gamma} \langle \mathrm{Tr}\hat{B}(\varkappa+\varkappa_1+\varepsilon) \rangle \langle \mathrm{Tr}\hat{B}(\varepsilon+\varkappa+\varkappa_1+\varkappa_2) \rangle^2$$

$$\times \sum_\omega \int \frac{d\mathbf{p}}{(2\pi)^3} \frac{\exp[i\mathbf{p}(\mathbf{x}-\mathbf{x}')]}{[\beta\varepsilon(\mathbf{p}|\,\varkappa+\varkappa_1|\,\langle p^2_{o\nu}[\varkappa+\varkappa_1]\rangle) + i\beta\omega][\beta^2\varepsilon^2(\mathbf{p}|\varkappa+\varkappa_1+\varkappa_2|\,\langle p^2_{o\nu}[\varkappa+\varkappa_1+\varkappa_2]\rangle)+\beta^2\omega^2]}$$

$$(\nabla_{x_\nu'} - \nabla_{x_\nu}) \, . \qquad (9.5.22)$$

The integral in (9.5.22) can be evaluated and is found to be a function of $\varkappa$, $\varkappa_1$, and $\varkappa_2$:

$$\frac{1}{8}\frac{\beta^2}{\pi^2}\langle N_{\varkappa+\varkappa_1}\rangle \langle \mathrm{Tr}\hat{B}(\varkappa+\varkappa_1+\varkappa_2) \rangle^2 \delta_{\gamma\beta'}\delta_{\beta''\beta_1'}\delta_{\beta_1'\gamma} , \qquad (9.5.23)$$

where

$$\langle N_{\varkappa+\varkappa_1} \rangle = \frac{\mu_e^{3/2} (\varepsilon_F + \mu_{\varkappa+\varkappa_1})^{3/2} v_o}{\sqrt{2}\,\pi^2\hbar^3} \langle \mathrm{Tr}\hat{\mathbf{B}}(\varkappa+\varkappa_1) \rangle . \qquad (9.5.24)$$

Using (9.5.23,24). we can introduce the function

$$\Psi_{\alpha\beta}(\varkappa|\Gamma_{\mathbf{x},\tau}) = N\left[\frac{7\zeta(3)\langle N_\varkappa\rangle}{8\pi^2 T_c^2[\varkappa|\Gamma_{\mathbf{x},\tau}]}\right]^{\frac{1}{2}} \mathrm{Tr}\hat{B}(\varkappa)\rangle\,\Phi_{\alpha\beta}(\Gamma_{\mathbf{x},\tau})\ . \tag{9.5.25}$$

This function is nonzero only in the superconducting phase, and its physical significance is determined by the equation that is satisfied by this function. If it is a wave equation, (9.5.25) may determine the wave function of bound electrons. In this case, the density of the superconducting electron pairs will be given by

$$\Psi^*_{\alpha\beta}(\varkappa|\Gamma_{\mathbf{x},\tau})\Psi_{\beta\alpha}(\varkappa|\Gamma_{\mathbf{x},\tau}) = N^2\left[\frac{7\zeta(3)\langle N_\varkappa\rangle}{8\pi^2 T_c^2[\varkappa|\Gamma_{\mathbf{x},\tau}]}\right]\langle \mathrm{Tr}\hat{B}(\varkappa)\rangle^2|\Phi(\Gamma_{\mathbf{x},\tau})|^2. \tag{9.5.26}$$

In (9.5.25,26), $T_c[\varkappa|\Gamma_{\mathbf{x},\tau}]$ is a slowly varying function of the contour variable $\Gamma_{\mathbf{x},\tau}$, and its mean value $T_c[n]$, i.e., the temperature of phase transition to the superconducting state T_{cn}, will now be defined by the expression

$$T_{cn} = \Big\langle \Big\{\mu_\varkappa^2 + (\langle\omega_D\rangle^2_{\rho(\varkappa)} - \mu_\varkappa^2) \times \exp\left[-\frac{2}{\nu_\varkappa(\varepsilon_F)\,\langle\lambda_{eph}(\mathbf{p}_o|\Gamma_{\mathbf{x},\tau}|\bar{p}_{\perp o})\rangle_{\rho(\varkappa,\bar{p}_{\perp o})}}\right]\Big\}^{\frac{1}{2}}\Big\rangle_{(\Gamma_V)} \tag{9.5.27}$$

At $\mu_\varkappa \to 0$, $T_c \to \frac{2\gamma}{\pi}\langle\omega_D\rangle \exp\left\{-\frac{1}{K_{enh}(x)(\lambda_{eph}-\tilde{\mu}^*)}\right\}$, where

$\nu_\varkappa(\varepsilon_F)\langle\lambda_{eph}(p_o|\Gamma_{\mathbf{x},\tau}|\bar{p}_{\perp o})\rangle_{\rho(\varkappa,\bar{p}_{\perp o})} \equiv \langle\tilde{\lambda}_{eph}\rangle = K_{enh}(x)(\lambda_{eph}-\tilde{\mu}^*)$,

$x = 2\pi^2 A(v_o^{1/3}/\langle r_c\rangle) < 10$.

Considering that $x \to \zeta(A = \frac{\sqrt{3}\,g\hbar}{\pi v_o^{1/3}\sqrt{J_o sM}})$ then

$$J_o s > \left(\frac{2\pi\sqrt{3}}{10}\right)^2\left(\frac{g^2\hbar^2}{\langle r_c\rangle^2 M}\right),$$

which coincides with the inequality (1.1.12). Away from the phase transition point, $|\chi/\chi_{os}| \approx 1$. In this case, the above expression coincides with (9.3.17) for

$\langle \mathrm{Tr}\hat{B}(\varkappa) \to 1$ and $\mu_\varkappa/\langle\omega_D\rangle_{\rho(\varkappa)} \ll 1$.

Let us write the expression for the current of bound electrons by taking into consideration (9.5.23,24):

$$\langle \hat{J}_\nu(\mathbf{x}) \rangle_S = -\frac{i}{2} g_1 \lim_{\substack{\mathbf{x},\tau\to\mathbf{x}',\tau' \\ \varepsilon\to 0}} \sum_{\alpha=1}^{N^2-1} \int_{-\varkappa,-\varkappa_1,-\varkappa_2\in\Gamma} \frac{D\varkappa(\mathbf{x}'',\tau'')}{2\pi C_\varkappa} \frac{D\varkappa_1(\mathbf{x}_1'',\tau_1'')}{2\pi C_\varkappa}$$

$$\times \frac{D\varkappa_2(\mathbf{x}_2'',\tau_2'')}{2\pi C_\varkappa} J^{\varepsilon+\varkappa+\varkappa_1}_{\varepsilon+\varkappa+\varkappa_1+\varkappa_2} (\nabla_{x'_\nu} - \nabla_{x_\nu})\Phi_{\gamma\beta''}(\mathbf{x}|\varkappa_1)\, \Phi^*_{\beta''\gamma}(\mathbf{x}'|\varkappa_2)$$

$$\times \exp(\int_\varkappa dx''_\nu p_{o\nu}[\varkappa])\ \langle \mathrm{Tr}[\hat{\tau}^\alpha, \hat{\tau}^\alpha\ \hat{B}(\varkappa)]\hat{B}(\Gamma)\rangle\ . \tag{9.5.28}$$

where

$$J^{\varepsilon+\varkappa+\varkappa_1}_{\varepsilon+\varkappa+\varkappa_1+\varkappa_2} = N^2 \frac{7\,\zeta(3)}{16\pi^2 T^2} \langle N_{\varepsilon+\varkappa+\varkappa_1}\rangle\ \langle \mathrm{Tr}\hat{B}(\varepsilon+\varkappa+\varkappa_1+\varkappa_2)\rangle^2\ .$$

Hence, by using (9.5.25,26) and integrating (9.5.28) with respect to $\varkappa$, $\varkappa_1$, and $\varkappa_2$, we can write the final expression for the function $\Psi_{\alpha\beta}(\mathbf{x},\tau|\Gamma)$, as well as the complete expression for the current $\langle \hat{J}_\nu(\mathbf{x})\rangle$, which includes both superconducting and normal components in the approximation of slowly varying functions of contour variables $\varkappa$ and $\Gamma_{\mathbf{x},\tau}$:

$$\frac{\hbar^2}{4\mu_e} \mathrm{Tr}[\hat{\tilde{\mathbb{D}}}^2_\nu\, \Psi_{\alpha\beta}(\mathbf{x},\tau|\Gamma)] + \left(\frac{3\pi^2 T^2}{\varepsilon_F A}\right)\left\{ \ln \left[\frac{T_c^2[\varkappa|\Gamma_{\mathbf{x},\tau}] - \mu_\varkappa^2}{T^2 - \mu_\varkappa^2}\right]^{\frac{1}{2}}\right.$$

$$\left. - \frac{1}{N\langle N_\varkappa\rangle} |\Psi(\mathbf{x},\tau|\Gamma)|^2 \right\}\Psi_{\alpha\beta}(\mathbf{x},\tau|\Gamma) = 0\ , \tag{9.5.29}$$

$$\langle \hat{J}_\nu(\mathbf{x})\rangle = -i\frac{2eh}{4\mu_e}\langle \mathrm{Tr}\Big\{[\hat{\tilde{\mathbb{D}}}_\nu \Psi^\alpha_{\gamma\delta}(\mathbf{x},\tau|\Gamma))^+ \Psi^\alpha_{\gamma\delta}(\mathbf{x},\tau|\Gamma)]$$

$$-[(\hat{\tilde{\mathbb{D}}}_\nu \Psi^\alpha_{\gamma\delta}(\mathbf{x},\tau|\Gamma))^+,\ \Psi^\alpha_{\gamma\delta}(\mathbf{x},\tau|\Gamma)]^+\Big\}\rangle$$

$$+\frac{i}{2}\frac{2e\hbar}{2\mu_e}\oint_\Gamma dx'_\nu\ \delta(\mathbf{x}-\mathbf{x}')\langle \mathrm{Tr}\,\hat{\tau}^\alpha\ \hat{B}(C_{\mathbf{x},\mathbf{x}'})\hat{\tau}^\alpha\hat{B}(C_{\mathbf{x}',\mathbf{x}})\rangle\ . \tag{9.5.30}$$

It can be easily seen that (9.5.29) is a wave equation, and hence $\Psi_{\alpha\beta}(\mathbf{x},\tau|\Gamma)$ is the wave function of the superconducting electrons. The quantity

$$\xi_\varkappa = \langle \left(\frac{\hbar^2 \varepsilon_F A}{12\pi^2 \mu_e\ T_c^2[\varkappa|\Gamma_{\mathbf{x},\tau}]}\right)^{\frac{1}{2}}\rangle_{(\Gamma_V)}$$

will be called the effective size of a bound electron pair in the superconducting phase, or the effective coherence length.

The operator wave functions $\hat{\Psi}^{\alpha}_{\gamma\delta}(\mathbf{x},\tau|\Gamma)$ in the expression (9.5.30) for the current are defined as follows:

$$\hat{\Psi}^{\alpha}_{\gamma\delta}(\mathbf{x},\tau|\Gamma) = \frac{1}{N}\left[\frac{7\zeta(3)\langle N_{\varkappa}\rangle N^2}{16\pi^2 T_c^2[\varkappa|\Gamma_{\mathbf{x},\tau}]}\right]^{\frac{1}{2}} \langle Tr\hat{B}(\varkappa)\rangle \hat{\tau}^{\alpha}\hat{T}$$

$$\times \exp\left[2\oint_{\Gamma} dx_{\nu}\hat{A}_{\nu} + \hat{O}(\mathbf{x},\tau)\right] \Phi_{\gamma\delta}(\mathbf{x},\tau|\Gamma) . \tag{9.5.31}$$

$$\hat{\Psi}^{\alpha+}_{\gamma\delta}(\mathbf{x},\tau|\Gamma) = [\hat{\Psi}^{\alpha}_{\gamma\delta}(\mathbf{x},\tau|\Gamma)]^{+} \tag{9.5.32}$$

The matrices O and O^+ belong to the group SU(N), and their explicit form is determined by the form of the equivalent wave function appearing in the expression for the effective Hamiltonian or free energy (Chap. 6). We can explain this statement as follows. In (9.5.30) we have carried out a convolution with respect to the upper spin index α. If we are interested in the current tensor $\langle \hat{J}^{\alpha}_{\nu}\rangle_S$, it is sufficient to decompose the matrix appearing under Tr sign in (9.5.30) into a product of two matrices, viz., $\hat{\tau}^{\alpha}\hat{J}^{\alpha}_{\nu}$. Taking the average over $\hat{J}^{\alpha}_{\nu}$, we obtain

$$\langle J^{\alpha}_{\nu}(\mathbf{x})\rangle_S = i\,\frac{2e\hbar}{4\mu_e}\,\langle Tr\hat{\tau}^{\alpha}\Big\{[(\hat{\tilde{\mathbb{D}}}_{\nu}\Psi_{\gamma\delta}(\mathbf{x},\tau|\Gamma))^{+},\Psi_{\gamma\delta}(\mathbf{x},\tau|\Gamma)]$$

$$-[(\hat{\tilde{\mathbb{D}}}_{\nu}\Psi_{\gamma\delta}(\mathbf{x},\tau|\Gamma))^{+},\Psi_{\gamma\delta}(\mathbf{x},\tau|\Gamma)]^{+}\Big\}\rangle , \tag{9.5.33}$$

where

$$\Psi_{\gamma\delta}(\mathbf{x},\tau|\Gamma) = \hat{\tau}^{\alpha}\hat{\Psi}^{\alpha}_{\gamma\delta}(\mathbf{x},\tau|\Gamma) ; \Psi^{+}_{\gamma\delta}(\mathbf{x},\tau|\Gamma) = \hat{\Psi}^{\alpha+}_{\gamma\delta}(\mathbf{x},\tau|\Gamma)\hat{\tau}^{\alpha} .$$

If the function $\Psi_{\gamma\delta}(\mathbf{x},\tau|\Gamma)$ in (9.5.33) is defined by a column matrix

$$\Psi_{\gamma\delta}(\mathbf{x},\tau|\Gamma) = \Psi^{0}_{\gamma\delta}(\mathbf{x},\tau|\Gamma)\hat{a} ; \quad \hat{a} = \begin{Vmatrix} 0 \\ 1 \end{Vmatrix}$$

(Sect. 6.3), the commutators in (9.5.33) are transformed into a product and we arrive at the following expression for the superconducting current:

$$\langle \hat{J}^{\alpha}_{\nu}(\mathbf{x}) \rangle_S = -i \frac{2e\hbar}{4\mu_e} \langle \mathrm{Tr}[(\hat{\tilde{\mathbb{D}}}^{o}_{\nu} \Phi_{\gamma\delta}(\mathbf{x},\tau|\Gamma))^{+} \hat{\tau}^{\alpha} \Phi_{\gamma\delta}(\mathbf{x},\tau|\Gamma)$$

$$- \Phi^{+}_{\gamma\delta}(\mathbf{x},\tau|\Gamma) \hat{\tau}^{\alpha} (\hat{\tilde{\mathbb{D}}}^{o}_{\nu} \Phi_{\gamma\delta}(\mathbf{x},\tau|\Gamma))] \rangle \quad . \tag{9.5.34}$$

This is identical to the phenomenological expression (6.3.19). Thus, we have obtained an equivalent expression for the current of superconducting electrons. The wave function $\Phi^{o}_{\gamma\delta}(\mathbf{x},\tau|\Gamma)$ is now chosen in such a way that the free energy of the system obtained with its help gives the results of the phenomenological theory. This is done by using $\hat{O}(\mathbf{x},\tau)$. Let us now suppose that $\Phi_{\gamma\delta}(\mathbf{x},\tau|\Gamma) = \sum_{\lambda=1}^{N^2-1} \hat{\tau}^{\lambda} \Phi^{\lambda}_{\gamma\delta}(\mathbf{x},\tau)$. In this case, a convolution over the upper spin indices followed by a computation of the trace of the product $\hat{\tau}^{\alpha} \hat{\tau}^{\beta}$ of matrices gives

$$\mathbf{J}_{\nu}(\mathbf{x},\tau) = -\frac{2e\hbar}{4\mu_e} \langle [(D_{\nu}\mathbf{\Psi}(\mathbf{x},\tau))^{*}, \mathbf{\Psi}(\mathbf{x},\tau)] + [(D_{\nu}\mathbf{\Psi}(\mathbf{x},\tau))^{*}, \mathbf{\Psi}(\mathbf{x},\tau)]^{*} \rangle , \tag{9.5.35}$$

where $D_{\nu} = \nabla_{\nu} - 2g_1 f^{\alpha}_{\mu\lambda} A^{\alpha}_{\nu}$. Hence the expression we have obtained is identical to the current found in Sect. 6.2, except for the averaging sign $\langle \dots \rangle_{(\Gamma_V)}$.

Thus, we have carried out a comparison with the phenomenological theory of the superconducting phase in rare-earth metal compounds. All the results are in good agreement. The equation for the wave function $\Psi_{\alpha\beta}(\mathbf{x},\tau|\Gamma)$ contains a cubic nonlinearity. However, it was shown in Chaps. 4 and 6 that the nonlinearity in all equations is described by a component of the form $|\Psi|^2 \Phi \ln(|\Psi|^2/g(\tau))$. The emergence of a nonlinear term of this type is associated with the exchange enhancement effect. Suppose that we consider a homogeneous superconducting phase n = 0 (Sect. 9.4). Let $\Psi_{\alpha\beta}(\mathbf{x},\tau|\Gamma)$ be an arbitrary function. The system contains thermodynamic magnetic fluctuations associated with a nonequilibrium distribution of magnetic moments, which may be uniform in view of the large value of the correlation radius $\langle r_c \rangle$.

The nonlinear term in (9.5.29) increases due to an increase in the correlation function $\mu_{\varkappa}$ near the phase transition point, and we must take into consideration the higher orders of the order parameter amplitude in the perturbation theory. The perturbation series can be summed selectively and we arrive at

$$\frac{\hbar^2}{4\mu_e} \mathrm{Tr}[\hat{\tilde{\mathbb{D}}}_\nu^2 \Phi_{\gamma\delta}(\mathbf{x},\tau|\Gamma)] + \left(\frac{3\pi^2 T^2}{\varepsilon_F A}\right)\Big\{\tau_n[\varkappa|\Gamma_{\mathbf{x},\tau}]$$

$$+ \frac{1}{N^2\langle N_\varkappa\rangle} |\Psi(\mathbf{x},\tau|\Gamma)|^2 \ln \frac{|\Psi(\mathbf{x},\tau|\Gamma)|^2}{N^2\langle N_\varkappa\rangle}\Big\}\Psi_{\gamma\delta}(\mathbf{x},\tau|\Gamma) = 0 , \tag{9.5.36}$$

where

$$\tau_n[\varkappa|\Gamma_{\mathbf{x},\tau}] = \frac{T_{cn}[\varkappa|\Gamma_{\mathbf{x},\tau}] - T}{T_{cn}[\varkappa|\Gamma_{\mathbf{x},\tau}]}\left(1 + \frac{\mu_\varkappa^2}{T_{cn}^2[\varkappa|\Gamma_{\mathbf{x},\tau}]}\right) .$$

In this case, $\xi_{\varkappa_n}(\tau_n) = \tau_n^{-\frac{1}{2}}\xi_\varkappa$, i.e., the coherence length is also quantized within the framework of the microscopic fluctuation theory (Sect. 9.4).

It follows from (9.5.36) that near the phase transition line, the nonlinearity now has the same structure as that obtained from the fluctuation theory of phase transitions. This means that the phase transition to the superconducting state is a first-order phase transition, a conclusion identical to that obtained in the phenomenological theory.

10. Theory of High-Temperature Superconductivity of Polymer Systems

10.1 Quasi-One-Dimensional Organic Superconductors: Polyacetylene as a Possible Superconductor

It follows from the microscopic theory of high-temperature superconductivity (Chaps. 8 and 9) that the effective electron-phonon interaction parameter, which determines the critical temperature T_c, will be larger for larger spin-phonon coupling strengths ζ (1.1.10). It was shown in Chap. 8 that in order to produce a high-temperature superconductor the highest possible value of the parameter ζ must be ensured. For this purpose, we must choose elements forming covalent bonds and having minimum values of atomic radius, mass, and ionization potential. Therefore, a reasonable possibility might be quasi-one-dimensional organic or polymer systems such as the polyacetylenes based on the -CH- group. Here the carbon atoms form the covalently bonded chain. In this system, in comparison to oxide ceramics the exchange forces between electrons should be shorter-ranger forces since the atomic radius of carbon is slightly larger than the atomic radius of oxygen, while the mass of a carbon atom is smaller than that of an oxygen atom. The ionization potentials of carbon and oxygen are comparable in magnitude. At present, it is well known that quasi-one-dimensional organic systems have a high, and even anomalous, conductivity [10.1]. This is true for polymers, including polyacetylene $(\text{-CH-})_x$ [10.1,2]. One of the ways to increase the conductivity of polymers is to dope them with donor or acceptor impurities. This can vary the conductivity by 12 orders of magnitude [10.1]. The introduction of impurities results in a violation of the translation invariance of the polymer chain. Translation symmetry may also be violated because of the existence of nonlinear excitations (topological solitons) in the polymers [10.2]. such solitons in polyacetylene are domain walls connecting the polyacetylene chains in degenerate A and B phases [10.2]. It was shown by the authors of [10.2] that the existence of soliton excitations determines the conduction mechanism in polyacetylene. Indeed, soliton excitations lead to the emergence of local levels in the forbidden band, and the conductivity is due to the jumps of electrons (occupying the P_z-orbital) between local levels. In other words, the conductivity has a discontinuous nature. Naturally, the existence of superconductivity in polyacetylene is ruled out by such a model.

In this chapter, we shall explore the possibility of superconductivity in polyacetylene and discuss the conditions under which such a possibility exists. We shall proceed from the model of a quasi-one-dimensional conductor in which there are two types of interaction between charge carriers, viz., electron-electron, and electron-hole interactions. The former type of interaction may be responsible for superconductivity, while the latter causes an instability which leads to the emergence of a dielectric gap σ_{21} in the system. For polyacetylene, $\sigma_{21} = g_{ph}u_0$, where g_{ph} is the parameter of electron interaction with the elastic deformations of the chain, and u_0 is the relative displacement of adjacent sites in a monomer. We shall also assume that the translation symmetry of the system is violated due to the existence of topological solitons having a topological charge $Q = 0$ and a spin $s = \pm(\frac{1}{2})$ [10.2]. Moreover, we shall assume that the concentration of such solitons is quite high (10^{14} cm^{-2}) so that they form a disordered spin system (for polymer films, this will be a two-dimensional disordered spin system). Consequently, we can describe such a model as follows. The spectrum of conduction electrons is assumed to be one-dimensional, i.e., $\omega = \varepsilon(p) - \mu$, $p = p_x$, μ being the effective chemical potential of the system. The Fermi surface has two flat regions for $p_x = \pm p_0$ (two points on a sphere in the case of a three-dimensional spectrum). The state of an electron having a momentum p in the vicinity of p_0 will be denoted by ψ_1, while for p lying in the vicinity of the point $-p_0$, the state of the electron is denoted by ψ_2. Thus, we have a two-band conduction system. Electrons in such a system interact with one another, with deformations of the chain, and also with the disordered system of soliton spins. We shall also assume that the interaction between chains takes place not only through the intermediary atoms (say, of sulphur), but also through spins. This interaction is of electrostatic origin, and is based on the quantum-mechanical principle of indistinguishability of particles. In other words, this interaction is of exchange type [10.2]. Hence each chain is a random sequence of polyacetylene C-C links in A or B phase (phase A and phase B differ in the sign of displacement u_0 of the adjacent sites in a monomer and in the sequence of double and single bonds, phase A corresponding to u_0 and phase B to $-u_0$). An electron moving along the x-axis interacts with the system of soliton spins of this chain. This interaction is of quasi-one-dimensional type, i.e., the interaction of electrons in a chain is much stronger than between two adjacent chains. The interaction with the field of the disordered spin system can be treated as an interaction with an effective field whose potential is denoted by A^{α}_{ν}, ν being the spatial index, $\nu = 1, 2$, and α is the spin index, $\alpha = 1,2,3$. The spin of a soliton assumes values $\pm\frac{1}{2}$. However, an electron interacts with a whole group of spins occupying an isolated two-dimensional volume V whose size is much larger than $\ell^2_{\varkappa}$, $\ell_{\varkappa}$ being the length of one monomer. The total spin of the isolated microscopic volume will be different from $\frac{1}{2}$ and assumes a whole set of values, since its projection s^z on an isolated axis z assumes values -s, ..., s. Consequently, it can be treated as a vector in the spin space and characterized by two Euler angles θ^1 and θ^2. In its turn, the axis also does not have an isolated direction, since the system does not have a magnetic anisotropy or an external magnetic field. Hence the vector **S** can be defined in terms of the angle of rotation about the z-axis $\tilde{\theta}^3$, and angles $\tilde{\theta}^1$ and $\tilde{\theta}^2$ describing the orientation of the z-axis in the spin space. It was shown in [10.3] that the field A^{α}_{ν} is a functional of the quantity $\nabla_{\nu}\tilde{\theta}^{\alpha}$, $A^{\alpha}_{\nu} = F[\nabla_{\nu}\tilde{\theta}^{\alpha}]$ and coincides with the Yang-Mills field.

The system of soliton spins oscillates as a result of spontaneous spin-flips. Hence the field A^{α}_{ν} fluctuates as well: $A^{\alpha}_{\nu} = A^{\alpha(0)}_{\nu} + \delta A^{\alpha}_{\nu}$. In this case, the state 1 or 2 of an electron interacting with the field A^{α}_{ν} can be defined with the help of the operator

$$\hat{\psi}_j(x/\Gamma_y) = \frac{1}{N}\hat{T}\exp\left(\oint_{\Gamma_y} dx_{\nu}\hat{A}_{\nu}\right)\hat{\psi}_j(x),\quad j = 1,2 \,. \qquad (10.1.1)$$

In this case, the contour Γ_y describes the entanglement between chains since it intersects the subsystem of chains, a soliton spin being located at each point of intersection. Having defined the electron operator (10.1.1), we can write the effective Hamiltonian of the system

$$\hat{\mathbb{H}}_{\Gamma} = N\mathrm{Tr}\int_{\mathbf{x}\in\Gamma_y,\, x\in C_{\infty}} d\mathbf{x}dx \left\{\hat{B}(\Gamma_y)\,\hat{\psi}^{+}_{i\alpha}(x)\left(\frac{\hat{p}^2_i}{2\mu_e} - \hat{\mu}_i\right)\hat{\psi}_{i\alpha}(x)\right.$$

$$+ \sum_{k,\varkappa}\hat{b}^{+}_{k\varkappa}(x)\hat{\xi}_{\varkappa}(x)\hat{b}_{k\varkappa}(x) + \frac{i}{\sqrt{2}}\, g^{ij}_{ph}(x)\sum_{k,\varkappa} e_{k\varkappa}\hat{\eta}_{\varkappa}(x)[\hat{b}_{k\varkappa}(x)$$

$$- \hat{b}^{+}_{k\varkappa}(x)]\hat{\psi}^{+}_{i\alpha}(x)\hat{\psi}_{j\alpha}(x) + \hat{\psi}^{t}_{i\alpha}(x)\hat{\Delta}^{+}_{ij\alpha\gamma}(x,y)\hat{\psi}_{j\gamma}(x)$$

$$+ \hat{\psi}^{+}_{i\alpha}(x)\hat{\Delta}_{ij\alpha\gamma}(x,y)\hat{\psi}^{+t}_{j\gamma}(x) + Q_{ij}(x,y)\hat{\psi}^{+}_{i\alpha}(x)\hat{\psi}_{j\alpha}(x)\Big\} \hat{B}^{+}(\Gamma_y)$$

$$+ \frac{1}{L_d}\Big[\frac{1}{2}\Big(\frac{\delta}{\delta s_{\mu\nu}}\Big)^2 + \hat{I}(\Gamma_y)\Big]\hat{B}(\Gamma_y) \Big\}$$

$$+ NTr \int_{\mathbf{x},\mathbf{x}'\in\Gamma_y, x,x'\in C_\infty} d\mathbf{x}d\mathbf{x}'dxdx' \; \hat{B}(\Gamma_y)\hat{\psi}^{+}_{i\alpha}(x)Tr[\hat{\psi}^{+}_{j\beta}(x')V(x-x')\hat{\psi}_{j\beta}(x')]$$

$$\times \hat{\psi}_{i\alpha}(x)\hat{B}^{+}(\Gamma_y) \; . \qquad (10.1.2)$$

In this expression, $\hat{\mu}_i$ ($i = 1,2$) are the effective chemical potentials near the points $\pm p_0$, g_{ph} is the parameter of electron interaction with elastic vibrations, x is the index of the point of intersection of the contour Γ_y with a chain, and $\hat{\Delta}_{ij\alpha\gamma}(x,y)$ is a function defining the dielectric gap in the electron spectrum and the superconducting order parameter (σ_{21}, Δ_{21}).

$$Q_{ij}(x,y) = \delta_{1i}\delta_{2j}q(x,y),$$

$$q(x,y) = q(x) \sum_{i=1}^{N_d} \delta(y-y_i), \quad q(x) = g^{12}_{ph}u_0 \tanh(x/\ell_\varkappa),$$

$$\hat{B}(\Gamma_y) = (1/N)\hat{T} \exp(\oint_{\Gamma_y} dx_\nu \hat{A}_\nu) \; . \qquad (10.1.3)$$

Here, N_d is the number of chains in a film. We introduce the Green functions

$$G_{ij\alpha\beta}(x,\tau|\Gamma_y|x',\tau') = \langle \hat{T}\hat{\psi}_{i\alpha}(x,\tau)Tr\hat{B}(\Gamma_y)\hat{\psi}^{+}_{j\beta}(x',\tau') \rangle \; , \qquad (10.1.4)$$

$$F^{+}_{ij\alpha\beta}(x,\tau|\Gamma_y|x',\tau') = \langle \hat{T}\hat{\psi}^{+}_{i\alpha}(x,\tau)Tr\hat{B}(\Gamma_y)\hat{\psi}_{j\beta}(x',\tau') \rangle \; , \qquad (10.1.5)$$

$$D_{kk'\varkappa\varkappa'}(x,\tau|\Gamma_y|x',\tau') = \langle \hat{T}\hat{b}_{k\varkappa}(x,\tau)Tr\hat{B}(\Gamma_y)\hat{b}^{+}_{k\varkappa'}(x',\tau') \rangle \; . \qquad (10.1.6)$$

Using the technique developed in Chap. 9, we can write down the equation of motion for the Green functions (10.1.4-6), and for the mean value $\langle Tr\hat{B}(\Gamma_y) \rangle$ of the operator $\hat{B}(\Gamma_y)$, having carried out the averaging over the fields $Q_{ij}(x,y)$. The equations of motion have solutions of the form

$$G_{ij\alpha\beta}(x,\tau|\Gamma_y|x',\tau') = \delta_{\alpha\beta}G_{ij}(x,\tau|\Gamma_y|x',\tau') \; , \qquad (10.1.7)$$

$$F^{*}_{\alpha\beta ij}(x,\tau|\Gamma_y|x',\tau') = -i\sigma^{y}_{\alpha\beta}F^{*}_{ij}(x,\tau|\Gamma_y|x',\tau') \; . \qquad (10.1.8)$$

The Fourier components of the Green functions have the form

$$y_{ij\varkappa}(p,i\omega_n) = \frac{N}{N^2-1}\sum_{=1}^{N^2-1} G_{ijx}(p,i\omega_n)\ \langle Tr\hat{\tau}^{\alpha}\hat{B}^{(1)}(\Gamma_y)\hat{\tau}^{\alpha}\hat{B}(x)\rangle\ , \tag{10.1.9}$$

$$f^*_{ij}(p,i\omega_n) = \frac{N}{N^2-1}\sum_{\alpha=1}^{N_2-1} F^*_{ijx}(p,i\omega_n)\langle\ Tr\hat{\tau}^{\alpha}\hat{B}^{(2)}(\Gamma_y)\hat{\tau}^{\alpha}\hat{B}(x)\rangle\ . \tag{10.1.10}$$

In the following analysis, we shall require the functions y_{11}, y_{22}, y_{21}, f^*_{11}, f^*_{21}. These functions have the form

$$\mathrm{Det}[y_{11},y_{21},f^*_{11},f^*_{21}]y_{11} = \langle\ Tr\hat{B}(\Gamma_y)\ \rangle[(i\omega_n - \xi_2)(i\omega_n+\xi_2)(i\omega_n+\xi_1)$$
$$- \sigma^2_{21}(i\omega_n - \xi_2) + \Delta^2_{21}(i\omega_n + \xi_2)]\ , \tag{10.1.11}$$

$$\mathrm{Det}[y_{11},y_{21},f^*_{11},f^*_{21}]y_{22} = \langle\ Tr\hat{B}(\Gamma_y)\ \rangle[(i\omega_n- \xi_1)(i\omega_n+ \xi_1)(i\omega_n+ \xi_2)$$
$$- \sigma^2_{21}(i\omega_n - \xi_1) + \Delta^2_{21}(i\omega_n + \xi_1)]\ , \tag{10.1.12}$$

$$\mathrm{Det}[y_{11},y_{21},f^*_{11},f^*_{21}]y_{21} = -\ \langle Tr\hat{B}(\Gamma_y)\ \rangle\sigma_{21}[(i\omega_n+\xi_1)(i\omega_n+\xi_2)$$
$$- \sigma^2_{21} + \Delta^2_{21}]\ , \tag{10.1.13}$$

$$\mathrm{Det}[y_{11},y_{21},f^*_{11},f^*_{21}]f^*_{11} = 2i\omega_n\sigma_{21}\varphi_{21}\langle\ Tr\hat{B}(\Gamma_y)\ \rangle\ , \tag{10.1.14}$$

$$\mathrm{Det}[y_{11},y_{21},f^*_{11},f^*_{21}]f^*_{21} = -\ \langle\ Tr\hat{B}(\Gamma_y)\ \rangle\varphi_{21}[(i\omega_n - \xi_2)(i\omega_n+\xi_1)$$
$$+ \sigma^2_{21} - \Delta^2_{21}]\ . \tag{10.1.15}$$

The following notation has been used in (10.1.11-15):

$$\mathrm{Det}[y_{11},y_{21},f^*_{11},f^*_{21}] = (\omega^2_+ + \omega^2_n)(\omega^2_- + \omega^2_n)\ ,$$

$$\omega_\pm = E\pm\mu_\varkappa\ , \quad E = \sqrt{\xi^2+\sigma^2_{21}}\ , \quad \tilde{\mu}_\varkappa = \sqrt{\mu^2_\varkappa + \Delta^2_{21}}\ ,$$

$$\xi_{1,2} = \pm\ \xi - \mu_\varkappa\ , \quad \xi = v_0|(p - p_0)|\ ,$$

where v_0 is the velocity of an electron along the chain near the point

$\pm p_0$, and $p-p_0$ is the momentum relative to p_0. The quantity $\mu_\varkappa$ is the contribution to the chemical potential of the system due to fluctuations of the disordered system of soliton spins $\hat{A}_\nu$, and is equal to

$$\mu_\varkappa = \frac{1}{2\mu_e} \langle p_0^2 [\varkappa] \rangle_{\rho(\varkappa)} . \qquad (10.1.16)$$

Here, $p_{0\nu}[\varkappa]$ is the departure of the electron momentum due to fluctuations during a one-dimensional motion. The quantity $\mu_\varkappa$ may be expressed in terms of the paramagnetic susceptibility of the system:

$$\mu_\varkappa = (p_f^2 /2\mu_e)(\chi/\chi_{os}) , \qquad (10.1.17)$$

where χ_{os} is the magnetic susceptibility of noninteracting electrons in the field A (the analog of Pauli's susceptibility), $p_f = 2\pi/\langle r_c\rangle$, $\langle r_c\rangle$ being the effective correlation radius of the fluctuations of a system of soliton spins, expressed in terms of the exchange interaction parameter $J(\mathbf{x})$ between spins, σ_{21} is the dielectric gap in the electron spectrum, and Δ_{21} is the superconducting gap. Carrying out the analytic continuation $i\omega_n \to \omega$ in the expression for $\mathrm{Det}[y_{11}, y_{21}, f_{11}^*, f_{21}^*]$, we find that the system's spectrum is defined by the zeros of the determinant and the poles of the functions y_{ij} and f_{ij}^*. Each of the two bands is split in this case into two subbands $\omega_\pm$ and $-\omega_\pm$. The difference $\delta N_\pm = N_+ - N_-$ in the population densities of each subband is nonzero, and if $\Delta_{21} = 0$, such a splitting is caused by fluctuations. Parameters σ_{21} and Δ_{21} are expressed in terms of the Green function as follows:

$$\sigma_{21}(\Gamma_y) = \langle\langle\lambda_{e-ph21}(2p_0|\Gamma_y|\bar{p}_{\perp 021}\rangle\rangle_{\rho(\varkappa,\bar{p}_{\perp 021})} \times G_{21}(x,\tau|\Gamma_y|x,\tau)\rangle_{\rho(\varkappa,\bar{p}_{\perp 0})} , \qquad (10.1.18)$$

$$\Delta_{21}(\Gamma_y) = \langle\langle\lambda_{e-ph21}(2p_0|\Gamma_y|\bar{\bar{p}}_{\perp 021})\rangle_{\rho(\varkappa,\tilde{\bar{p}}_{\perp 021})} \times F_{21}(x,\tau|\Gamma_y|x,\tau)\rangle_{\rho(\varkappa,\bar{p}_0)} . \qquad (10.1.19)$$

The parameters $\langle\lambda_{e-ph}(2p_0|\Gamma_y|\tilde{\bar{p}}_{\perp 021})\rangle_{\rho(\varkappa,\tilde{\bar{p}}_{\perp 021})}$, $\langle\tilde{\lambda}_{e-ph21}(2p_0|\Gamma_y|\tilde{\bar{p}}_{\perp 021})\rangle_{\rho(\varkappa,\tilde{\bar{p}}_{\perp 021})}$ are the effective electron-phonon interaction parameters and have the form:

$$\langle\lambda_{e\text{-}ph21}(2p_0|\Gamma_y|\bar{p}_{\perp 021})\rangle_{\rho(\varkappa,\bar{p}_{\perp 021})} = \langle \mathrm{Tr}\hat{B}(\Gamma_y)\, \nu(\varepsilon_0)$$

$$\times [\tilde{g}^{112}_{ph} \; \mathrm{Tr}\hat{B}(\Gamma_y)\,\rangle \langle\, \cosh(\pi \ell_\varkappa \bar{p}_{\perp 021}\,\rangle_{\rho(\varkappa,\bar{p}_{\perp 021})} - \langle V_\varkappa(2p_0)\rangle_{\rho(\varkappa)}\,] \qquad (10.1.20)$$

$$\langle\, \tilde{\lambda}_{e\text{-}ph21}(2p_0|\Gamma_y|\tilde{\bar{p}}_{\perp 021})\rangle_{\rho(\varkappa,\tilde{\bar{p}}_{\perp 021})} = \langle\, \mathrm{Tr}\hat{B}(\Gamma_y)\, \nu(\varepsilon_0)$$

$$\times [\tilde{g}^{112}_{ph} \langle\, \mathrm{Tr}\hat{B}(\Gamma_y)\,\rangle\; \langle\, \cosh(\pi\ell_\varkappa \tilde{\bar{p}}_{\perp 021})\rangle_{\rho(\varkappa,\tilde{\bar{p}}_{\perp 021})} - \langle\, V_\varkappa(2p_0)\rangle_{\rho(\varkappa)}\,],$$

$$\varepsilon_0 \to \mu \quad . \qquad (10.1.21)$$

In (10.1.20,21), $\tilde{g}^{11}_{ph} = g^{11}_{ph} + 2{g^{21}_{ph}}^2/\langle\omega_D\rangle_{\rho(\varkappa)}$. The term $2{g^{21}_{ph}}^2/\langle\omega_D\rangle_{\rho(\varkappa)}$ appears on account of the interaction of electrons with longwave deformations whose existence is responsible for the appearance of a dielectric gap in the system, and $\langle\omega_D\rangle_{\rho(\varkappa)}$ is the mean Debye energy.

$$\langle\cosh(\pi\ell_\varkappa\bar{p}_{\perp 0})\rangle_{\rho(\varkappa,\bar{p}_{\perp 021})} = (1/\pi\ell_\varkappa\bar{p}_{\perp 021})\sinh(\pi\ell_\varkappa\bar{p}_{\perp 021}) \;;$$

$$\bar{p}_{\perp 021} = 2\pi/\langle r_{c21}\rangle \; . \qquad (10.1.22)$$

$$\langle\cosh(\pi\ell_\varkappa\tilde{\bar{p}}_{\perp 0})\rangle_{\rho(\varkappa,\tilde{\bar{p}}_{\perp 021})} = (1/\pi\ell_\varkappa\tilde{\bar{p}}_{\perp 021})\sinh(\pi\ell_\varkappa\tilde{\bar{p}}_{\perp 021}) \;;$$

$$\tilde{\bar{p}}_{\perp 021} = 2\pi/\langle\tilde{r}_{c21}\rangle \; . \qquad (10.1.23)$$

Here, $\langle\, V_\varkappa(2p_0)\rangle_{\rho(\varkappa)}$ is the mean strength of the Coulomb interaction between electrons. The functions (10.1.22,23) appear because the electrons in different chains also exchange virtual transverse phonons with a momentum $\bar{p}_{\perp 021}(\tilde{\bar{p}}_{\perp 021})$ (Sect. 9.3). The magnitude of this momentum depends on whether this interaction is of electron-electron type or of electron-hole type. The former interaction is responsible for the appearance of superconductivity in the system, while the latter accounts for the emergence of a forbidden gap in the electron spectrum. In both cases, the interaction is determined by the fluctuation correlation radius, $\langle\tilde{r}_{c21}\rangle$ and $\langle r_{c21}\rangle$ respectively. There is no exchange interaction between electrons and holes. Hence $\langle r_{c21}\rangle \to \infty$, and consequently, $\langle\, \cosh(\pi\ell_\varkappa\bar{p}_{\perp 021})\rangle_{\rho(\varkappa,\tilde{\bar{p}}_{\perp 021})} \to 1$. For electron-electron interaction, $\langle\, r_{c21}\,\rangle \approx L_\varkappa$, where $L_\varkappa$ is the mean separation between the chains. Considering that Green functions are functionals of the order parameters $\sigma_{21}(\Gamma_y)$ and $\Delta_{21}(\Gamma_y)$, we can easily show that (10.1.18,19) are the equations for the functions $\sigma_{21}(\Gamma_y)$ and $\varphi_{21}(\Gamma_y)$ by equations for the parameter $\delta N_\pm$ describing the imbalance between the population densities of the two subbands $\omega_\pm$ in each band, we obtain the system of equations

$$\frac{\sigma_{21}(\Gamma_y)}{\langle\lambda_{e-ph21}(2p_0|\Gamma_y|)\rangle_{\rho(\varkappa)}} = \Big\langle \int_{-\varkappa\in\Gamma_y} \frac{D\varkappa(x'',y'')}{2\pi C_\varkappa} \exp\{\textstyle\int_\varkappa dx''_\nu p_{0\nu'}[\varkappa]\}\langle \mathrm{Tr}\hat{B}(\Gamma_{y''})\rangle$$

$$\times\, \sigma_{21}(\Gamma_{y''}) \int_0^{v_0|(p-p_0)|} \nu(\varepsilon_0)\,\frac{d\xi}{E}\Big(\tanh\frac{\beta\omega+}{2} + \tanh\frac{\beta\omega-}{2}\Big)\Big\rangle_{\rho(\varkappa,\tilde{\tilde{p}}_{\perp 021})} , \tag{10.1.24}$$

$$\frac{\Delta_{21}(\Gamma_y)}{\langle\lambda e\text{-}ph(2p_0|\Gamma y|\tilde{p}_{\perp 021})\rangle_{\rho(\varkappa,\tilde{\tilde{p}}_{\perp 021})}} = \Big\langle \int_{-\varkappa\in\Gamma_y} \frac{D\varkappa(x'',y'')}{2\pi C_\varkappa}$$

$$\exp\{\textstyle\int_\varkappa dx''_{\nu'} p_{0\nu'}[\varkappa]\}\ \langle \mathrm{Tr}\hat{B}(\Gamma_{y''})\rangle\ \Delta_{21}(\Gamma_{y''}) \int_0^{v_0|(p-p_0)|} \nu(\varepsilon_0)\,\frac{d\xi}{2\tilde{\mu}_\varkappa}$$

$$\times\,\Big(\tanh\frac{\beta\omega+}{2} - \tanh\frac{\beta\omega-}{2}\Big)\Big\rangle_{\rho(\varkappa,\tilde{\tilde{p}}_{\perp 021})} , \tag{10.1.25}$$

$$\delta N_\pm = \Big\langle \int_{-\varkappa\in\Gamma_y} \frac{D\varkappa(x'',y'')}{2\pi C_\varkappa} \exp\{\textstyle\int_\varkappa dx''_{\nu'} p_{0\nu'}[\varkappa]\}\ \langle \mathrm{Tr}\hat{B}(\Gamma_{y''})\rangle$$

$$\times \int_0^{v_0(|p-p_0|)} \nu(\varepsilon_0)d\xi\,\frac{\mu_\varkappa}{\tilde{\tilde{\mu}}_\varkappa}\Big(\tanh\frac{\beta\omega+}{2} - \tanh\frac{\beta\omega-}{2}\Big)\Big\rangle_{\rho(\varkappa,\tilde{\tilde{p}}_{\perp 021})} . \tag{10.1.26}$$

10.2 Criterion for the Emergence of Superconductivity in Polyacetylene

Equations (10.1.23-25) are quite complicated. Hence we shall consider the simplest case when the field $\hat{A}_\nu^{(0)}$ satisfies the gauge equation $\oint_{\Gamma_y} dx_\nu \hat{A}_\nu^{(0)} = 0$. Physically, this corresponds to the absence of vortex-type excitations in the spin system [10.4]. In this case, (10.1.23,24) are simplified considerably:

$$\frac{1}{\langle \lambda_{e-ph21}(2p_0|\Gamma_y|)\rangle_{\rho(\varkappa)}} = \nu(\varepsilon_0) \int_0^{v_0|(p-p_0)|} \frac{d\xi}{2E}\left(\tanh\frac{\beta\omega+}{2} + \tanh\frac{\beta\omega-}{2}\right), \tag{10.2.1}$$

$$\frac{1}{\langle \tilde{\lambda}_{e-ph21}(2p_0|\Gamma_y|\tilde{\tilde{p}}_{\perp 021})\rangle_{\rho(\varkappa,\tilde{\tilde{p}}_{\perp 021})}} = \nu(\varepsilon_0) \int_0^{v_0|(p-p_0)|} \frac{d\xi}{2\tilde{\mu}_\varkappa}$$

$$\times \left(\tanh\frac{\beta\omega+}{2} - \tanh\frac{\beta\omega-}{2}\right), \tag{10.2.2}$$

$$\delta N_\pm\Big|_{\beta\to\infty} = 2\,\frac{\mu_\varkappa}{\tilde{\mu}_\varkappa}\sqrt{\tilde{\mu}_\varkappa^2 - \sigma_{21}^2} \quad . \tag{10.2.3}$$

We introduce the notation

$$\langle \lambda_{e-ph}(2p_0|\Gamma_y|)\rangle_{\rho(\varkappa)} = \lambda_{21}, \quad \langle \tilde{\lambda}_{e-ph}(2p_0|\Gamma_y|\tilde{\tilde{p}}_{\perp 021})\rangle_{\rho(\varkappa,\tilde{\tilde{p}}_{\perp 021})} = \tilde{\lambda}_{21}.$$

In the simplest case, when $\Delta_{21} = 0$, the dielectric transition temperature is

$$T_s = (2\gamma/\pi)\varepsilon_0 \exp\left[-\frac{1}{\nu(\varepsilon_0)\lambda_{21}}\right] \quad , \tag{10.2.4}$$

where $\gamma = e^C$ (C is Euler's constant). We assume that after the phase transition at a temperature T_s, and following the emergence of a dielectric gap σ_{21} in the system, a subsequent lowering of temperature results in a phase transition to the superconducting state, which is observed down to zero temperature. In this case, we can determine the superconducting gap $\Delta_{21}(0)$ at zero temperature from (10.2.1-3). This gap is found to be

$$\Delta_{21}(0) = \left[\frac{\sigma_{21}^2(0)}{1 - \tilde{\lambda}_{21}^{-2}} - \mu_\varkappa^2\right]^{\frac{1}{2}} \quad . \tag{10.2.5}$$

It follows from (10.2.5) that the condition $\tilde{\lambda}_{21} > 1$ must be satisfied for the radical to be nonzero and real. This is the condition necessary for the emergence of superconductivity in polyacetylene. It can be shown that for T = 0, (10.2.1) can have a solution for $\tilde{\lambda}_{21} > 1$ only if the following condition is satisfied:

$$\tilde{\mu}_\varkappa > \left[\langle \omega_D\rangle^2_{\rho(\varkappa)} + \sigma_{21}^2\right]^{\frac{1}{2}} \quad . \tag{10.2.6}$$

Taking (10.2.5) into account, we obtain

$$\sigma_{21}(0) > \langle \omega_D \rangle_{\rho(\varkappa)} (\tilde{\lambda}_{21}^2 - 1)^{\frac{1}{2}} . \tag{10.2.7}$$

However, considering that the dielectric gap at T = 0 is associated with the phase transition temperature T_s through the relation,

$$\sigma_{21}(0) = \frac{\pi}{\gamma} T_s .$$

We obtain the condition for the dielectric instability temperature:

$$T_s > \frac{\gamma}{\pi} \langle \omega_D \rangle_{\rho(\varkappa)} (\tilde{\lambda}_{21}^2 - 1)^{\frac{1}{2}} , \tag{10.2.8}$$

By using (10.2.4) we obtain

$$\nu(\varepsilon_0)\lambda_{21} > \ln^{-1} \left(\frac{2\varepsilon_0}{\langle \omega_D \rangle_{\rho(\varkappa)} \sqrt{\tilde{\lambda}_{21}^2 - 1}} \right) . \tag{10.2.9}$$

Together with the condition $\tilde{\lambda}_{21} > 1$, (10.2.7) is the necessary and sufficient condition for the emergence of superconductivity in polyacetylene. The condition $\tilde{\lambda}_{21} > 1$ can be satisfied by increasing the parameter $\langle \cosh(\pi \ell_\varkappa \tilde{\tilde{p}}_{\perp 021}) \rangle_{\rho(\varkappa_0, \tilde{\tilde{p}}_{\perp 021})}$. This becomes possible by decreasing the separation between the chains, i.e., by applying external pressure. The critical pressure can be estimated from (10.2.5,9), as well as the data of [10.5-9] under the condition that $L_\varkappa = 10^{-7}$ cm, and $\sigma_{21} = g_{ph} u_0$. This quantity is found to be $p_c \approx$ 30-100 kbar. For a small gap, the superconducting transition temperature is found to be

$$T_c = \langle \omega_D \rangle_{\rho(\varkappa)} \ln^{-1} \left(\frac{\tilde{\lambda}_{21} + 1}{\tilde{\lambda}_{21} - 1} \right) , \quad \sigma_{21} \to 0 . \tag{10.2.10}$$

By applying an external pressure, we can increase the effective electron-phonon interaction parameter as a result of a decrease in the separation between the chains, and hence a decrease in the effective correlation radius of fluctuations in the system of soliton spins. However, a decrease in the separation between chains causes an increase in the overlap integral t_{ij} and an increase in the probability of electron jumps between chains, which means a decrease in the effective interaction of superconducting electrons. Moreover, polyacetylene is a metastable compound, and hence the application of pressure is hardly the best way of inducing it to be superconductive. There is, however, an alternative technique. Let us consider a system of polyacetylene chains in a solution whose atoms and ions have a nonzero spin ($\pm\frac{1}{2}$). In addition to the electron spin interactions, as the conduction electrons move along a chain, they interact with the elastic vibrations (phonons) and with the phonons of the solution. The spin component of the solution can be described by introducing a fluctuating field $\hat{A}_\nu$ in the same way as before. The potential of this field is identical to the Yang-Mills field potential for SU(2)-symmetry. Carrying out the analysis as before, we can write the strengths of the effective electron-electron and electron-hole interactions $\tilde{\lambda}_{21}$ and λ_{21} responsible for superconducting and dielectric phase transitions

$$\tilde{\lambda}_{21} = \langle \mathrm{Tr}\hat{B}(\Gamma_{\mathbf{x}})\rangle \; \nu(\varepsilon_0)[2(g^2_{ph\parallel} + g^2_{ph\perp}) \; \langle \mathrm{Tr}\hat{B}(\Gamma_{\mathbf{x}}) \times \cosh(\pi\ell_\varkappa \tilde{\tilde{p}}_{\perp 021})\rangle_{\rho(\varkappa,\tilde{\tilde{p}}_{\perp 021})} - \langle V_\varkappa(2p_0)\rangle_{\rho(\varkappa)}] \,, \tag{10.2.11}$$

$$\lambda_{21} = \langle \mathrm{Tr}\hat{B}(\Gamma_{\mathbf{x}})\rangle \; \nu(\varepsilon_0)[2(g^2_{ph\parallel} + g^2_{ph\perp}) \; \langle \mathrm{Tr}\hat{B}(\Gamma_{\mathbf{x}}) - \langle V_\varkappa(2p_0)\rangle_{\rho(\varkappa)}] \,, \tag{10.2.12}$$

where $\nu(\varepsilon_0)$ is the density of states at the Fermi level (points $\pm p_0$), and $\tilde{\tilde{p}}_{\perp 021} = 2\pi / \langle \tilde{r}_{c21}\rangle$. The parameter $\langle \tilde{r}_{c21}\rangle$ is determined by the mean correlation radius of the spins in the solution, but is always smaller than the separation $L_\varkappa$ between chains; consequently, $\tilde{\lambda}_{21}$ in this system is found to be larger than for the polymer (polyacetylene) film in which $\langle \tilde{r}_{c21}\rangle \gtrsim L_\varkappa$. Since $\tilde{\lambda}_{21}$ in this system is found to be larger than in the film (if we take into consideration the gauge equation $\oint_{\Gamma y} dx_\nu \hat{A}_\nu = 0$), the superconducting transition temperature will also be higher for the system. However, it should be noted that the superconducting phase will exist in the interval $T_{c1} \le T \le T_{c2}$, since in this case we must account for the stability condition for the system [10.10]:

$$\frac{e^2}{\langle\varepsilon_\varkappa\rangle_{\rho(\varkappa)} \, \ell_\varkappa T} < 1 \,. \tag{10.2.13}$$

Here, $\langle \varepsilon_\varkappa\rangle_{\rho(\varkappa)}$ is the permittivity ($\langle \varepsilon_\varkappa\rangle_{\rho(\varkappa)} \gg 1$). The quantity T_{c1} is defined by (10.2.1). As the temperature is decreased, the inequality (10.2.13) will be violated and the ions in the solution will be condensed on the chains. But this will entail a decrease in the charge density on the polyacetylene chains, i.e., a decrease in T_s and a violation of the condition (10.2.9). Consequently, superconductivity disappears and the lower critical temperature is $T_{c2} \approx e^2/\langle\varepsilon_\varkappa\rangle_{\rho(\varkappa)}\ell_\varkappa$. The phase transition to the nonsuperconducting state as a result of fluctuations may turn out to be first-order but close to second-order. In this case, hysteresis is possible in the vicinity of the point T_{c2}. Hence, a system of polyacetylene chains immersed in a solution of a spin liquid is analogous to a reversible superconductor based on rare-earth metal compounds. In such a system, superconductivity can be created more easily than in a polyacetylene film, since the condition $\tilde{\lambda}_{21} > 1$ can be satisfied by increasing the spin-density in the solution. For crystalline polyacetylene, T_c may reach values up to several hundred Kelvin for $\tilde{\lambda}_{21} > 1$.

10.3 High-Frequency Properties of Polyacetylene in the Nonsuperconducting State

Results on the experimental investigation of superconductivity in polyacetylene are not yet available. The dependence of the conductivity of polyacetylene on frequency ω is described in [10.2]. This dependence shows two peaks at $\omega \simeq 2$ and 8 eV. The existence of the peak at $\omega \approx 2$ eV can be explained by **Schrieffer**'s theory [10.2], but no explanation is available for the second peak at $\omega \approx 8$ eV. For this purpose, we can use our theory developed for calculating the frequency dependence of the conductivity of polyacetylene. In order to do so, we must find the permittivity $\varepsilon_{\varkappa\varkappa}(\omega)$ of the system and consider its imaginary part $\mathrm{Im}\{\varepsilon_{\varkappa\varkappa}(\omega)\}$. Using the results obtained in Sect. 9.5, we can show that the reciprocal permittivity tensor satisfies the equation

$$\begin{aligned}
\langle\varepsilon^{-1}_{\varkappa ij}(p,\omega|\Gamma_y)\rangle_{\rho(\varkappa,\tilde{\tilde{p}}_{\perp 021})} &= \delta_{ij} + \frac{4}{v_0 p_0}\langle\cosh(\pi\ell_\varkappa\tilde{\tilde{p}}_{\perp 021})\rangle_{\rho(\varkappa,\tilde{\tilde{p}}_{\perp 021})} \\
&\times \langle\omega G_{ij\varkappa}(p,\omega|\varepsilon)\rangle_{\rho(\varkappa,\tilde{\tilde{p}}_{\perp 021})} - \frac{4}{v_0 p_0}\langle\cosh(\pi\ell_\varkappa\tilde{\tilde{p}}_{\perp 021})\rangle_{\rho(\varkappa,\tilde{\tilde{p}}_{\perp 021})} \\
&\times\Big\langle\omega\int_{-\varkappa_1\in\Gamma_y}\frac{D\varkappa_1(x'',y'')}{2\pi C_\varkappa}G_{ij'\varkappa}(p,\omega|\varepsilon+\varkappa_1|)\,\mathrm{Im}\{\varepsilon^{-1}_{j'j''}(p,\omega|\varkappa_1)\} \\
&\times G_{j''j\varkappa}(p,\omega|\varepsilon+\varkappa_1|)\Big\rangle_{\rho(\varkappa,\tilde{\tilde{p}}_{\perp 021})} \quad .
\end{aligned} \tag{10.3.1}$$

Taking the imaginary part $\langle \mathrm{Im}\{\varepsilon^{-1}_{ij\varkappa}(p,\omega|\Gamma_y)\}\rangle_{\rho(\varkappa,\tilde{\bar{p}}_{\perp 021})}$ of the tensor, we obtain the resistance tensor $\langle (R_{ij\varkappa}(p,\omega|\Gamma_y)\rangle_{\rho(\varkappa,\tilde{\bar{p}}_{\perp 021})}$, whose reciprocal will be the conductivity tensor

$$\langle G_{ij}(p,\omega|\Gamma_y)\rangle_{\rho(\varkappa,\tilde{\bar{p}}_{\perp 021})} = \langle R^{-1}_{ij\varkappa}(p,\omega|\Gamma_y)\rangle_{\rho(\varkappa,\tilde{\bar{p}}_{\perp 021})} .$$

Using (10.3.1), we can show that the tensor

$$\langle R_{ij\varkappa}(p,\omega|\Gamma_y)\rangle_{\rho(\varkappa,\tilde{\bar{p}}_{\perp 021})}$$

satisfies the following equation:

$$\langle R_{ij}(p,\omega|\Gamma_y)\rangle_{\rho(\varkappa,\tilde{\bar{p}}_{\perp 021})} = \frac{4}{v_0 p_0}\langle \cosh(\pi \ell_\varkappa \tilde{\bar{p}}_{\perp 021})\rangle_{\rho(\varkappa,\tilde{\bar{p}}_{\perp 021})}$$

$$\left|\langle \mathrm{Im}\{\omega G_{ij\varkappa}(p,\omega)\}\rangle_{\rho(\varkappa,\tilde{\bar{p}}_{\perp 021})}\right| - \frac{4}{v_0 p_0}\langle \cosh(\pi \ell_\varkappa \tilde{\bar{p}}_{\perp 021})\rangle_{\rho(\varkappa,\tilde{\bar{p}}_{\perp 021})}$$

$$\times \left\langle \int_{-\varkappa\in\Gamma_y} \frac{D\varkappa_1(x'',y'')}{2\pi C_\varkappa} \mathrm{Re}\{\omega\, G_{ij'\varkappa}(p,\omega|\varkappa_1)\}\, R_{j'j''}(p,\omega|\varkappa_1)\right.$$

$$\left.\times\ \mathrm{Re}\,\{G_{j''j\varkappa}(p,\omega|\varkappa_1)\}\right\rangle_{\rho(\varkappa,\tilde{\bar{p}}_{\perp 021})} - \frac{4}{v_0 p_0}\langle \cosh(\pi \ell_\varkappa \tilde{\bar{p}}_{\perp 021})\rangle_{\rho(\varkappa,\tilde{\bar{p}}_{\perp 021})}$$

$$\times \left\langle \int_{-\varkappa_1\in\Gamma_y} \frac{D\varkappa_1(x'',y'')}{2\pi C_\varkappa} \left|\mathrm{Im}\{\omega\, G_{ij'\varkappa}(p,\omega|\varkappa_1)\}\right| \left|R_{ij\varkappa}(p,\omega|\varkappa_1)\right|\right.$$

$$\left.\times\left|\mathrm{Im}\,\{G_{j''j\varkappa}(p,\omega|\varkappa_1)\}\right|\right\rangle_{(\varkappa,\tilde{\bar{p}}_{\perp 021})} . \qquad (10.3.2)$$

Using the gauge equation $\oint_{\Gamma_y} dx_\nu \hat{A}^{(0)}_\nu = 0$, we can evaluate the components of the conductivity tensor for $p - p_0 = 0$ which satisfy the conditions

$$G_{11}(\omega) = G_{22}(\omega) , \quad G_{12}(\omega) = G_{21}(\omega) .$$

$$G_{11}(\omega) = [(\omega + \mu_\varkappa)(\omega - \mu_\varkappa)^2 - \sigma^2_{21}(\omega + \mu_\varkappa) - \Delta^2_{21}(\omega - \mu_\varkappa)]$$

$$\times \left[\pi \frac{\delta(\omega - \omega_+) - \delta(\omega - \omega_-)}{(\omega+\omega_+)(\omega+\omega_-)(\omega_+ - \omega_-)} + \frac{1}{a\omega(\omega + \tilde{\omega}_-)(\omega - \tilde{\omega}_+)} \right.$$

$$\left. + \operatorname{Re}\left\{ \frac{\tilde{\varepsilon}_0}{(\omega^2-\omega_+^2)(\omega^2-\omega_-^2)} \right\} \right] \quad , \Delta_{21} \to +0 \; .$$

$$G_{12}(\omega) = \sigma_{21}[(\omega - \mu_\varkappa)^2 - \sigma_{21}^2 + \Delta_{21}^2]$$

$$\times \left[\pi \frac{\delta(\omega - \omega_+) - \delta(\omega-\omega_-)}{(\omega+\omega_+)(\omega+\omega_-)(\omega_+-\omega_-)} - \frac{1}{a\omega(\omega+\tilde{\omega}_-)(\omega-\tilde{\omega}_+)} \right.$$

$$\left. - \operatorname{Re}\left\{ \frac{\varepsilon_0}{(\omega^2-\omega_+^2)(\omega^2 - \omega_-^2)} \right\} \right] \quad , \Delta_{21} \to +0$$

$$a = \frac{4\tilde{\varepsilon}_0}{v_0 p_0} \langle \cosh(\pi \ell_\varkappa \tilde{\tilde{p}}_{\perp 021}) \rangle_{\rho(\varkappa, \tilde{\tilde{p}}_{\perp 021})} \; , \qquad (10.3.3)$$

where $\tilde{\varepsilon}_0$ is the zero point energy in the system, $\tilde{\varepsilon}_0 \approx \sigma_{21}$, $\tilde{\omega}_\pm = \sigma_{21} \pm \mu_\varkappa$, $\omega_\pm = \sigma_{21} \pm \tilde{\mu}_\varkappa$. In this case, the conductivity is defined by the trace of the tensor:

$$G(\omega) = \frac{1}{2} \left| \operatorname{Tr} G_{ij}(\omega) \right| \; .$$

Figure 10.1 shows the dependence of the conductivity of the system on frequency ω . This dependence clearly shows the peak at $\omega = \sigma_{21} - \tilde{\mu}_\varkappa$ and two peaks in the vicinity of the point $\omega = \sigma_{21} + \tilde{\mu}_\varkappa$. If the parameter Δ_{21} is exactly equal to zero, the frequency dependence of the conductivity is

$$G(\omega) = \left| -\pi \frac{(\omega+\mu_\varkappa)(\omega - \tilde{\omega}_+)}{(\omega + \tilde{\omega}_+)(\tilde{\omega}_+ - \tilde{\omega}_-)} \delta(\omega - \tilde{\omega}_-) + \frac{\omega + \mu_\varkappa}{a\omega} \right.$$

$$\left. + \operatorname{Re}\left\{ \frac{\tilde{\varepsilon}_0(\omega + \mu_\varkappa)}{(\omega - \tilde{\omega}_-)(\omega + \tilde{\omega}_+)} \right\} \right| \; .$$

In this case, we obtain one peak at $\omega = \tilde{\omega}_-$, as did **Schrieffer** [10.2]. Since the peaks at $\omega \simeq \sigma_{21} + \tilde{\mu}_\varkappa$ and $\Delta_{21} \to +0$ are very close to each other, it is quite possible that only their envelope (dashed curve in Fig. 10.1) is observed in the experiment. But if this were the case, the theory concludes that the second conductivity peak

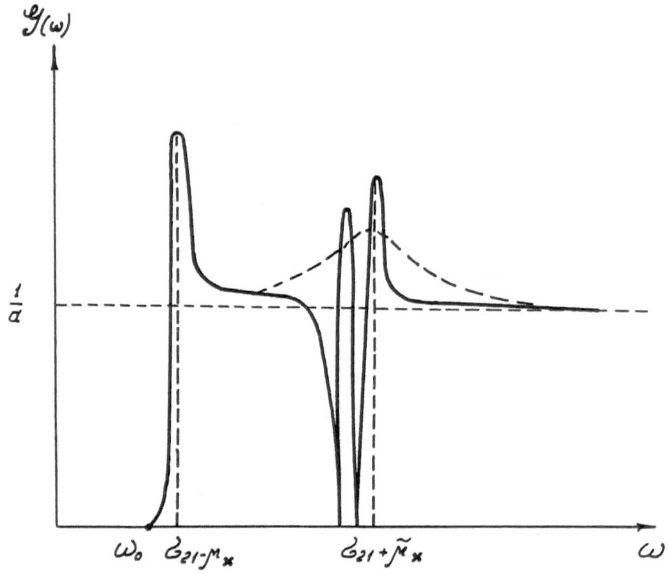

Fig. 10.1. Imaginary part of the permittivity (conductivity) of polyacetylene as a function of frequency in a system containing superconducting fluctuations:

$$\omega_0 = \sqrt{\sigma_{21}^2 + a\mu_\varkappa + \frac{a^2}{4}} - \left(\mu_\varkappa + \frac{a}{2}\right); \ \omega_0 \leq \omega_{s_2}\varepsilon_F; \ \Delta_{21} \to +0$$

in the system is caused by the narrow superconducting gap Δ_{21}. However, the system has a finite conductivity in spite of the presence of a superconducting order parameter. As was mentioned above, this is due to the fact that the superconductivity in polyacetylene is fluctuational, and the effective electron-phonon interaction parameter $\tilde{\lambda}_{21}$ generally depends on the spatial variable Γ_y due to the fact that vortex-type metastable excited states satisfying the condition $\oint_{\Gamma_y} dx_\nu \hat{A}_\nu^{(0)} \neq 0$ may emerge in the soliton spin system. In this case, the function $\langle \mathrm{Tr}\hat{B}(\Gamma_y) \rangle$ varies in space. Consequently, the superconducting state may turn out to be spatially inhomogeneous, i.e., it may occur in the form of superconducting fluctuations. Then in this case, the resistance of the system does not vanish. Hence it can be concluded that the second peak at $\omega = \omega_+$ is due to the existence of superconducting fluctuations [10.11].

References

Chapter 1

1.1 H. C. Ku, F. Acker, B. T. Matthias: Phys. Lett. **A76**, 399 (1980)
1.2 T. D. Thanh, A. Koma, S. Tanaka: J. Appl. Phys. **22**, 205 (1980)
1.3 J. G. Bednorz, K. A. Müller: Z. Phys. **B64**, 189 (1986)
1.4 C. W. Chu, P. H. Hor, R. L. Meng, et al: Phys. Rev. Lett. **58**, 405 (1987)
1.5 C. W. Chu, P. H. Hor, R. L. Meng, et al: Science **235**, 567 (1987)
1.6 P. H. Hor, R. L. Meng, L. Gao, et al: Phys. Rev. Lett. **58**, 908 (1987)
1.7 F. Izumi, H. Asano, T. Ishigaki, et al: Jap. J. Appl. Phys. **26**, L649 (1987)
1.8 Z. Z. Sheng, A. M. Hermann, A. El Ali, et al: Phys. Rev. Lett. **60**, 937 (1988)
1.9 C. W. Chu, J. Bechtold, L. Gao, et al: Phys. Rev. Lett. **60**, 941 (1988)
1.10 J. H. Van Vleck: Phys. Rev. **52**, 1178 (1937)
1.11 M. A. Savchenko: Fiz. Tverd. Tela (Leningrad) **6**, 864 (1964)
1.12 V. I. Ozhogin, P. P. Maximenkov: IEEE Trans. Magn. **8**, 645 (1972)
1.13 M. H. Seavy: Solid State Commun. **10**, 219 (1972)
1.14 M. A. Savchenko, L. A. Shishkin: Izv. Vuzov., Radiotekhnika **5**, 454 (1962)
1.15 M. A. Savchenko, A. V. Stefanovich: Phys. Status Solidi (b) **122**, 367 (1984)
1.16 V. G. Vaks, A. I. Larkin, S. A. Pikin: Preprint No. 1312, Kurchatov Inst. Atom. Energ., Moscow (1967) (Russ.)
1.17 M. A. Savchenko, A. V. Stefanovich: Solid State Commun. **47**, 863 (1983)
1.18 H. A. Mook, W. C. Koehler, S. K. Sinha, et al: J. Appl. Phys. **53**, 2614 (1982)
1.19 N. F. Berk, J. R. Schrieffer: Phys. Rev. Lett. **17**, 433 (1966)
1.20 K. Okida, S. Noguchi, A. Yamagishi et al.: Jpn. J. Appl. Phys. **26**, L822 (1987)
1.21 A. Junod, A. Bezinge, J. Müller: Physica **C152**, 65 (1988)
1.22 J. Bardeen, L. N. Cooper, J. R. Schrieffer: Phys. Rev. **108**, 1175 (1957)
1.23 W. L. McMillan, J. M. Rowell: in *Superconductivity*, ed. by R. D. Parks, Vol. I (Marcell Dekker, New York 1969)
1.24 M. B. Maple, F. Fisher: *Superconductivity in Ternary Compounds* (Springer, Berlin-Heidelberg 1982)

Chapter 2

2.1 E. F. Bertaut: Ann. Phys. **7**, 203 (1972)
2.2 J. H. Van Vleck: Phys. Rev. **52**, 1178 (1937)
2.3 M. A. Savchenko, B. V. Moshchinskii, A. V. Stefanovich: Fiz. Metal. Metalloved. **47**, 711 (1979)
2.4 M. Gell-Mann, F. E. Low: Phys. Rev. **95**, 1300 (1954)

2.5 A. I. Larkin, D. E. Khmel'nitskii: Zh. Eksp. Teor. Fiz. **56**, 2087 (1969)
2.6 K. G. Wilson: Phys. Rev. **B4**, 3174 (1971); ibid., 3184
2.7 K. G. Wilson, M. E. Fisher: Phys. Rev. Lett. **28**, 240 (1972)
2.8 K. G. Wilson, F. Kogut: Phys. Rep. **12C**, 75 (1974)
2.9 N. N. Bogolyubov, D. V. Shirkov: *Introduction to the Quantum Field Theory* (Russ.), (Nauka, Moscow 1976)
2.10 A. A. Vladimirov, D. V. Shirkov: Usp. Fiz. Nauk **129**, 407 (1979)
2.11 A. I. Larkin, S. A. Pikin: Zh. Eksp. Teor. Fiz. **56**, 1664 (1969)
2.12 P. M. Levy: Phys. Rev. **177**, 509 (1969)
2.13 M. A. Savchenko: D.Sc. Thesis (Russ.), Inst. Space Research, Acad. Sci. USSR, Moscow (1968)
2.14 I. K. Kamilov: D.Sc. Thesis (Russ.), Kazan State Univ., Kazan (1975)
2.15 K. S. Aleksandrov, A. T. Ansitratov, B. V. Beznosikov, N. V. Fedoseeva: *Phase Transitions in Halide Crystals* (Russ.), (Nauka, Novosibirsk 1981)
2.16 A. V. Stefanovich: Ph.D. Dissertation (Russ.), Inst. Radio-Eng., Electronics and Automation, Moscow (1978)
2.17 K. G. Wilson: Phys. Rev. Lett. **28**, 548 (1972)
2.18 M. A. Savchenko, A. V. Stefanovich: Fiz. Metal. Metalloved. **45**, 926 (1978)
2.19 I. E. Chupis: Fiz. Nizk. Temp. **1**, 183 (1975)
2.20 M. A. Savchenko, A. V. Stefanovich, M. A. Khabakhpashev: Fiz. Metal. Metalloved. **47**, 1320 (1979)
2.21 F. Iona, G. Shirane: *Ferroelectric Crystals* (Pergamon, Oxford 1962)
2.22 M. A. Savchenko: Fiz. Tverd. Tela (Leningrad) **6**, 864 (1964)
2.23 M. A. Savchenko: in *Magnetic Resonance and Related Phenomena Proc. XVI Congr., Ampere (1970)*, p. 498; ibid., p. 484

Chapter 3

3.1 N. P. Grazhdankina: Usp. Fiz. Nauk **96**, 291 (1968)
3.2 D. Mukamel, S. Krinsky: Phys. Rev. **B13**, 5065 (1976)
3.3 D. Mukamel, S. Krinsky: Phys. Rev. **B13**, 5078 (1976)
3.4 P. Bak, D. Mukamel: Phys. Rev. **B13**, 5086 (1976)
3.5 K. G. Wilson: Phys. Rev. **B4**, 3174 (1971); ibid., 3184
3.6 K. G. Wilson, M. E. Fisher: Phys. Rev. Lett. **28**, 240 (1972)
3.7 K. G. Wilson, F. Kogut: Phys. Rep. **12C**, 75 (1974)
3.8 D. Sherrington: J. Phys. **C6**, 1037 (1973)
3.9 A. W. Overhauser: Phys. Rev. **128**, 1437 (1962)
3.10 A. D. B. Woods, T. M. Holden, B. M. Powell, M. W. Stringfellow: Phys. Rev. Lett. **23**, 81 (1969)
3.11 T. M. Holden, B. M. Powell, M. W. Stringfellow, A. D. B. Woods: J. Appl. Phys. **41**, 1176 (1970)
3.12 A. H. Milhouse, W. C. Koehler: Int. J. Magn. **2**, 389 (1971)
3.13 K. P. Belov, M. A. Belyanchikova, R. Z. Levitin, S. A. Nikitin: *Rare-Earth Ferro- and Antiferromagnets* (Russ.), (Nauka, Moscow 1963)
3.14 M. A. Savchenko, A. V. Stefanovich: Zh. Eksp. Teor. Fiz. **74**, 2300 (1978)
3.15 V. E. Naish: Fiz. Metal. Metalloved. **14**, 315 (1962)
3.16 V. A. Buldovskii, M. A. Savchenko, A. V. Stefanovich: Izv. Akad. Nauk SSSR, ser. Fiz. **44**, 1367 (1980)
3.17 D. Husemoller: *Fibre Bundles* (McGraw-Hill, New York 1966)
3.18 G. A. Smolenskii, P. A. Krainik: Usp. Fiz. Nauk **97**, 657 (1969)
3.19 A. V. Stefanovich, V. A. Buldovskii: Fiz. Metal. Metalloved. **49**, 1299 (1980)

Chapter 4

4.1 D. E. Monston, D. B. McWhan, J. Eckert, et al: Phys. Rev. Lett. **39**, 1164 (1977)
4.2 R. W. McCallum, D. C. Johnston, R. N. Shelton, et al: Solid State Commun. **24**, 501 (1977)
4.3 M. Ishikawa, F. Fisher: Solid State Commun. **24**, 747 (1977)
4.4 W. A. Fertig, D. C. Johnston, L. E. DeLong, et al: Phys. Rev. Lett. **38**, 987 (1977)
4.5 D. C. Johnston, W. A. Fertig, M. B. Maple, B. T. Matthias: Solid State Commun. **26**, 141 (1978)
4.6 H. R. Ott, W. A. Fertig, D. C. Johnston, et al: J. Phys. (France) **39**, C375 (1978)
4.7 H. R. Ott, W. A. Fertig, D. C. Johnston, et al: J. Low Temp. Phys. **33**, 159 (1978)
4.8 J. W. Lynn, D. E. Moncton, W. Thomlinson, et al: Solid State Commun. **28**, 493 (1978)
4.9 M. B. Maple: J. Phys. (France) **39**, C1374 (1978)
4.10 H. B. MacKay, L. D. Wolf, M. B. Maple, D. C. Johnston: Phys. Rev. Lett. **42**, 918 (1979)
4.11 D. E. Moncton: J. Appl. Phys. **50**, 1880 (1979)
4.12 B. Chevalier, P. Lejay, J. Etourneau, et al: Mater. Res. Bull. **17**, 1211 (1982)
4.13 H. Adrian, K. Müller, G. Saemann-Ishenko: in *Ternary Superconductors, Proc. Intl. Conf., Lake Geneva, Wis.* (New York 1981)
4.14 H. Adrian, K. Müller: Phys. Rev. **B26**, 2450 (1982)
4.15 R. Tournier, R. Chevrel, M. Sergent: in *Ternary Superconductors, Proc. Intl. Conf., Lake Geneva, Wis.* (New York 1981)
4.16 I. Felner, I. Nowik: Solid State Commun. **47**, 831 (1983)
4.17 J. L. Holdeau, M. Marezino, J. P. Remeika: Acta Crystall. **B40**, 26 (1984)
4.18 J. W. Lynn, J. A. Gataas, R. W. Erwin, et al: Phys. Rev. Lett. **52**, 133 (1984)
4.19 B. Chevalier, P. Lejay, J. Etoruneau, P. Hagenmuller: Solid State Commun. **49**, 753 (1984)
4.20 S. Miraglia, J. L. Holdeau, M. Marezino, et al: Solid State Commun. **52**, 135 (1984)
4.21 D. Sherrington: J. Phys. **C6**, 1037 (1973)
4.22 H. A. Mook, W. C. Koehler, S. K. Sinha, et al: J. Appl. Phys. **53**, 2614 (1982)
4.23 K. G. Wilson, M. E. Fisher: Phys. Rev. Lett. **28**, 240 (1972)
4.24 K. G. Wilson, F. Kogut: Phys. Rep. **12C**, 75 (1974)
4.25 M. A. Savchenko, A. V. Stefanovich: Pis'ma Zh. Eksp. Teor. Fiz. **29**, 132 (1979)
4.26 M. A. Savchenko, A. V. Stefanovich: Fiz. Metal. Metalloved. **50**, 471 (1980)
4.27 H. R. Ott, H. Rudieger, Z. Fisk, J. L. Smith: Phys. Rev. Lett. **50**, 1595 (1983)
4.28 H. R. Ott, H. Rudieger, T. M. Rice, et al: Phys. Rev. Lett. **53**, 1915 (1984)
4.29 V. Korenman, J. L. Murray, R. E. Prange: Phys. Rev. **B16**, 4032 (1977)
4.30 L. V. Panina, M. A. Savchenko, A. V. Stefanovich: Phys. Status Solidi (b) **109**, 37 (1982)
4.31 L. V. Panina, M. A. Savchenko, A. V. Stefanovich: in *Proc. II Intl. Conf. on Selected Topics in Stat. Mech., Dubna (1981)*
4.32 M. A. Savchenko, A. V. Stefanovich: Solid State Commun. **44**, 1031 (1982)

Chapter 5

5.1 A. P. Cracknell, K. C. Wong: *Fermi Surface* (Clarendon, Oxford 1973)
5.2 D. E. Moncton: J. Appl. Phys. **50**, 1880 (1979)
5.3 J. W. Lynn, J. A. Gataas, R. W. Erwin, et al: Phys. Rev. Lett. **52**, 133 (1984)
5.4 R. W. McCallum, D. C. Johnston, R. N. Shelton, et al: Solid State Commun. **24**, 501 (1977)
5.5 M. Ishikawa, F. Fisher: Solid State Commun. **24**, 747 (1977)
5.6 M. A. Savchenko, A. V. Stefanovich: Pis'ma Zh. Eksp. Teor. Fiz. **29**, 661 (1979)
5.7 M. A. Savchenko, A. V. Stefanovich: Fiz. Metal. Metalloved. **50**, 471 (1980)
5.8 T. Moriya: J. Magn. Magn. Mater. **14**, 1 (1979)
5.9 L. V. Panina, M. A. Savchenko, A. V. Stefanovich: Phys. Status Solidi (b) **109**, 37 (1982)
5.10 L. V. Panina, M. A. Savchenko, A. V. Stefanovich: in *Proc. II Intl. Conf. on Selected Topics in Stat. Mech., Dubna (1981)*
5.11 N. F. Berk, J. R. Schrieffer: Phys. Rev. Lett. **17**, 433 (1966)
5.12 H. R. Ott, W. A. Fertig, D. C. Johnston, et al: J. Phys. (France) **39**, C375 (1978)
5.13 H. R. Ott, W. A. Fertig, D. C. Johnston, et al: J. Low Temp. Phys. **33**, 159 (1978)
5.14 M. A. Savchenko, A. V. Stefanovich: Solid State Commun. **47**, 863 (1983)
5.15 M. A. Savchenko, A. V. Stefanovich: Pis'ma Zh. Eksp. Teor. Fiz. **29**, 132 (1979)
5.16 H. C. Ku, F. Acker, B. T. Matthias: Phys. Lett. **A76**, 399 (1980)
5.17 H. C. Ku, H. F. Braun, F. Acker: Physica B+C **108**, 1231 (1981)
5.18 I. Felner, I. Nowik: Solid State Commun. **47**, 831 (1983)
5.19 D. J. Kim: Jap. J. Appl. Phys. **26**, L741 (1987)

Chapter 6

6.1 D. E. Moncton: J. Appl. Phys. **50**, 1880 (1979)
6.2 M. A. Savchenko, A. V. Stefanovich: Pis'ma Zh. Eksp. Teor. Fiz. **29**, 132 (1979)
6.3 M. A. Savchenko, A. V. Stefanovich: Teor. Mat. Fiz. **46**, 139 (1980)
6.4 C. N. Yang, R. G. Mills: Phys. Rev. **96**, 191 (1954)
6.5 N. D. Mermine, G. Stare: Phys. Rev. Lett. **30**, 1135 (1973)
6.6 M. A. Savchenko, A. V. Stefanovich: Fiz Metal. Metalloved. **50**, 471 (1980)
6.7 L. V. Panina, M. A. Savchenko, A. V. Stefanovich: Phys. Status Solidi (b) **109**, 37 (1982)
6.8 L. V. Panina, M. A. Savchenko, A. V. Stefanovich: in *Proc. II Intl. Conf. on Selected Topics in Stat. Mech., Dubna (1981)*
6.9 M. A. Savchenko, A. V. Stefanovich: Solid State Commun. **44**, 1031 (1982)
6.10 M. A. Savchenko, A. V. Stefanovich: Pis'ma Zh. Eksp. Teor. Fiz. **29**, 661 (1979)
6.11 J. W. Lynn, G. Shirane, W. Thomlinson, et al: Phys. Rev. **B24**, 3817 (1983)
6.12 C. F. Majkrzak, D. E. Cox, G. Shirane, et al: Phys. Rev. **B26**, 245 (1982)
6.13 M. A. Savchenko, A. V. Stefanovich: Solid State Commun. **33**, 725 (1980)
6.14 G. T'Hooft: Phys. Rev. Lett. **37**, 8 (1976)
6.15 H. A. Mook, W. C. Koehler, S. K. Sinha, et al: J. Appl. Phys. **53**, 2614 (1982)

Chapter 7

7.1 D. Sherrington: J. Phys. **C6**, 1037 (1973)
7.2 D. B. Richards, S. Legvold: Phys. Rev. **186**, 508 (1969)
7.3 M. A. Savchenko, A. V. Stefanovich: Pis'ma Zh. Eksp. Teor. Fiz. **28**, 337 (1978)
7.4 T. O. Brun, S. K. Sinha, N. Wakabayashi, et al: Phys. Rev. **B1**, 1251 ((1974)
7.5 G. S. Cargill: Solid State Phys. **30**, 227 (1975)
7.6 V. Canella, J. A. Mydosh: in *Proc. Intl. Conf. Magn. ICM-73* (Nauka, Moscow 1974)
7.7 V. Canella, J. A. Mydosh: in *Magn. and Magn. Mater., Proc. 19th Ann. Conf. AIP, Boston, Mass. 1973* (New York 1974)
7.8 G. S. Cargill: in *Magn. and Magn. Mater., Proc. 20th Ann. Conf. AIP, San Francisco 1974* (New York 1975)
7.9 S. F. Edwards, P. W. Anderson: J. Phys. **F5**, 965 (1975)
7.10 D. Sherrington, S. Kirkpatrick: Phys. Rev. Lett. **35**, 1792 (1975)
7.11 D. I. Thouless, P. W. Anderson, R. G. Palmer: Phil. Mag. **35**, 593 (1977)
7.12 A. B. Harris, S. Kirkpatrick: Phys. Rev. **B16**, 542 (1977)
7.13 C. M. Soukoulis: Phys. Rev. **B18**, 3757 (1978)
7.14 R. W. Tustison, P. A. Bek: Solid State Commun. **21**, 517 (1977)
7.15 A. K. Mukhopadhyay, R. D. suhl, P. A. Bek: J. Less Common Met. **43**, 69 (1975)
7.16 D. J. Elderfield, A. J. McKane: Phys. Rev. **B18**, 3730 (1978)
7.17 R. A. Pelcovits: Phys. Rev. **B19**, 465 (1979)
7.18 B. I. Halperin, W. M. Saslow: Phys. Rev. **B16**, 2154 (1977)
7.19 J. Vanuimenus, G. Toulouse: J. Phys. **C10**, L537 (1977)
7.20 J. Villain: J. Phys. **C10**, 1717 (1977)
7.21 P. G. de Gennes: *The Physics of Liquid Crystals* (Oxford University Press, New York 1974)
7.22 M. A. Savchenko, A. V. Stefanovich: Solid State Commun. **33**, 725 (1980)
7.23 E. H. Spanier: *Algebraic Topology* (McGraw-Hill, New York 1966)
7.24 M. W. Hirsch: *Differential Topology* (Springer, New York 1976)
7.25 V. A. Buldovskii, M. A. Savchenko, A. V. Stefanovich: Izv. Akad. Nauk SSSR, ser. Fiz. **44**, 1367 (1980)
7.26 M. A. Savchenko, A. V. Stefanovich: Fiz. Metal. Metalloved. **50**, 269 (1980)
7.27 J. A. Simmons, R. de Wit, R.K.Bullough (Eds.): *Fundamental Aspects of Dislocations*, Nat. Bur. Stand (USA) Special Publ. # 317, Vol. I (1970)
7.28 C. N. Yang, R. G. Mills: Phys. Rev. **96**, 191 (1954)

Chapter 8

8.1 M. A. Savchenko, A. V. Stefanovich: *Proc. Conf. Neodnorodnye electronnye sostoyaniya, 14-16 March 1984, USSR, Novosibirsk*, pp. 64,65
8.2 M. A. Savchenko, A. V. Stefanovich: *Proc. Conf. Poluchenie i primenenie segneto- i pyezoelectricheskikh materialov, 20-23 October 1984, Moscow*, p. 172
8.3 J. G. Bednorz, K. A. Müller: Z. Phys. **B64**, 189 (1986)
8.4 C. W. Chu, P. H. Hor, R. L. Meng, et al: Phys. Rev. Lett. **58**, 405 (1987)
8.5 C. W. Chu, P. H. Hor, R. L. Meng, et al: Science **235**, 567 (1987)
8.6 P. H. Hor, R. L. Meng, L. Gao, et al: Phys. Rev. Lett. **58**, 908 (1987)
8.7 F. Izumi, H. Asano, T. Ishigaki, et al: Jap. J. Appl. Phys. **26**, L649 (1987)

8.8 Z. Z. Sheng, A. M. Hermann, A. El Ali, et al: Phys. Rev. Lett. **69**, 937 (1988)
8.9 C. W. Chu, J. Bechtold, L. Gao, et al: Phys. Rev. Lett. **60**, 941 (1988)
8.10 P. P. Edwards, M. R. Harrison, R. Jones: Chemistry in Britain, October, 962 (1987)
8.11 B. Batlogg, G. Kourouklis, W. Weber, et al: Phys. Rev. Lett. **59**, 912 (1987)
8.12 T. A. Faltens, W. K. Ham, S. W. Keller, et al: Phys. Rev. Lett. **59**, 915 (1987)
8.13 J. Kevin, et al: Phys. Rev. Lett. **59**, 1236 (1987)
8.14 P. P. Tveitas, C. C. Tsuei, T. S. Plaskett: Phys. Rev. **B36**, 833 (1987)
8.15 J. W. Lynn, W. H. Li, Q. Li, et al: Phys. Rev. **B36**, 2374 (1987)
8.16 C. J. Lobb: Phys. Rev. **B36**, 3990 (1987)
8.17 M. A. Dubson, J. J. Calabuse, S. T. Herbett, et al: in *Növel Superconductivity*, ed. by S. A. Wolf, V. Kresin (Plenum, New York 1987), p. 981
8.18 G. Shirane, Y. Dndoh, R. J. Birgeneau, et al: Phys. Rev. Lett. **59**, 1613 (1987)
8.19 V. I. Ilyichev, M. A. Savchenko, A. V. Stefanovich: Solid State Commun. **69**,605 (1989)
8.20 M. B. Maple, F. Fisher: *Superconductivity in Ternary Compounds* (Springer, Berlin-Heidelberg 1982)
8.21 C.Rettori, D. Davidov, I. Belaish, et al: Phys. Rev. **B36**, 4028 (1987)
8.22 D. C. Larbalestier, M. Daeumling, X. Cai, et al: J. Appl. Phys. **62**, 3308 (1987)
8.23 M. A. Dubson, S. T. Herbert, J. J. Calabrese, et al: Phys. Rev. Lett. **60**, 1061 (1988)
8.24 D. K. Finnemore, R. N. Shelton, J. R. Clem, et al: Phys. Rev. **B35**, 5319 (1987)
8.25 M. A. Savchenko, A. V. Stefanovich: Solid State Commun. **44**, 1031 (1982)
8.26 M. Oussena, S. Senoussi, G. Collin, et al: Phys. Rev. **B36**, 4014 (1987)
8.27 R. Caspary, C. D. Bredle, E. Spille, M. Winkelmann, F. Steglich, H. Schmidt, T. Wolf, R. Flukiger: Physica **C153–155**,Pt. 1. 876 (1988)
8.28 R. A. Fisher, J. E. Gordon, S. Kim, N. E. Phillips. A. M. Stacy: Physica **C153–155**, Pt. 2, 1092 (1988)
8.29 M. E. Reeves, D. S. Citrin, B. G. Pazol, et al: Phys. Rev. **B36**, 6915 (1987)
8.30 T. Tamaki, T. Komai, A. Ito, et al: Solid State Commun. **65**, 43 (1988)
8.31 H. Kotayama-Yoshida, T. Harioska, A. Oyamada, et al: Physica, **C156**, 481 (1988)

Chapter 9

9.1 L. H. Ryder: *Quantum Field Theory* (Cambridge, London 1985); (Mir, USSR 1987), p. 128
9.2 S. Mandelstam: Phys. Rev. **175**, 1580 (1968)
9.3 S. Seiler: *Gauge Theories as a Problem of Constructive Quantum Field Theory and Statistical Mechanics* (Springer, Berlin-Heidelberg, 1982); (Mir, USSR 1985), p. 11
9.4 M. A. Savchenko, A. V. Stefanovich: Solid State Commun. **39**, 725 (1981)
9.5 J. Bardeen, L. N. Cooper, J. R. Schrieffer: Phys. Rev. **108**, 1175 (1957)

9.6 L. D. Landau, E. M. Lifshitz: *Quantum Mechanics Nonrelativistic Theory* (Russ.), (Nauka, Moscow 1974)
9.7 A. A. Abrikosov, L. P. Gorkov, I. E. Dzyaloshinskii: Metodi Kvantovoi Teorii Polya v Statisticheskoi Fizike (G. Ed. Phys.-Math. Let., Moscow 1962),p. 388
9.8 V. Heine, A. B. Pippard: Phil. Mag. **3**, 1046 (1958)
9.9 M. B. Maple, F. Fisher: *Superconductivity in Ternary Compounds* (Springer, Berlin-Heidelberg 1982)
9.10 M. A. Savchenko, A. V. Stefanovich: *Proc. Conf. Poluchenie i primenenie segneto- i Pyezoelektricheskikh materialov, 20-23 October 1984, Moscow*, p. 172
9.11 R. N. Bhargava, S. P. Herko, W. N. Osborne: Phys. Rev. Lett. **59**, 1468 (1987)

Chapter 10

10.1 I. I. Frankevich, I. A. Sokolik, D. I. Kadirov, V. M. Kobryanskii: Pisma v JETE **36**, 401 (1982)
10.2 W. P. Su, J. R. Schrieffer, A. J. Heeger: Phys. Rev. **B22**, 2099 (1980)
10.3 M. A. Savchenko, A. V. Stefanovich: Solid State Commun. **52**, 527 (1984)
10.4 M. A. Savchenko, A. V. Stefanovich: Doklady Akad. Nauk, SSSR, **277**, 419 (1984)
10.5 Y. W. Park, M. A. Druy, C. K. Chiang, et al: J. Polym. Sci. Polym. Lett. **17**, 195-201 (1979)
10.6 C. A. Coulson, A. Golebiewski: Proc. Phys. Soc. **78**, N 6, Pt. 2, 1310 (1961)
10.7 M. Tsuji, S. Hazinaga, T. Hasino: Rev. Mod. Phys. **32**, 425 (1960)
10.8 J. A. Pople, S. H. Walmsley: Trans. Faraday Soc. **18**, 441 (1962)
10.9 G. M. Holob, P. Enrlich, R. D. Allendoerfer: Macromolecules **5**, 569
10.10 P. G. de Gennes: *Scaling Concepts in Polymer Physics* (Cornell University Press, Ithaca 1979); (Mir, USSR 1982)
10.11 G. Appli: Physica **B148**, 163 (1987)

Subject Index